McGraw-Hill Dictionary of
MECHANICAL
AND DESIGN
ENGINEERING

McGraw-Hill Dictionary of
MECHANICAL AND DESIGN ENGINEERING

Sybil P. Parker
EDITOR IN CHIEF

McGraw-Hill Book Company

New York   St. Louis   San Francisco

Auckland   Bogotá   Guatemala   Hamburg
Johannesburg   Lisbon   London   Madrid   Mexico
Montreal   New Delhi   Panama   Paris   San Juan
São Paulo   Singapore   Sydney   Tokyo   Toronto

On the cover: A computer-generated graphical representation of a waveguide divided into finite elements for structural analysis. (Courtesy of Peter Marks, Design Insight, Milford, Ohio)

ISBN 0-07-045414-0

Library of Congress Cataloging in Publication Data

McGraw-Hill dictionary of mechanical and design engineering.

1. Mechanical engineering—Dictionaries.
2. Engineering design—Dictionaries.   I. Parker, Sybil P.   II. McGraw-Hill Book Company.   III. Title: Dictionary of mechanical and design engineering.
TJ9.M395   1984   620.1'003'21   84-28853
ISBN 0-07-045414-0 (pbk.)

Editorial Staff

Sybil P. Parker, Editor in Chief

Jonathan Weil, Editor
Betty Richman, Editor
Edward J. Fox, Art director
Joe Faulk, Editing manager
Frank Kotowski, Jr., Editing supervisor

Consulting and Contributing Editors
from the McGraw-Hill Dictionary of Scientific and Technical Terms

How to Use the Dictionary

I. ALPHABETIZATION

The terms in the *McGraw-Hill Dictionary of Mechanical and Design Engineering* are alphabetized on a letter-by-letter basis; word spacing, hyphen, comma, solidus, and apostrophe in a term are ignored in the sequencing. For example, an ordering of terms would be:

> **air belt**
> **air-breathing**
> **aircraft**
> **air hp**
> **airspeed**
> **APU**

II. FORMAT

The basic format for a defining entry provides the term in boldface, the field in small capitals, and the single definition in lightface:

> **term** [FIELD] Definition.

A field may be followed by multiple definitions, each introduced by a boldface number:

> **term** [FIELD] **1.** Definition. **2.** Definition. **3.** Definition.

A term may have definitions in two or more fields:

> **term** [AERO ENG] Definition. [MECH] Definition.

A simple cross-reference entry appears as:

> **term** *See* another term.

A cross-reference may also appear in combination with definitions:

> **term** [AERO ENG] Definition. [MECH] *See* another term.

III. CROSS-REFERENCING

A cross-reference entry directs the user to the defining entry. For example, the user looking up "astatic governor" finds:

> **astatic governor** *See* isochronous governor.

The user then turns to the "I" terms for the definition.

Cross-references are also made from variant spellings, acronyms, abbreviations, and symbols.

A-5 *See* Vigilante.
aerofoil *See* airfoil.
atm *See* atmosphere.
ATS *See* applications technology satellite.

IV. ALSO KNOWN AS . . . , etc.

A definition may conclude with a mention of a synonym of the term, a variant spelling, an abbreviation for the term, or other such information, introduced by "Also known as . . . ," "Also spelled . . . ," "Abbreviated . . . ," "Symbolized . . . ," "Derived from . . . ." When a term has more than one definition, the positioning of any of these phrases conveys the extent of applicability. For example:

term [AERO ENG] **1.** Definition. Also known as synonym.
2. Definition. Symbolized T.

In the above arrangement, "Also known as . . ." applies only to the first definition. "Symbolized . . ." applies only to the second definition.

term [AERO ENG] **1.** Definition. **2.** Definition. [MECH] Definition. Also known as synonym.

In the above arrangement, "Also known as . . ." applies only to the second field.

term [AERO ENG] Also known as synonym. **1.** Definition. **2.** Definition. [MECH] Definition.

In the above arrangement, "Also known as . . ." applies to both definitions in the first field.

term Also known as synonym. [AERO ENG] **1.** Definition. **2.** Definition. [MECH] Definition.

In the above arrangement, "Also known as . . ." applies to all definitions in both fields.

Field Abbreviations and Scope

aerospace engineering—Engineering pertaining to the design and construction of aircraft and space vehicles and of power units, and dealing with the special problems of flight in both the earth's atmosphere and space, such as in the flight of air vehicles and the launching, guidance, and control of missiles, earth satellites, and space vehicles and probes.

control systems—The study of those systems in which one or more outputs are forced to change in a desired manner as time progresses.

design engineering—A branch of engineering concerned with the design of a product or facility according to generally accepted uniform standards and procedures, such as the specification of a linear dimension, or a manufacturing practice, such as the consistent use of a particular size of screw to fasten covers.

fluid mechanics—The science concerned with fluids, either at rest or in motion, and dealing with pressures, velocities, and accelerations in the fluid, including fluid deformation and compression or expansion.

mechanical engineering—The branch of engineering concerned with the generation, transmission, and utilization of heat and mechanical power, and with the production and operation of tools, machinery, and their products.

mechanics—The branch of physics which seeks to formulate general rules for predicting the behavior of a physical system under the influence of any type of interaction with its environment.

thermodynamics—The branch of physics which seeks to derive, from a few basic postulates, relations between properties of substances, especially those which are affected by changes in temperature, and a description of the conversion of energy from one form to another.

McGraw-Hill Dictionary of
MECHANICAL
AND DESIGN
ENGINEERING

Å *See* angstrom.

A-1 *See* Skyraider.

A-3 *See* Skywarrior.

A-5 *See* Vigilante.

ablation [AERO ENG] The carrying away of heat, generated by aerodynamic heating, from a vital part by arranging for its absorption in a nonvital part, which may melt or vaporize and then pass away, taking the heat with it. Also known as ablative cooling.

ablative cooling *See* ablation.

ablative shielding [AERO ENG] A covering of material designed to reduce heat transfer to the internal structure through sublimation and loss of mass.

abort [AERO ENG] **1.** To cut short or break off an action, operation, or procedure with an aircraft, space vehicle, or the like, especially because of equipment failure. **2.** An aircraft, space vehicle, or the like which aborts. **3.** An act or instance of aborting.

abort zone [AERO ENG] The area surrounding the launch within which malperforming missiles will be contained with known and acceptable probability.

abrasion test [MECH ENG] The measurement of abrasion resistance, usually by the weighing of a material sample before and after subjecting it to a known abrasive stress throughout a known time period, or by reflectance or surface finish comparisons, or by dimensional comparisons.

abrasive belt [MECH ENG] A cloth, leather, or paper band impregnated with grit and rotated as an endless loop to abrade materials through continuous friction.

abrasive blasting [MECH ENG] The cleaning or finishing of surfaces by the use of an abrasive entrained in a blast of air.

abrasive cloth [MECH ENG] Tough cloth to whose surface an abrasive such as sand or emery has been bonded for use in grinding or polishing.

abrasive cone [MECH ENG] An abrasive sintered or shaped into a solid cone to be rotated by an arbor for abrasive machining.

abrasive disk [MECH ENG] An abrasive sintered or shaped into a disk to be rotated by an arbor for abrasive machining.

abrasive machining [MECH ENG] Grinding, drilling, shaping, or polishing by abrasion.

abreast milling [MECH ENG] A milling method in which parts are placed in a row parallel to the axis of the cutting tool and are milled simultaneously.

absolute angle of attack [AERO ENG] The acute angle between the chord of an airfoil at any instant in flight and the chord of that airfoil at zero lift.

absolute ceiling [AERO ENG] The greatest altitude at which an aircraft can maintain level flight in a standard atmosphere and under specified conditions.

absolute expansion [THERMO] The true expansion of a liquid with temperature, as calculated when the expansion of the container in which the volume of the liquid is measured is taken into account; in contrast with apparent expansion.

absolute potential vorticity *See* potential vorticity.

absolute scale *See* absolute temperature scale.

absolute specific gravity [MECH] The ratio of the weight of a given volume of a substance in a vacuum at a given temperature to the weight of an equal volume of water in a vacuum at a given temperature.

absolute temperature [THERMO] 1. The temperature measurable in theory on the thermodynamic temperature scale. 2. The temperature in Celsius degrees relative to the absolute zero at $-273.16°C$ (the Kelvin scale) or in Fahrenheit degrees relative to the absolute zero at $-459.69°F$ (the Rankine scale).

absolute temperature scale [THERMO] A scale with which temperatures are measured relative to absolute zero. Also known as absolute scale.

absolute viscosity [FL MECH] The tangential force per unit area of two parallel planes at unit distance apart when the space between them is filled with a fluid and one plane moves with unit velocity in its own plane relative to the other.

absolute vorticity [FL MECH] The vorticity of a fluid relative to an absolute coordinate system; especially, the vorticity of the atmosphere relative to axes not rotating with the earth.

absolute zero [THERMO] The temperature of $-273.16°C$, or $-459.69°F$, or 0 K, thought to be the temperature at which molecular motion vanishes and a body would have no heat energy.

absorber [MECH ENG] 1. A device which holds liquid for the absorption of refrigerant vapor or other vapors. 2. That part of the low-pressure side of an absorption system used for absorbing refrigerant vapor.

absorption cycle [MECH ENG] In refrigeration, the process whereby a circulating refrigerant, for example, ammonia, is evaporated by heat from an aqueous solution at elevated pressure and subsequently reabsorbed at low pressure, displacing the need for a compressor.

absorption refrigeration [MECH ENG] Refrigeration in which cooling is effected by the expansion of liquid ammonia into gas and absorption of the gas by water; the ammonia is reused after the water evaporates.

absorption system [MECH ENG] A refrigeration system in which the refrigerant gas in the evaporator is taken up by an absorber and is then, with the application of heat, released in a generator.

abutting joint [DES ENG] A joint which connects two pieces of wood in such a way that the direction of the grain in one piece is angled (usually at 90°) with respect to the grain in the other.

abutting tenons [DES ENG] Two tenons inserted into a common mortise from opposite sides so that they contact.

acceleration [MECH] The rate of change of velocity with respect to time.

acceleration analysis [MECH ENG] A mathematical technique, often done graphically, by which accelerations of parts of a mechanism are determined.

acceleration error constant [CONT SYS] The ratio of the acceleration of a controlled variable of a servomechanism to the actuating error when the actuating error is constant.

acceleration feedback [AERO ENG] The use of accelerometers strategically located within the body of a missile so that they sense body accelerations during flight and interact with another device on board the missile or with a control center on the ground or in an airplane to keep the missile's speed within design limits.

acceleration measurement [MECH] The technique of determining the magnitude and direction of acceleration, including translational and angular acceleration.

acceleration of free fall *See* acceleration of gravity.

acceleration of gravity [MECH] The acceleration imparted to bodies by the attractive force of the earth; has an international standard value of 980.665 cm/sec^2 but varies with latitude and elevation. Also known as acceleration of free fall; apparent gravity.

acceleration potential [FL MECH] The sum of the potential of the force field acting on a fluid and the ratio of the pressure to the fluid density; the negative of its gradient gives the acceleration of a point in the fluid.

accelerator [MECH ENG] A device for varying the speed of an automotive vehicle by varying the supply of fuel.

accelerator jet [MECH ENG] The jet through which the fuel is injected into the incoming air in the carburetor of an automotive vehicle with rapid demand for increased power output.

accelerator linkage [MECH ENG] The linkage connecting the accelerator pedal of an automotive vehicle to the carburetor throttle valve or fuel injection control.

accelerator pedal [MECH ENG] A pedal that operates the carburetor throttle valve or fuel injection control of an automotive vehicle.

accelerator pump [MECH ENG] A small cylinder and piston controlled by the throttle of an automotive vehicle so as to provide an enriched air-fuel mixture during acceleration.

accessory [MECH ENG] A part, subassembly, or assembly that contributes to the effectiveness of a piece of equipment without changing its basic function; may be used for testing, adjusting, calibrating, recording, or other purposes.

accordion roller conveyor [MECH ENG] A conveyor with a flexible latticed frame which permits variation in length.

accumulator [AERO ENG] A device sometimes incorporated in the fuel system of a gas-turbine engine to store fuel and release it under pressure as an aid in starting. [MECH ENG] 1. A device, such as a bag containing pressurized gas, which acts upon hydraulic fluid in a vessel, discharging it rapidly to give high hydraulic power, after which the fluid is returned to the vessel with the use of low hydraulic power. 2. A device connected to a steam boiler to enable a uniform boiler output to meet an irregular steam demand. 3. A chamber for storing low-side liquid refrigerant in a refrigeration system. Also known as surge drum; surge header.

Ackeret method [FL MECH] A method of studying the behavior of an airfoil in a supersonic airstream based on the hypothesis that the disturbance caused by the airfoil consists of two plane waves, at the leading and trailing edges, which propagate outward like sound waves and each makes an angle equal to the Mach angle with the direction of flow.

Ackerman linkage *See* Ackerman steering gear.

Ackerman steering gear [MECH ENG] Differential gear or linkage that turns the two steered road wheels of a self-propelled vehicle so that all wheels roll on circles with a common center. Also known as Ackerman linkage.

acme screw thread [DES ENG] A standard thread having a profile angle of 29° and a flat crest; used on power screws in such devices as automobile jacks, presses, and lead screws on lathes. Also known as acme thread.

acme thread   See acme screw thread.

acoustic approximation   [FL MECH] The approximation that leads from the nonlinear hydrodynamic equations of a gas to the linear wave equation for sound wave propagation.

acoustic fatigue   [MECH] The tendency of a material, such as a metal, to lose strength after acoustic stress.

acoustic Mach meter   [AERO ENG] A device which registers data on sound propagation for the calculation of Mach number.

acoustic streaming   [FL MECH] Unidirectional flow currents in a fluid that are due to the presence of sound waves.

acoustic theory   [AERO ENG] The linearized small-disturbance theory used to predict the approximate airflow past an airfoil when the disturbance velocities caused by the flow are small compared to the flight speed and to the speed of sound.

acre   [MECH] A unit of area, equal to 43,560 square feet, or to 4046.8564224 square meters.

action   [MECH] An integral associated with the trajectory of a system in configuration space, equal to the sum of the integrals of the generalized momenta of the system over their canonically conjugate coordinates. Also known as phase integral.

active communications satellite   [AERO ENG] Satellite which receives, regenerates, and retransmits signals between stations.

active controls technology   [AERO ENG] The development of special forms of augmentation systems to stabilize airplane configurations and to limit, or tailor, the design loads that the airplane structure must support.

active satellite   [AERO ENG] A satellite which transmits a signal.

actual exhaust velocity   [AERO ENG] 1. The real velocity of the exhaust gas leaving a nozzle as determined by accurately measuring at a specified point in the nozzle exit plane. 2. The velocity obtained when the kinetic energy of the gas flow produces actual thrust.

actual horsepower   See actual power.

actual power   [MECH ENG] The power delivered at the output shaft of a source of power. Also known as actual horsepower.

actuate   [MECH ENG] To put into motion or mechanical action, as by an actuator.

actuated roller switch   [MECH ENG] A centrifugal sequence-control switch that is placed in contact with a belt conveyor, immediately preceding the conveyor it controls.

actuating system   [CONT SYS] An electric, hydraulic, or other system that supplies and transmits energy for the operation of other mechanisms or systems.

actuator   [CONT SYS] A mechanism to activate process control equipment by use of pneumatic, hydraulic, or electronic signals; for example, a valve actuator for opening or closing a valve to control the rate of fluid flow.   [MECH ENG] A device that produces mechanical force by means of pressurized fluid.

acyclic motion   See irrotational flow.

adamantine drill   [MECH ENG] A core drill with hardened steel shot pellets that revolve under the rim of the rotating tube; employed in rotary drilling in very hard ground.

adapter skirt   [AERO ENG] A flange or extension of a space vehicle that provides a ready means for fitting some object to a stage or section.

adaptive control   [CONT SYS] A control method in which one or more parameters are sensed and used to vary the feedback control signals in order to satisfy the performance criteria.

adaptive control function [CONT SYS] That level in the functional decomposition of a large-scale control system which updates parameters of the optimizing control function to achieve a best fit to current plant behavior, and updates parameters of the direct control function to achieve good dynamic response of the closed-loop system.

addendum [DES ENG] The radial distance between two concentric circles on a gear, one being that whose radius extends to the top of a gear tooth (addendum circle) and the other being that which will roll without slipping on a circle on a mating gear (pitch line).

addendum circle [DES ENG] The circle on a gear passing through the tops of the teeth.

adhesion [MECH] The force of static friction between two bodies, or the effects of this force.

adhesional work [THERMO] The work required to separate a unit area of a surface at which two substances are in contact. Also known as work of adhesion.

adhesive bond [MECH] The forces such as dipole bonds which attract adhesives and base materials to each other.

adiabatic [THERMO] Referring to any change in which there is no gain or loss of heat.

adiabatic compression [THERMO] A reduction in volume of a substance without heat flow, in or out.

adiabatic cooling [THERMO] A process in which the temperature of a system is reduced without any heat being exchanged between the system and its surroundings.

adiabatic ellipse [FL MECH] A plot of the speed of sound as a function of the speed of flow for the adiabatic flow of a gas, which forms one quadrant of an ellipse.

adiabatic engine [MECH ENG] A heat engine or thermodynamic system in which there is no gain or loss of heat.

adiabatic envelope [THERMO] A surface enclosing a thermodynamic system in an equilibrium which can be disturbed only by long-range forces or by motion of part of the envelope; intuitively, this means that no heat can flow through the surface.

adiabatic expansion [THERMO] Increase in volume without heat flow, in or out.

adiabatic flow [FL MECH] Movement of a fluid without heat transfer.

adiabatic process [THERMO] Any thermodynamic procedure which takes place in a system without the exchange of heat with the surroundings.

adiabatic recovery temperature [FL MECH] 1. The temperature reached by a moving fluid when brought to rest through an adiabatic process. Also known as recovery temperature; stagnation temperature. 2. The final and initial temperature in an adiabatic, Carnot cycle.

adiabatic vaporization [THERMO] Vaporization of a liquid with virtually no heat exchange between it and its surroundings.

adiabatic wall temperature [FL MECH] The temperature assumed by a wall in a moving fluid stream when there is no heat transfer between the wall and the stream.

adsorption system [MECH ENG] A device that dehumidifies air by bringing it into contact with a solid adsorbing substance.

advance [MECH ENG] To effect the earlier occurrence of an event, for example, spark advance or injection advance.

adz [DES ENG] A cutting tool with a thin arched blade, sharpened on the concave side, at right angles on the handle; used for rough dressing of timber.

adz block [MECH ENG] The part of a machine for wood planing that carries the cutters.

aerator [DES ENG] A tool having a roller equipped with hollow fins; used to remove cores of soil from turf. [MECH ENG] Equipment used to inject compressed air into sewage in the treatment process.

aerial cableway *See* aerial tramway.

aerial ropeway *See* aerial tramway.

aerial sound ranging [AERO ENG] The process of locating an aircraft by means of the sounds it emits.

aerial spud [MECH ENG] A cable for moving and anchoring a dredge.

aerial tramway [MECH ENG] A system for transporting bulk materials that consists of one or more cables supported by steel towers and is capable of carrying a traveling carriage from which loaded buckets can be lowered or raised. Also known as aerial cableway; aerial ropeway.

aeroballistics [MECH] The study of the interaction of projectiles or high-speed vehicles with the atmosphere.

aeroduct [AERO ENG] A ramjet type of engine designed to scoop up ions and electrons freely available in the outer reaches of the atmosphere or in the atmospheres of other spatial bodies and, by a metachemical process within the engine duct, to expel particles derived from the ions and electrons as a propulsive jetstream.

aerodynamic [FL MECH] Pertaining to forces acting upon any solid or liquid body moving relative to a gas (especially air).

aerodynamically rough surface [FL MECH] A surface whose irregularities are sufficiently high that the turbulent boundary layer reaches right down to the surface.

aerodynamically smooth surface [FL MECH] A surface whose irregularities are sufficiently small to be entirely embedded in the laminar sublayer.

aerodynamic center [AERO ENG] A point on a cross section of a wing or rotor blade through which the forces of drag and lift are acting and about which the pitching moment coefficient is practically constant.

aerodynamic characteristics [AERO ENG] The performance of a given airfoil profile as related to lift and drag, to angle of attack, and to velocity, density, viscosity, compressibility, and so on.

aerodynamic chord [AERO ENG] A straight line intersecting or touching an airfoil profile at two points; specifically, that part of such a line between two points of intersection.

aerodynamic coefficient [FL MECH] Any nondimensional coefficient relating to aerodynamic forces or moments, such as a coefficient of drag or a coefficient of lift.

aerodynamic configuration [AERO ENG] The form of an aircraft, incorporating desirable aerodynamic qualities.

aerodynamic control [AERO ENG] A control surface whose use causes local aerodynamic forces.

aerodynamic drag [FL MECH] A retarding force that acts upon a body moving through a gaseous fluid and that is parallel to the direction of motion of the body; it is a component of the total fluid forces acting on the body. Also known as aerodynamic resistance.

aerodynamic force [FL MECH] The force between a body and a gaseous fluid caused by their relative motion. Also known as aerodynamic load.

aerodynamic heating [FL MECH] The heating of a body produced by passage of air or other gases over its surface; caused by friction and by compression processes and significant chiefly at high speeds.

aerodynamic instability [AERO ENG] An unstable state caused by oscillations of a structure that are generated by spontaneous and more or less periodic fluctuations in the flow, particularly in the wake of the structure.

aerodynamic lift [FL MECH] That component of the total aerodynamic force acting on a body perpendicular to the undisturbed airflow relative to the body. Also known as lift.

aerodynamic load *See* aerodynamic force.

aerodynamic missile [AERO ENG] A missile with surfaces which produce lift during flight.

aerodynamic moment [AERO ENG] The torque about the center of gravity of a missile or projectile moving through the atmosphere, produced by any aerodynamic force which does not act through the center of gravity.

aerodynamic phenomena [FL MECH] Acoustic, thermal, electrical, and mechanical effects, among others, that result from the flow of air over a body.

aerodynamic resistance *See* aerodynamic drag.

aerodynamics [FL MECH] The science that deals with the motion of air and other gaseous fluids and with the forces acting on bodies when they move through such fluids or when such fluids move against or around the bodies.

aerodynamic stability [AERO ENG] The property of a body in the air, such as an aircraft or rocket, to maintain its attitude, or to resist displacement, and if displaced, to develop aerodynamic forces and moments tending to restore the original condition.

aerodynamic trail [FL MECH] A condensation trail formed by adiabatic cooling to saturation (or slight supersaturation) of air passing over the surfaces of high-speed aircraft.

aerodynamic trajectory [MECH] A trajectory or part of a trajectory in which the missile or vehicle encounters sufficient air resistance to stabilize its flight or to modify its course significantly.

aerodynamic turbulence [FL MECH] A state of fluid flow in which the instantaneous velocities exhibit irregular and apparently random fluctuations.

aerodynamic vehicle [AERO ENG] A device, such as an airplane or glider, capable of flight only within a sensible atmosphere and relying on aerodynamic forces to maintain flight.

aerodynamic wave drag [FL MECH] The force retarding an airplane, especially in supersonic flight, as a consequence of the formation of shock waves ahead of it.

aerodyne [AERO ENG] Any heavier-than-air craft that derives its lift in flight chiefly from aerodynamic forces, such as the conventional airplane, glider, or helicopter.

aeroelasticity [MECH] The deformation of structurally elastic bodies in response to aerodynamic loads.

aerofall mill [MECH ENG] A grinding mill of large diameter with either lumps of ore, pebbles, or steel balls as crushing bodies; the dry load is airswept to remove mesh material.

aerofoil *See* airfoil.

aeromechanics [FL MECH] The science of air and other gases in motion or equilibrium; has two branches, aerostatics and aerodynamics.

aeromotor [AERO ENG] An engine designed to provide motive power for an aircraft.

aeronaut [AERO ENG] A person who operates or travels in an airship or balloon.

aeronautical engineering [AERO ENG] The branch of engineering concerned primarily with the design and construction of aircraft structures and power units, and with the special problems of flight in the atmosphere.

aeronautical flutter [FL MECH] An aeroelastic, self-excited vibration in which the external source of energy is the airstream and which depends on the elastic, inertial, and dissipative forces of the system in addition to the aerodynamic forces. Also known as flutter.

aeronautics [FL MECH] The science that deals with flight through the air.

aerophysics [AERO ENG] The physics dealing with the design, construction, and operation of aerodynamic devices.

aeropulse engine *See* pulsejet engine.

aeroservoelasticity [AERO ENG] The study of the interaction of automatic flight controls on aircraft and aeroelastic response and stability.

aerosol generator [MECH ENG] A mechanical means of producing a system of dispersed phase and dispersing medium, that is, an aerosol.

aerospace ground equipment [AERO ENG] Support equipment for air and space vehicles. Abbreviated AGE.

aerospace vehicle [AERO ENG] A vehicle capable of flight both within and outside the sensible atmosphere.

aerostat [AERO ENG] Any aircraft that derives its buoyancy or lift from a lighter-than-air gas contained within its envelope or one of its compartments; for example, ships and balloons.

aerostatics [FL MECH] The science of the equilibrium of gases and of solid bodies immersed in them when under the influence only of natural gravitational forces.

aerothermochemistry [FL MECH] The study of gases which takes into account the effect of motion, heat, and chemical changes.

aerothermodynamics [FL MECH] The study of aerodynamic phenomena at sufficiently high gas velocities that thermodynamic properties of the gas are important.

aerothermoelasticity [FL MECH] The study of the response of elastic structures to the combined effects of aerodynamic heating and loading.

afterbody [AERO ENG] **1.** A companion body that trails a satellite. **2.** A section or piece of a rocket or spacecraft that enters the atmosphere behind the nose cone or other body that is protected for entry. **3.** The afterpart of a vehicle.

afterburner [AERO ENG] A device for augmenting the thrust of a jet engine by burning additional fuel in the uncombined oxygen in the gases from the turbine.

afterburning [AERO ENG] The function of an afterburner. [MECH ENG] Combustion in an internal combustion engine following the maximum pressure of explosion.

afterburnt [AERO ENG] Descriptive of the condition following the complete transformation of the solid propellant to gaseous form.

aftercondenser [MECH ENG] A condenser in the second stage of a two-stage ejector; used in steam power plants, refrigeration systems, and air conditioning systems.

aftercooler [MECH ENG] A heat exchanger which cools air that has been compressed; used on turbocharged engines.

aftercooling [MECH ENG] The cooling of a gas after its compression.

afterfilter [MECH ENG] In an air-conditioning system, a high-efficiency filter located near a terminal unit. Also known as final filter.

afterflaming [AERO ENG] With liquid- or solid-propellant rocket thrust chambers, a characteristic low-grade combustion that takes place in the thrust chamber assembly and around its nozzle exit after the main propellant flow has been stopped.

AGE *See* aerospace ground equipment.

Agena rocket [AERO ENG] A liquid-fuel, upper-stage rocket usually used with a first-stage Atlas or Thor booster in certain space satellite projects.

agitating speed [MECH ENG] The rate of rotation of the drum or blades of a truck mixer or other device used for agitation of mixed concrete.

agitating truck [MECH ENG] A vehicle carrying a drum or agitator body, in which freshly mixed concrete can be conveyed from the point of mixing to that of placing, the drum being rotated continuously to agitate the contents.

agitator [MECH ENG] A device for keeping liquids and solids in liquids in motion by mixing, stirring, or shaking.

agitator body [MECH ENG] A truck-mounted drum for transporting freshly mixed concrete; rotation of internal paddles or of the drum prevents the setting of the mixture prior to delivery.

aileron [AERO ENG] The hinged rear portion of an aircraft wing moved differentially on each side of the aircraft to obtain lateral or roll control moments. [ARCH] A half gable, such as that which closes the end of a penthouse roof or of a church aisle.

air base [AERO ENG] 1. In the U.S. Air Force, an establishment, comprising an airfield, its installations, facilities, personnel, and activities, for the flight operation, maintenance, and supply of aircraft and air organizations. 2. A similar establishment belonging to any other air force. 3. In a restricted sense, only the physical installation.

air belt [MECH ENG] The chamber which equalizes the pressure that is blasted into the cupola at the tuyeres.

airboat *See* seaplane.

airborne [AERO ENG] Of equipment and material, carried or transported by aircraft.

air brake [MECH ENG] An energy-conversion mechanism activated by air pressure and used to retard, stop, or hold a vehicle or, generally, any moving element.

air breakup [AERO ENG] The breakup of a test reentry body within the atmosphere.

air-breathing [MECH ENG] Of an engine or aerodynamic vehicle, required to take in air for the purpose of combustion.

air cap [MECH ENG] A device used in thermal spraying which directs the air pattern for purposes of atomization.

air chamber [MECH ENG] A pressure vessel, partially filled with air, for converting pulsating flow to steady flow of water in a pipeline, as with a reciprocating pump.

air classifier [MECH ENG] A device to separate particles by size through the action of a stream of air.

air compressor [MECH ENG] A machine that increases the pressure of air by increasing its density and delivering the fluid against the connected system resistance on the discharge side.

air-compressor unloader [MECH ENG] A device for control of air volume flowing through an air compressor.

air-compressor valve [MECH ENG] A device for controlling the flow into or out of the cylinder of a compressor.

air condenser [MECH ENG] 1. A steam condenser in which the heat exchange occurs through metal walls separating the steam from cooling air. Also known as air-cooled condenser. 2. A device that removes vapors, such as of oil or water, from the airstream in a compressed-air line.

air conditioner [MECH ENG] A mechanism primarily for comfort cooling that lowers the temperature and reduces the humidity of air in buildings.

air conditioning [MECH ENG] The maintenance of certain aspects of the environment within a defined space to facilitate the function of that space; aspects controlled include air temperature and motion, radiant heat level, moisture, and concentration of pollutants such as dust, microorganisms, and gases. Also known as climate control.

air conveyor *See* pneumatic conveyor.

air-cooled condenser *See* air condenser.

air-cooled engine [MECH ENG] An engine cooled directly by a stream of air without the interposition of a liquid medium.

air-cooled heat exchanger [MECH ENG] A finned-tube (extended-surface) heat exchanger with hot fluids inside the tubes, and cooling air that is fan-blown (forced draft) or fan-pulled (induced draft) across the tube bank.

air cooling [MECH ENG] Lowering of air temperature for comfort, process control, or food preservation.

aircraft [AERO ENG] Any structure, machine, or contrivance, especially a vehicle, designed to be supported by the air, either by the dynamic action of the air upon the surfaces of the structure or object or by its own buoyancy. Also known as air vehicle.

aircraft axes *See* axes of an aircraft.

aircraft bonding [AERO ENG] Electrically connecting together all of the metal structure of the aircraft, including the engine and metal covering on the wiring.

aircraft engine [AERO ENG] A component of an aircraft that develops either shaft horsepower or thrust and incorporates design features advantageous for aircraft propulsion.

aircraft instrumentation [AERO ENG] Electronic, gyroscopic, and other instruments for detecting, measuring, recording, telemetering, processing, or analyzing different values or quantities in the flight of an aircraft.

aircraft instrument panel [AERO ENG] A coordinated instrument display arranged to provide the pilot and flight crew with information about the aircraft's speed, altitude, attitude, heading, and condition; also advises the pilot of the aircraft's response to his control efforts.

aircraft propeller [AERO ENG] A hub-and-multiblade device for transforming the rotational power of an aircraft engine into thrust power for the purpose of moving an aircraft through the air.

aircraft propulsion [AERO ENG] The means, other than gliding, whereby an aircraft moves through the air; effected by the rearward acceleration of matter through the use of a jet engine or by the reactive thrust of air on a propeller.

aircraft pylon [AERO ENG] A suspension device externally installed under the wing or fuselage of an aircraft; it is aerodynamically designed to fit the configuration of specific aircraft so as to create the least amount of drag; it provides a means of attaching fuel tanks, bombs, rockets, torpedoes, rocket motors, or machine-gun pods.

aircraft testing [AERO ENG] The subjecting of an aircraft or its components to simulated or actual flight conditions while measuring and recording pertinent physical phenomena that indicate operating characteristics.

air current [FL MECH] Very generally, any moving stream of air.

air curtain [MECH ENG] A stream of high-velocity temperature-controlled air which is directed downward across an opening; it excludes insects, exterior drafts, and so forth, prevents the transfer of heat across it, and permits air-conditioning of a space with an open entrance.

air cushion [MECH ENG] A mechanical device using trapped air to arrest motion without shock.

air-cushion vehicle [MECH ENG] A transportation device supported by low-pressure, low-velocity air capable of traveling equally well over water, ice, marsh, or relatively level land. Also known as ground-effect machine (GEM); hovercraft.

air cycle [MECH ENG] A refrigeration cycle characterized by the working fluid, air, remaining as a gas throughout the cycle rather than being condensed to a liquid; used primarily in airplane air conditioning.

air cylinder [MECH ENG] A cylinder in which air is compressed by a piston, compressed air is stored, or air drives a piston.

air density [MECH] The mass per unit volume of air.

airdraulic [MECH ENG] Combining pneumatic and hydraulic action for operation.

air drill [MECH ENG] A drill powered by compressed air.

air ejector [MECH ENG] A device that uses a fluid jet to remove air or other gases, as from a steam condenser.

air eliminator [MECH ENG] In a piping system, a device used to remove air from water, steam, or refrigerant.

air engine [MECH ENG] An engine in which compressed air is the actuating fluid.

air-exhaust ventilator [MECH ENG] Any air-exhaust unit used to carry away dirt particles, odors, or fumes.

airflow [FL MECH] 1. A flow or stream of air which may take place in a wind tunnel or, as a relative airflow, past the wing or other parts of a moving craft. Also known as airstream. 2. A rate of flow, measured by mass or volume per unit of time.

airflow stack effect [FL MECH] The variation of pressure with height in air flowing in a vertical duct due to a difference in temperature between the flowing air and the air outside the duct.

airfoil [AERO ENG] A body of such shape that the force exerted on it by its motion through a fluid has a larger component normal to the direction of motion than along the direction of motion; examples are the wing of an airplane and the blade of a propeller. Also known as aerofoil.

airfoil profile [AERO ENG] The closed curve defining the external shape of the cross section of an airfoil. Also known as airfoil section; airfoil shape; wing section.

airfoil section *See* airfoil profile.

airfoil shape *See* airfoil profile.

airframe [AERO ENG] The basic assembled structure of any aircraft or rocket vehicle, except lighter-than-air craft, necessary to support aerodynamic forces and inertia loads imposed by the weight of the vehicle and contents.

air-handling system [MECH ENG] An air-conditioning system in which an air-handling unit provides part of the treatment of the air.

air-handling unit [MECH ENG] A packaged assembly of air-conditioning components (coils, filters, fan humidifier, and so forth) which provides for the treatment of air before it is distributed.

air heater *See* air preheater.

air-heating system *See* air preheater.

air hoist [MECH ENG] A lifting tackle or tugger constructed with cylinders and pistons for reciprocating motion and air motors for rotary motion, all powered by compressed air. Also known as pneumatic hoist.

air horsepower [MECH ENG] The theoretical (minimum) power required to deliver the specified quantity of air under the specified pressure conditions in a fan, blower, compressor, or vacuum pump. Abbreviated air hp.

air hp *See* air horsepower.

air-injection system [MECH ENG] A device that uses compressed air to inject the fuel into the cylinder of an internal combustion engine.

air inlet [MECH ENG] In an air-conditioning system, a device through which air is exhausted from a room or building.

air intake [AERO ENG] An open end of an air duct or similar projecting structure so that the motion of the aircraft is utilized in capturing air to be conducted to an engine or ventilator.

air launch [AERO ENG] Launching from an aircraft in the air.

air leakage [MECH ENG] 1. In ductwork, air which escapes from a joint, coupling, and such. 2. The undesired leakage or uncontrolled passage of air from a ventilation system.

air lift [MECH ENG] 1. Equipment for lifting slurry or dry powder through pipes by means of compressed air. 2. *See* air-lift pump.

airlift [AERO ENG] 1. To transport passengers and cargo by the use of aircraft. 2. The total weight of personnel or cargo carried by air.

air-lift hammer [MECH ENG] A gravity drop hammer used in closed die forging in which the ram is raised to its starting point by means of an air cylinder.

air-lift pump [MECH ENG] A device composed of two pipes, one inside the other, used to extract water from a well; the lower end of the pipes is submerged, and air is delivered through the inner pipe to form a mixture of air and water which rises in the outer pipe above the water in the well; also used to move corrosive liquids, mill tailings, and sand. Also known as air lift.

air-line lubricator *See* line oiler.

air log [AERO ENG] A distance-measuring device used especially in certain guided missiles to control range.

air-mixing plenum [MECH ENG] In an air-conditioning system, a chamber in which the recirculating air is mixed with air from outdoors.

air motor [MECH ENG] A device in which the pressure of confined air causes the rotation of a rotor or the movement of a piston.

air outlet [MECH ENG] In an air-conditioning system, a device at the end of a duct through which air is supplied to a space.

airplane [AERO ENG] A heavier-than-air vehicle designed to use the pressures created by its motion through the air to lift and transport useful loads.

air preheater [MECH ENG] A device used in steam boilers to transfer heat from the flue gases to the combustion air before the latter enters the furnace. Also known as air heater; air-heating system.

air-pressure drop [FL MECH] The pressure lost in overcoming friction along an airway.

air propeller [AERO ENG] A hub-and-multiblade device for changing rotational power of an aircraft engine into thrust power for the purpose of propelling an aircraft through the air. [MECH ENG] A rotating fan for moving air.

air pump [MECH ENG] A device for removing air from an enclosed space or for adding air to an enclosed space.

air purge [MECH ENG] Removal of particulate matter from air within an enclosed vessel by means of air displacement.

air regulator [MECH ENG] A device for regulating airflow, as in the burner of a furnace.

air reheater [MECH ENG] In a heating system, any device used to add heat to the air circulating in the system.

air release valve [MECH ENG] A valve, usually manually operated, which is used to release air from a water pipe or fitting.

air resistance [MECH] Wind drag giving rise to forces and wear on buildings and other structures.

air scoop [DES ENG] An air-duct cowl projecting from the outer surface of an aircraft or automobile, which is designed to utilize the dynamic pressure of the airstream to maintain a flow of air.

air screw [MECH ENG] A screw propeller that operates in air.

air separator [MECH ENG] A device that uses an air current to separate a material from another of greater density or particles from others of greater size.

airship [AERO ENG] A propelled and steered aerial vehicle, dependent on gases for flotation.

Airslide conveyor [MECH ENG] An air-activated gravity-type conveyor, of the Fuller Company, using low-pressure air to aerate or fluidize pulverized material to a degree which will permit it to flow on a slight incline by the force of gravity.

airspace [AERO ENG] 1. The space occupied by an aircraft formation or used in a maneuver. 2. The area around an airplane in flight that is considered an integral part of the plane in order to prevent collision with another plane; the space depends on the speed of the plane.

airspeed [AERO ENG] The speed of an airborne object relative to the atmosphere; in a calm atmosphere, airspeed equals ground speed.

air spring [MECH ENG] A spring in which the energy storage element is air confined in a container that includes an elastomeric bellows or diaphragm.

air stack [AERO ENG] A group of planes flying at prescribed heights while waiting to land at an airport.

air-standard cycle [THERMO] A thermodynamic cycle in which the working fluid is considered to be a perfect gas with such properties of air as a volume of 12.4 ft³/lb at 14.7 psi (approximately 0.7756 m³/kg at 101.36 kPa) and 492°R and a ratio of specific heats of 1:4.

air-standard engine [MECH ENG] A heat engine operated in an air-standard cycle.

airstart [AERO ENG] An act or instance of starting an aircraft's engine while in flight, especially a jet engine after flameout.

air starting valve [MECH ENG] A device that admits compressed air to an air starter.

airstream *See* airflow.

air strip *See* landing strip.

air-suspension system [MECH ENG] Parts of an automotive vehicle that are intermediate between the wheels and the frame, and support the car body and frame by means of a cushion of air to absorb road shock caused by passage of the wheels over irregularities.

air system [MECH ENG] A mechanical refrigeration system in which air serves as the refrigerant in a cycle comprising compressor, heat exchanger, expander, and refrigerating core.

air taxi [AERO ENG] A carrier of passengers and cargo engaged in charter flights, feeder air services to large airline facilities, or contract airmail transportation.

air transportation [AERO ENG] The use of aircraft, predominantly airplanes, to move passengers and cargo.

air valve [MECH ENG] A valve that automatically lets air out of or into a liquid-carrying pipe when the internal pressure drops below atmospheric.

air vane [AERO ENG] A vane that acts in the air, as contrasted to a jet vane which acts within a jetstream.

air vehicle *See* aircraft.

air-velocity measurement [FL MECH] The measurement of the rate of displacement of air or gas at a specific location, as when ascertaining wind speed or airspeed of an aircraft.

air washer [MECH ENG] 1. A device for cooling and cleaning air in which the entering warm, moist air is cooled below its dew point by refrigerated water so that although the air leaves close to saturation with water, it has less moisture per unit volume than when it entered. 2. Apparatus to wash particulates and soluble impurities from air by passing the airstream through a liquid bath or spray.

Airy stress function [MECH] A biharmonic function of two variables whose second partial derivatives give the stress components of a body subject to a plane strain.

alignment pin [DES ENG] Pin in the center of the base of an octal, loctal, or other tube having a single vertical projecting rib that aids in correctly inserting the tube in its socket.

all-burnt time [AERO ENG] The point in time at which a rocket has consumed its propellants.

all-burnt velocity *See* burnout velocity.

Allen screw [DES ENG] A screw or bolt which has an axial hexagonal socket in its head.

Allen wrench [DES ENG] A wrench made from a straight or bent hexagonal rod, used to turn an Allen screw.

alligator shears *See* lever shears.

alligator wrench [DES ENG] A wrench having fixed jaws forming a V, with teeth on one or both jaws.

allowable load [MECH] The maximum force that may be safely applied to a solid, or is permitted by applicable regulators.

allowable stress [MECH] The maximum force per unit area that may be safely applied to a solid.

allowance [DES ENG] An intentional difference in sizes of two mating parts, allowing clearance usually for a film of oil, for running or sliding fits.

all-weather aircraft [AERO ENG] Aircraft that are designed or equipped to perform by day or night under any weather conditions.

all-weather fighter [AERO ENG] A fighter aircraft equipped with radar and other special devices which enable it to intercept its target in the dark, or in daylight weather conditions that do not permit visual interception; it is usually a multiplace (pilot plus navigator-observer) airplane.

alternating stress [MECH] A stress produced in a material by forces which are such that each force alternately acts in opposite directions.

altitude valve [AERO ENG] A valve that adjusts the composition of the air-fuel mixture admitted into an airplane carburetor as the air density varies with altitude.

altitude wind tunnel [AERO ENG] A wind tunnel in which the air pressure, temperature, and humidity can be varied to simulate conditions at different altitudes.

ambient pressure [FL MECH] The pressure of the surrounding medium, such as a gas or liquid, which comes into contact with an apparatus or with a reaction.

American standard pipe thread [DES ENG] Taper, straight, or dryseal pipe thread whose dimensions conform to those of a particular series of specified sizes established as a standard in the United States. Also known as Briggs pipe thread.

American standard screw thread [DES ENG] Screw thread whose dimensions conform to those of a particular series of specified sizes established as a standard in the United States; used for bolts, nuts, and machine screws.

American system drill *See* churn drill.

ammonia absorption refrigerator [MECH ENG] An absorption-cycle refrigerator which uses ammonia as the circulating refrigerant.

ammonia compressor [MECH ENG] A device that decreases the volume of a quantity of gaseous ammonia by the amplification of pressure; used in refrigeration systems.

ammonia condenser [MECH ENG] A device in an ammonia refrigerating system that raises the pressure of the ammonia gas in the evaporating coil, conditions the ammonia, and delivers it to the condensing system.

amphibious [MECH ENG] Said of vehicles or equipment designed to be operated or used on either land or water.

analog output [CONT SYS] Transducer output in which the amplitude is continuously proportional to a function of the stimulus.

analytic mechanics [MECH] The application of differential and integral calculus to classical (nonquantum) mechanics.

analyzer [MECH ENG] The component of an absorption refrigeration system where the mixture of water vapor and ammonia vapor leaving the generator meets the relatively cool solution of ammonia in water entering the generator and loses some of its vapor content.

anchor [MECH ENG] A vehicle used in steam plowing and located on the side of the field opposite that of the engine while maintaining the tension on the endless wire by means of a pulley.

anchor and collar [DES ENG] A door or gate hinge whose socket is attached to an anchor embedded in the masonry.

anchor nut [DES ENG] A nut in the form of a tapped insert forced under steady pressure into a hole in sheet metal.

Andrade's creep law [MECH] A law which states that creep exhibits a transient state in which strain is proportional to the cube root of time and then a steady state in which strain is proportional to time.

anelasticity [MECH] Deviation from a proportional relationship between stress and strain.

aneroid valve [MECH ENG] A valve actuated or controlled by an aneroid capsule.

angle back-pressure valve [MECH ENG] A back-pressure valve with its outlet opening at right angles to its inlet opening.

angle board [DES ENG] A board whose surface is cut at a desired angle; serves as a guide for cutting or planing other boards at the same angle.

angle collar [DES ENG] A cast-iron pipe fitting which has a socket at each end for joining with the spigot ends of two pipes that are not in alignment.

angle divider [DES ENG] A square for setting or bisecting angles; one side is an adjustable hinged blade.

Angledozer [MECH ENG] A power-operated machine fitted with a blade, adjustable in height and angle, for pushing, sidecasting, and spreading loose excavated material as for opencast pits, clearing land, or leveling runways. Also known as angling dozer.

angle gear *See* angular gear.

angle of action [MECH ENG] The angle of revolution of either of two wheels in gear during which any particular tooth remains in contact.

angle of advance *See* angular advance.

angle of approach [MECH ENG] The angle that is turned through by either of paired wheels in gear from the first contact between a pair of teeth until the pitch points of these teeth fall together.

angle of attack [AERO ENG] The angle between a reference line which is fixed with respect to an airframe (usually the longitudinal axis) and the direction of movement of the body.

angle of bite *See* angle of nip.

angle of cant [AERO ENG] In a spin-stabilized rocket, the angle formed by the axis of a venturi tube and the longitudinal axis of the rocket.

angle of climb [AERO ENG] The angle between the flight path of a climbing vehicle and the local horizontal.

angle of contact [FL MECH] The angle between the surface of a liquid and the surface of a partially submerged object or of the container at the line of contact. Also known as contact angle.

angle of departure [AERO ENG] The vertical angle, at the origin, between the line of site and the line of departure.

angle of descent [AERO ENG] The angle between the flight path of a descending vehicle and the local horizontal.

angle of fall [MECH] The vertical angle at the level point, between the line of fall and the base of the trajectory.

angle of friction *See* angle of repose.

angle of glide [AERO ENG] Angle of descent for an airplane or missile in a glide.

angle of impact [MECH] The acute angle between the tangent to the trajectory at the point of impact of a projectile and the plane tangent to the surface of the ground or target at the point of impact.

angle of nip [MECH ENG] The largest angle that will just grip a lump between the jaws, rolls, or mantle and ring of a crusher. Also known as angle of bite; nip.

angle of obliquity *See* angle of pressure.

angle of orientation [MECH] Of a projectile in flight, the angle between the plane determined by the axis of the projectile and the tangent to the trajectory (direction of motion), and the vertical plane including the tangent to the trajectory.

angle of pitch [AERO ENG] The angle, as seen from the side, between the longitudinal body axis of an aircraft or similar body and a chosen reference line or plane, usually the horizontal plane.

angle of pressure [DES ENG] The angle between the profile of a gear tooth and a radial line at its pitch point. Also known as angle of obliquity.

angle of recess [MECH ENG] The angle that is turned through by either of two wheels in gear, from the coincidence of the pitch points of a pair of teeth until the last point of contact of the teeth.

angle of repose [MECH] The angle between the horizontal and the plane of contact between two bodies when the upper body is just about to slide over the lower. Also known as angle of friction.

angle of roll [AERO ENG] The angle that the lateral body axis of an aircraft or similar body makes with a chosen reference plane in rolling; usually, the angle between the lateral axis and a horizontal plane.

angle of stall [AERO ENG] The angle of attack at which the flow of air begins to break away from the airfoil, the lift begins to decrease, and the drag begins to increase. Also known as stalling angle.

angle of thread [DES ENG] The angle occurring between the sides of a screw thread, measured in an axial plane.

angle of torsion [MECH] The angle through which a part of an object such as a shaft or wire is rotated from its normal position when a torque is applied. Also known as angle of twist.

angle of twist See angle of torsion.

angle of yaw [AERO ENG] The angle, as seen from above, between the longitudinal body axis of an aircraft, a rocket, or the like and a chosen reference direction. Also known as yaw angle.

angle plate [DES ENG] An L-shaped plate or a plate having an angular section.

angle press [MECH ENG] A hydraulic plastics-molding press with both horizontal and vertical rams; used to produce complex moldings with deep undercuts.

angle valve [DES ENG] A manually operated valve with its outlet opening oriented at right angles to its inlet opening; used for regulating the flow of a fluid in a pipe.

angle variable [MECH] The dynamical variable w conjugate to the action variable J, defined only for periodic motion.

angling dozer See Angledozer.

angstrom [MECH] A unit of length, 10^{-10} meter, used primarily to express wavelengths of optical spectra. Abbreviated A; Å. Also known as tenthmeter.

angular acceleration [MECH] The time rate of change of angular velocity.

angular advance [MECH ENG] The amount by which the angle between the crank of a steam engine and the virtual crank radius of the eccentric exceeds a right angle. Also known as angle of advance; angular lead.

angular bitstalk See angular bitstock.

angular bitstock [MECH ENG] A bitstock whose handles are positioned to permit its use in corners and other cramped areas. Also known as angular bitstalk.

angular clearance [DES ENG] The relieved space located below the straight of a die, to permit passage of blanks or slugs.

angular-contact bearing [MECH ENG] A rolling-contact antifriction bearing designed to carry heavy thrust loads and also radial loads.

angular cutter [MECH ENG] A tool-steel cutter used for finishing surfaces at angles greater or less than 90° with its axis of rotation.

angular gear [MECH ENG] A gear that transmits motion between two rotating shafts that are not parallel. Also known as angle gear.

angular impulse [MECH] The integral of the torque applied to a body over time.

angular lead See angular advance.

angular length [MECH] A length expressed in the unit of the length per radian or degree of a specified wave.

angular milling [MECH ENG] Milling surfaces that are flat and at an angle to the axis of the spindle of the milling machine.

angular momentum [MECH] 1. The cross product of a vector from a specified reference point to a particle, with the particle's linear momentum. Also known as moment of momentum. 2. For a system of particles, the vector sum of the angular momenta (first definition) of the particles.

angular pitch [DES ENG] The angle determined by the length along the pitch circle of a gear between successive teeth.

angular rate *See* angular speed.

angular shear [MECH ENG] A shear effected by two cutting edges inclined to each other to reduce the force needed for shearing.

angular speed [MECH] Change of direction per unit time, as of a target on a radar screen, without regard to the direction of the rotation axis; in other words, the magnitude of the angular velocity vector. Also known as angular rate.

angular travel error [MECH] The error which is introduced into a predicted angle obtained by multiplying an instantaneous angular velocity by a time of flight.

angular velocity [MECH] The time rate of change of angular displacement.

animal power [MECH ENG] The time rate at which muscular work is done by a work animal, such as a horse, bullock, or elephant.

anker [MECH] A unit of capacity equal to 10 U.S. gallons (37.854 liters); used to measure liquids, especially honey, oil, vinegar, spirits, and wine.

annealing point [THERMO] The temperature at which the viscosity of a glass is $10^{13.0}$ poises. Also known as annealing temperature; 13.0 temperature.

annealing temperature *See* annealing point.

annular auger [DES ENG] A ring-shaped boring tool which cuts an annular channel, leaving the core intact.

annular effect [FL MECH] A phenomenon observed in the flow of fluid in a tube when its motion is alternating rapidly, as in the propagation of sound waves, in which the mean velocity rises progressing from the center of the tube toward the walls and then falls within a thin laminar boundary layer to zero at the wall itself.

annular gear [DES ENG] A gear having a cylindrical form.

annular nozzle [DES ENG] A nozzle with a ring-shaped orifice.

annulus [MECH ENG] A plate serving to protect or cover a machine.

anomalous viscosity *See* non-Newtonian viscosity.

antidrag [AERO ENG] 1. Describing structural members in an aircraft or missile that are designed or built to resist the effects of drag. 2. Referring to a force acting against the force of drag.

antifriction [MECH] Making friction smaller in magnitude. [MECH ENG] Employing a rolling contact instead of a sliding contact.

antifriction bearing [MECH ENG] Any bearing having the capability of reducing friction effectively.

anti-icing [AERO ENG] The prevention of the formation of ice upon any object, especially aircraft, by means of windshield sprayer, addition of antifreeze materials to the carburetor, or heating the wings and tail.

apical angle [MECH] The angle between the tangents to the curve outlining the contour of a projectile at its tip.

Apollo program [AERO ENG] The scientific and technical program of the United States that involved placing men on the moon and returning them safely to earth.

apothecaries' dram *See* dram.

apothecaries' ounce *See* ounce.

apothecaries' pound *See* pound.

apparent additional mass [FL MECH] A fictitious mass of fluid added to the mass of the body to represent the force required to accelerate the body through the fluid.

apparent expansion [THERMO] The expansion of a liquid with temperature, as measured in a graduated container without taking into account the container's expansion.

apparent force [MECH] A force introduced in a relative coordinate system in order that Newton's laws be satisfied in the system; examples are the Coriolis force and the centrifugal force incorporated in gravity.

apparent gravity *See* acceleration of gravity.

apparent motion *See* relative motion.

apparent weight [MECH] For a body immersed in a fluid (such as air), the resultant of the gravitational force and the buoyant force of the fluid acting on the body; equal in magnitude to the true weight minus the weight of the displaced fluid.

applications technology satellite [AERO ENG] Any artificial satellite in the National Aeronautics and Space Administration program for the evaluation of advanced techniques and equipment for communications, meteorological, and navigation satellites. Abbreviated ATS.

approach [MECH ENG] The difference between the temperature of the water leaving a cooling tower and the wet-bulb temperature of the surrounding air.

apron [AERO ENG] A protective device specially designed to cover an area surrounding the fuel inlet on a rocket or spacecraft.

apron conveyor [MECH ENG] A conveyor used for carrying granular or lumpy material and consisting of two strands of roller chain separated by overlapping plates, forming the carrying surface, with sides 2–6 inches high.

apron feeder [MECH ENG] A limited-length version of apron conveyor used for controlled-rate feeding of pulverized materials to a process or packaging unit. Also known as plate-belt feeder; plate feeder.

APU *See* auxiliary power unit.

arbor hole [DES ENG] A hole in a revolving cutter or grinding wheel for mounting it on an arbor.

arbor press [MECH ENG] A machine used for forcing an arbor or a mandrel into drilled or bored parts preparatory to turning or grinding. Also known as mandrel press.

arc force [MECH] The force of a plasma arc through a nozzle or opening.

Archimedes number [FL MECH] One of a dimensionless group of numbers denoting the ratio of gravitational force to viscous force.

Archimedes' screw [MECH ENG] A device for raising water by means of a rotating broad-threaded screw or spirally bent tube within an inclined hollow cylinder.

arch press [MECH ENG] A punch press having an arch-shaped frame to permit operations on wide work.

arc jet engine [AERO ENG] An electromagnetic propulsion engine used to supply motive power for flight; hydrogen and ammonia are used as the propellant, and some plasma is formed as the result of electric-arc heating.

arc of action *See* arc of contact.

arc of approach [DES ENG] In toothed gearing, the part of the arc of contact along which the flank of the driving wheel contacts the face of the driven wheel.

arc of contact [MECH ENG] 1. The angular distance over which a gear tooth travels while it is in contact with its mating tooth. Also known as arc of action. 2. The angular distance a pulley travels while in contact with a belt or rope.

arc of recess [DES ENG] In toothed gearing, the part of the arc of contact wherein the face of the driving wheel touches the flank of the driven wheel.

are [MECH] A unit of area, used mainly in agriculture, equal to 100 square meters.

area rule [AERO ENG] A prescribed method of design for obtaining minimum zero-lift drag for a given aerodynamic configuration, such as a wing-body configuration, at a given speed.

Ariel [AERO ENG] A series of artificial satellites launched for Britain by the United States.

arm conveyor [MECH ENG] A conveyor in the form of an endless belt or chain to which are attached projecting arms or shelves which carry the materials.

arm elevator [MECH ENG] A chain elevator with protruding arms to cradle fixed-shape objects, such as drums or barrels, as they are moved upward.

armored faceplate [DES ENG] A tamper-proof faceplate or lock front, mortised in the edge of a door to cover the lock mechanism.

armored front [DES ENG] A lock front used on mortise locks that consists of two plates, the underplate and the finish plate.

arrester hook [AERO ENG] A hook in the tail section of an airplane; used to engage the arrester wires on an aircraft carrier's deck.

arrow wing [AERO ENG] An aircraft wing of V-shaped planform, either tapering or of constant chord, suggesting a stylized arrowhead.

articulated leader [MECH ENG] A wheel-mounted transport unit with a pivotal loading element used in earth moving.

artificial asteroid [AERO ENG] An object made by humans and placed in orbit about the sun.

artificial feel [AERO ENG] A type of force feedback incorporated in the control system of an aircraft or spacecraft whereby a portion of the forces acting on the control surfaces are transmitted to the cockpit controls.

artificial gravity [AERO ENG] A simulated gravity established within a space vehicle by rotation or acceleration.

artificial satellite [AERO ENG] Any man-made object placed in a near-periodic orbit in which it moves mainly under the gravitational influence of one celestial body, such as the earth, sun, another planet, or a planet's moon.

ascending branch [MECH] The portion of the trajectory between the origin and the summit on which a projectile climbs and its altitude constantly increases.

ascent [AERO ENG] Motion of a craft in which the path is inclined upward with respect to the horizontal.

ash conveyor [MECH ENG] A device that transports refuse from a furnace by fluid or mechanical means.

aspect ratio [AERO ENG] The ratio of the square of the span of an airfoil to the total airfoil area, or the ratio of its span to its mean chord. [DES ENG] 1. The ratio of frame width to frame height in television; it is 4:3 in the United States and Britain. 2. In any rectangular configuration (such as the cross section of a rectangular duct), the ratio of the longer dimension to the shorter.

asphalt cutter [MECH ENG] A powered machine having a rotating abrasive blade; used to saw through bituminous surfacing material.

assault aircraft [AERO ENG] Powered aircraft, including helicopters, which move assault troops and cargo into an objective area and which provide for their resupply.

assembly [MECH ENG] A unit containing the component parts of a mechanism, machine, or similar device.

assisted takeoff [AERO ENG] A takeoff of an aircraft or a missile by using a supplementary source of power, usually rockets.

astatic governor *See* isochronous governor.

ASTM-CFR engine [MECH ENG] A special engine developed by the Coordinating Fuel and Equipment Research Committee of the Coordinating Research Council, Inc., to determine the knock tendency of gasolines.

astragal front [DES ENG] A lock front which is shaped to fit the edge of a door with an astragal molding.

astral dome *See* astrodome.

astroballistics [MECH] The study of phenomena arising out of the motion of a solid through a gas at speeds high enough to cause ablation; for example, the interaction of a meteoroid with the atmosphere.

astrodome [AERO ENG] A transparent dome in the fuselage or body of an aircraft or spacecraft intended primarily to permit taking celestial observations in navigating. Also known as astral dome; navigation dome.

astrodynamics [AERO ENG] The practical application of celestial mechanics, astroballistics, propulsion theory, and allied fields to the problem of planning and directing the trajectories of space vehicles.

astronaut [AERO ENG] In United States terminology, a person who rides in a space vehicle.

astronautical engineering [AERO ENG] The engineering aspects of flight in space.

astronautics [AERO ENG] 1. The art, skill, or activity of operating spacecraft. 2. The science of space flight.

asymmetric rotor [MECH ENG] A rotating element for which the axis (center of rotation) is not centered in the element.

asymmetric top [MECH] A system that has all three principal moments of inertia different.

asynchronous control [CONT SYS] A method of control in which the time allotted for performing an operation depends on the time actually required for the operation, rather than on a predetermined fraction of a fixed machine cycle.

asynchronous device [CONT SYS] A device in which the speed of operation is not related to any frequency in the system to which it is connected.

at *See* technical atmosphere.

ata [MECH] A unit of absolute pressure in the metric technical system equal to 1 technical atmosphere.

Athey wheel [MECH ENG] A crawler wheel assembly used on tractors for moving over soft terrain.

athodyd [AERO ENG] A type of jet engine, consisting essentially of a duct or tube of varying diameter and open at both ends, which admits air at one end, compresses it by the forward motion of the engine, adds heat to it by the combustion of fuel, and discharges the resulting gases at the other end to produce thrust.

Atlas-Centaur launch vehicle [AERO ENG] A two-stage rocket consisting of an Atlas first stage and a Centaur second stage; used for launching uncrewed spacecraft.

atm *See* atmosphere.

atmolysis [FL MECH] The separation of gas mixtures by using their relative diffusibility through a porous partition.

atmosphere [MECH] A unit of pressure equal to 1.013250×10^6 dynes/cm², which is the air pressure measured at mean sea level. Abbreviated atm. Also known as standard atmosphere.

atmospheric braking [AERO ENG] 1. Slowing down an object entering the atmosphere of the earth or other planet from space by using the drag exerted by air or other gas particles in the atmosphere. 2. The action of the drag so exerted.

atmospheric cooler [MECH ENG] A fluids cooler that utilizes the cooling effect of ambient air surrounding the hot, fluids-filled tubes.

atmospheric drag [FL MECH] A major perturbation of close artificial satellite orbits caused by the resistance of the atmosphere; the secular effects are decreasing eccentricity, semidiameter, and period.

atmospheric entry [AERO ENG] The penetration of any planetary atmosphere by any object from outer space; specifically, the penetration of the earth's atmosphere by a crewed or uncrewed capsule or spacecraft.

atomic rocket [AERO ENG] A rocket propelled by an engine in which the energy for the jetstream is to be generated by nuclear fission or fusion. Also known as nuclear rocket.

atomization [MECH ENG] The mechanical subdivision of a bulk liquid or meltable solid, such as certain metals, to produce drops, which vary in diameter depending on the process from under 10 to over 1000 micrometers.

atomizer [MECH ENG] A device that produces a mechanical subdivision of a bulk liquid, as by spraying, sprinkling, misting, or nebulizing.

atomizer burner [MECH ENG] A liquid-fuel burner that atomizes the unignited fuel into a fine spray as it enters the combustion zone.

atomizer mill [MECH ENG] A solids grinder, the product from which is a fine powder.

atomizing humidifier [MECH ENG] A humidifier in which tiny particles of water are introduced into a stream of air.

ATS *See* applications technology satellite.

attached shock *See* attached shock wave.

attached shock wave [FL MECH] An oblique or conical shock wave that appears to be in contact with the leading edge of an airfoil or the nose of a body in a supersonic flow field. Also known as attached shock.

attack plane [AERO ENG] A multiweapon carrier aircraft which can carry bombs, torpedoes, and rockets.

attemperation of steam [MECH ENG] The controlled cooling, in a steam boiler, of steam at the superheater outlet or between the primary and secondary stages of the superheater to regulate the final steam temperature.

attitude [AERO ENG] The position or orientation of an aircraft, spacecraft, and so on, either in motion or at rest, as determined by the relationship between its axes and some reference line or plane or some fixed system of reference axes.

attitude control [AERO ENG] 1. The regulation of the attitude of an aircraft, spacecraft, and so on. 2. A device or system that automatically regulates and corrects attitude, especially of a pilotless vehicle.

attitude gyro [AERO ENG] Also known as attitude indicator. 1. A gyro-operated flight instrument that indicates the attitude of an aircraft or spacecraft with respect to a reference coordinate system. **2.** Any gyro-operated instrument that indicates attitude.

attitude indicator   *See* attitude gyro.

attitude jet   [AERO ENG] **1.** A stream of gas from a jet used to correct or alter the attitude of a flying body either in the atmosphere or in space. **2.** The nozzle that directs this jetstream.

attrition mill   [MECH ENG] A machine in which materials are pulverized between two toothed metal disks rotating in opposite directions.

Atwood machine   [MECH ENG] A device comprising a pulley over which is passed a stretch-free cord with a weight hanging on each end.

auger   [DES ENG] **1.** A wood-boring tool that consists of a shank with spiral channels ending in two spurs, a central tapered feed screw, and a pair of cutting lips. **2.** A large augerlike tool for boring into soil.

auger bit   [DES ENG] A bit shaped like an auger but without a handle; used for wood boring and for earth drilling.

auger conveyor   *See* screw conveyor.

auger packer   [MECH ENG] A feed mechanism that uses a continuous auger or screw inside a cylindrical sleeve to feed hard-to-flow granulated solids into shipping containers, such as bags or drums.

augmentation system   [AERO ENG] An electronic servomechanism or feedback control system which provides improvements in aircraft performance or pilot handling characteristics over that of the basic unaugmented aircraft.

augmenter tube   [AERO ENG] A tube or pipe, usually one of several, through which the exhaust gases from an aircraft reciprocating engine are directed to provide additional thrust.

austausch coefficient   *See* exchange coefficient.

autogenous grinding   [MECH ENG] The secondary grinding of material by tumbling the material in a revolving cylinder, without balls or bars taking part in the operation.

autogenous mill   *See* autogenous tumbling mill.

autogenous tumbling mill   [MECH ENG] A type of ball-mill grinder utilizing as the grinding medium the coarse feed (incoming) material. Also known as autogenous mill.

autogiro   [AERO ENG] A type of aircraft which utilizes a rotating wing (rotor) to provide lift and a conventional engine-propeller combination to pull the vehicle through the air.

autoignition   [MECH ENG] Spontaneous ignition of some or all of the fuel-air mixture in the combustion chamber of an internal combustion engine.

automatic batcher   [MECH ENG] A batcher for concrete which is actuated by a single starter switch, opens automatically at the start of the weighing operations of each material, and closes automatically when the designated weight of each material has been reached.

automatic check-out system   [CONT SYS] A system utilizing test equipment capable of automatically and simultaneously providing actions and information which will ultimately result in the efficient operation of tested equipment while keeping time to a minimum.

automatic control   [CONT SYS] Control in which regulating and switching operations are performed automatically in response to predetermined conditions. Also known as automatic regulation.

automatic-control block diagram [CONT SYS] A diagrammatic representation of the mathematical relationships defining the flow of information and energy through the automatic control system, in which the components of the control system are represented as functional blocks in series and parallel arrangements according to their position in the actual control system.

automatic-control error coefficient [CONT SYS] Three numerical quantities that are used as a measure of the steady-state errors of an automatic control system when the system is subjected to constant, ramp, or parabolic inputs.

automatic-control frequency response [CONT SYS] The steady-state output of an automatic control system for sinusoidal inputs of varying frequency.

automatic controller [CONT SYS] An instrument that continuously measures the value of a variable quantity or condition and then automatically acts on the controlled equipment to correct any deviation from a desired preset value. Also known as automatic regulator; controller.

automatic-control servo valve [CONT SYS] A mechanically or electrically actuated servo valve controlling the direction and volume of fluid flow in a hydraulic automatic control system.

automatic-control stability [CONT SYS] The property of an automatic control system whose performance is such that the amplitude of transient oscillations decreases with time and the system reaches a steady state.

automatic control system [CONT SYS] A control system having one or more automatic controllers connected in closed loops with one or more processes. Also known as regulating system.

automatic-control transient analysis [CONT SYS] The analysis of the behavior of the output variable of an automatic control system as the system changes from one steady-state condition to another in terms of such quantities as maximum overshoot, rise time, and response time.

automatic coupling [MECH ENG] A device which couples rail cars when they are bumped together.

automatic drill [DES ENG] A straight brace for bits whose shank comprises a coarse-pitch screw sliding in a threaded tube with a handle at the end; the device is operated by pushing the handle.

automatic fire pump [MECH ENG] A pump which provides the required water pressure in a fire standpipe or sprinkler system; when the water pressure in the system drops below a preselected value, a sensor causes the pump to start.

automatic press [MECH ENG] A press in which mechanical feeding of the work is synchronized with the press action.

automatic regulation *See* automatic control.

automatic regulator *See* automatic controller.

automatic sampler [MECH ENG] A mechanical device to sample process streams (gas, liquid, or solid) either continuously or at preset time intervals.

automatic screw machine [MECH ENG] A machine designed to automatically produce finished parts from bar stock at high production rates; the term is not an exact, specific machine-tool classification.

automatic stability [AERO ENG] Stability achieved with the controls operated by automatic devices, as by an automatic pilot.

automatic stabilization equipment [AERO ENG] Apparatus which automatically operates control devices to maintain an aircraft in a stable condition.

automatic stoker [MECH ENG] A device that supplies fuel to a boiler furnace by mechanical means. Also known as mechanical stoker.

automatic tuning system   [CONT SYS] An electrical, mechanical, or electromechanical system that tunes a radio receiver or transmitter automatically to a predetermined frequency when a button or lever is pressed, a knob turned, or a telephone-type dial operated.

automatic-type belt tensioning device   [MECH ENG] Any device which maintains a predetermined tension in a conveyor belt.

automechanism   [CONT SYS] A machine or other device that operates automatically or under control of a servomechanism.

automobile   [MECH ENG] A four-wheeled, trackless, self-propelled vehicle for land transportation of as many as eight people. Also known as car.

automobile chassis   [MECH ENG] The automobile frame, together with the wheels, power train, brakes, engine, and steering system.

automotive air conditioning   [MECH ENG] A system for maintaining comfort of occupants of automobiles, buses, and trucks, limited to air cooling, air heating, ventilation, and occasionally dehumidification.

automotive brake   [MECH ENG] A friction mechanism that slows or stops the rotation of the wheels of an automotive vehicle, so that tire traction slows or stops the vehicle.

automotive engine   [MECH ENG] The fuel-consuming machine that provides the motive power for automobiles, airplanes, tractors, buses, and motorcycles and is carried in the vehicle.

automotive engineering   [MECH ENG] The branch of mechanical engineering concerned primarily with the special problems of land transportation by a four-wheeled, trackless, automotive vehicle.

automotive ignition system   [MECH ENG] A device in an automotive vehicle which initiates the chemical reaction between fuel and air in the cylinder charge.

automotive steering   [MECH ENG] Mechanical means by which a driver controls the course of a moving automobile, bus, truck, or tractor.

automotive suspension   [MECH ENG] The springs and related parts intermediate between the wheels and frame of an automotive vehicle that support the frame on the wheels and absorb road shock caused by passage of the wheels over irregularities.

automotive transmission   [MECH ENG] A device for providing different gear or drive ratios between the engine and drive wheels of an automotive vehicle, a principal function being to enable the vehicle to accelerate from rest through a wide speed range while the engine operates within its most effective range.

automotive vehicle   [MECH ENG] A self-propelled vehicle or machine for land transportation of people or commodities or for moving materials, such as a passenger car, bus, truck, motorcycle, tractor, airplane, motorboat, or earthmover.

autopatrol   [MECH ENG] A self-powered blade grader. Also known as motor grader.

autorail   [MECH ENG] A self-propelled vehicle having both flange wheels and pneumatic tires to permit operation on both rails and roadways.

autorotation   [MECH] **1.** Rotation about any axis of a body that is symmetrical and exposed to a uniform airstream and maintained only by aerodynamic moments. **2.** Rotation of a stalled symmetrical airfoil parallel to the direction of the wind.

autosled   [MECH ENG] A propeller-driven machine equipped with runners and wheels and adaptable to use on snow, ice, or bare roads.

autostability   [CONT SYS] The ability of a device (such as a servomechanism) to hold a steady position, either by virtue of its shape and proportions, or by control by a servomechanism.

auxiliary dead latch [DES ENG] A supplementary latch in a lock which automatically deadlocks the main latch bolt when the door is closed. Also known as auxiliary latch bolt; deadlocking latch bolt; trigger bolt.

auxiliary fluid ignition [AERO ENG] A method of ignition of a liquid-propellant rocket engine in which a liquid that is hypergolic with either the fuel or the oxidizer is injected into the combustion chamber to initiate combustion.

auxiliary landing gear [AERO ENG] The part or parts of a landing gear, such as an outboard wheel, which is intended to stabilize the craft on the surface but which bears no significant part of the weight.

auxiliary power plant [MECH ENG] Ancillary equipment, such as pumps, fans, and soot blowers, used with the main boiler, turbine, engine, waterwheel, or generator of a power-generating station.

auxiliary power unit [AERO ENG] A power unit carried on an aircraft or spacecraft which can be used in addition to the main sources of power. Abbreviated APU.

auxiliary rim lock [DES ENG] A secondary or extra lock that is surface-mounted on a door to provide additional security.

auxiliary rope-fastening device [MECH ENG] A device attached to an elevator car, to a counterweight, or to the overhead dead-end rope-hitch support, that automatically supports the car or counterweight in case the fastening for the wire rope (cable) fails.

available draft [MECH ENG] The usable differential pressure in the combustion air in a furnace, used to sustain combustion of fuel or to transport products of combustion.

available energy [MECH ENG] Energy which can in principle be converted to mechanical work.

available heat [MECH ENG] The heat per unit mass of a working substance that could be transformed into work in an engine under ideal conditions for a given amount of heat per unit mass furnished to the working substance.

aviation [AERO ENG] 1. The science and technology of flight through the air. 2. The world of airplane business and its allied industries.

avogram [MECH] A unit of mass, equal to 1 gram divided by the Avogadro number.

avoirdupois pound See pound.

avoirdupois weight [MECH] The system of units which has been commonly used in English-speaking countries for measurement of the mass of any substance except precious stones, precious metals, and drugs; it is based on the pound (approximately 453.6 grams) and includes the short ton (2000 pounds), long ton (2240 pounds), ounce (one-sixteenth pound), and dram (one-sixteenth ounce).

awl [DES ENG] A point tool with a short wooden handle used to mark surfaces and to make small holes, as in leather or wood.

ax [DES ENG] An implement consisting of a heavy metal wedge-shaped head with one or two cutting edges and a relatively long wooden handle; used for chopping wood and felling trees.

axes of an aircraft [AERO ENG] Three fixed lines of reference, usually centroidal and mutually perpendicular: the longitudinal axis, the normal or yaw axis, and the lateral or pitch axis. Also known as aircraft axes.

axhammer [DES ENG] An ax having one cutting edge and one hammer face.

axial fan [MECH ENG] A fan whose housing confines the gas flow to the direction along the rotating shaft at both the inlet and outlet.

axial flow [FL MECH] Flow of fluid through an axially symmetric device such that the direction of the flow is along the axis of symmetry. Also known as axisymmetric flow.

axial-flow compressor [MECH ENG] A fluid compressor that accelerates the fluid in a direction generally parallel to the rotating shaft.

axial-flow jet engine [AERO ENG] 1. A jet engine in which the general flow of air is along the longitudinal axis of the engine. 2. A turbojet engine that utilizes an axial-flow compressor and turbine.

axial-flow pump [MECH ENG] A pump having an axial-flow or propeller-type impeller; used when maximum capacity and minimum head are desired. Also known as propeller pump.

axial hydraulic thrust [MECH ENG] In single-stage and multistage pumps, the summation of unbalanced impeller forces acting in the axial direction.

axial jet [FL MECH] A flowing, turbulent stream which mixes with standing water in three dimensions.

axial load [MECH] A force with its resultant passing through the centroid of a particular section and being perpendicular to the plane of the section.

axial moment of inertia [MECH] For any object rotating about an axis, the sum of its component masses times the square of the distance to the axis.

axial rake [MECH ENG] The angle between the face of a blade of a milling cutter or reamer and a line parallel to its axis of rotation.

axial relief [MECH ENG] The relief behind the end cutting edge of a milling cutter.

axial runout [MECH ENG] The total amount, along the axis of rotation, by which the rotation of a cutting tool deviates from a plane.

axis [MECH] A line about which a body rotates.

axis of freedom [DES ENG] An axis in a gyro about which a gimbal provides a degree of freedom.

axis of rotation [MECH] A straight line passing through the points of a rotating rigid body that remain stationary, while the other points of the body move in circles about the axis.

axis of symmetry [MECH] An imaginary line about which a geometrical figure is symmetric. Also known as symmetry axis.

axisymmetric flow *See* axial flow.

axle [MECH ENG] A supporting member that carries a wheel and either rotates with the wheel to transmit mechanical power to or from it, or allows the wheel to rotate freely on it.

azimuth circle [DES ENG] A ring calibrated from 0 to 360 degrees over a compass, compass repeater, radar plan position indicator, direction finder, and so on, which provides means for observing compass bearings and azimuths.

B

B-47 *See* Stratojet.

B-52 *See* Stratofortress.

Babcock coefficient of friction [FL MECH] An approximation to the coefficient of friction for steam flowing in a circular pipe of diameter d inches, given by $0.0027(1 + 3.6/d)$.

backacter *See* backhoe.

back check [DES ENG] In a hydraulic door closer, a mechanism that slows the speed with which a door may be opened.

backdigger *See* backhoe.

back-draft damper [MECH ENG] A damper with blades actuated by gravity, permitting air to pass through them in one direction only.

back end *See* thrust yoke.

backflap hinge [DES ENG] A hinge having a flat plate or strap which is screwed to the face of a shutter or door. Also known as flap hinge.

backflow [FL MECH] Any flow in a direction opposite to the natural or intended direction of flow.

back gearing [MECH ENG] The technique of using gears on machine tools to obtain an increase in the number of speed changes that can be gotten with cone belt drives.

backhoe [MECH ENG] An excavator fitted with a hinged arm to which is rigidly attached a bucket that is drawn toward the machine in operation. Also known as backacter; backdigger; dragshovel; pullshovel.

backing pump [MECH ENG] A vacuum pump, in a vacuum system using two pumps in tandem, which works directly to the atmosphere and reduces the pressure to an intermediate value, usually between 100 and 0.1 pascals. Also known as fore pump.

backlash [DES ENG] The amount by which the tooth space of a gear exceeds the tooth thickness of the mating gear along the pitch circles.

back nut [DES ENG] **1.** A threaded nut, one side of which is dished to retain a grommet; used in forming a watertight pipe joint. **2.** A locking nut on the shank of a pipe fitting, tap, or valve.

backout [AERO ENG] An undoing of previous steps during a countdown, usually in reverse order.

backplate lamp holder [DES ENG] A lamp holder, integrally mounted on a plate, which is designed for screwing to a flat surface.

back pressure [MECH] Pressure due to a force that is operating in a direction opposite to that being considered, such as that of a fluid flow. [MECH ENG] Resistance transferred from rock into the drill stem when the bit is being fed at a faster rate than the bit can cut.

back rake [DES ENG] An angle on a single-point turning tool measured between the plane of the tool face and the reference plane.

backsaw [DES ENG] A fine-tooth saw with its upper edge stiffened by a metal rib to ensure straight cuts.

backspace [MECH ENG] To move a typewriter carriage back one space by depressing a back space key.

backward-bladed aerodynamic fan [MECH ENG] A fan that consists of several stream-lined blades mounted in a revolving casing.

badger [DES ENG] *See* badger plane.

badger plane [DES ENG] A hand plane whose mouth is cut obliquely from side to side, so that the plane can work close up to a corner. Also known as badger.

bailout [AERO ENG] The exiting from a flying aircraft and descending by parachute in an emergency.

bailout bottle [AERO ENG] A personal supply of oxygen usually contained in a cylinder under pressure and utilized when the individual has left the central oxygen system, as in a parachute jump.

Bairstow number [FL MECH] A term previously used for Mach number.

balance [AERO ENG] 1. The equilibrium attained by an aircraft, rocket, or the like when forces and moments are acting upon it so as to produce steady flight, especially without rotation about its axes. 2. The equilibrium about any specified axis that counterbalances something, especially on an aircraft control surface, such as a weight installed forward of the hinge axis to counterbalance the surface aft of the hinge axis.

balanced surface [AERO ENG] A control surface that extends on both sides of the axis of the hinge or pivot, or that has auxiliary devices or extensions connected with it, in such a manner as to effect a small or zero resultant moment of the air forces about the hinge axis.

balance tool [MECH ENG] A tool designed for taking the first cuts when the external surface of a piece in a lathe is being machined; it is supported in the tool holder at an unvarying angle.

balance wheel [MECH ENG] 1. A wheel which governs or stabilizes the movement of a mechanism. 2. *See* flywheel.

baler [MECH ENG] A machine which takes large quantities of raw or finished materials and binds them with rope or metal straps or wires into a large package.

ball [MECH ENG] In fine grinding, one of the crushing bodies used in a ball mill.

ball-and-race-type pulverizer [MECH ENG] A grinding machine in which balls rotate under an applied force between two races to crush materials, such as coal, to fine consistency. Also known as ball-bearing pulverizer.

ball-and-socket joint [MECH ENG] A joint in which a member ending in a ball is joined to a member ending in a socket so that relative movement is permitted within a certain angle in all planes passing through a line. Also known as ball joint.

ballast [AERO ENG] A relatively dense substance that is placed in the car of a vehicle and can be thrown out to reduce the load or can be shifted to change the center of gravity.

ball bearing [MECH ENG] An antifriction bearing permitting free motion between moving and fixed parts by means of balls confined between outer and inner rings.

ball-bearing hinge   [MECH ENG] A hinge which is equipped with ball bearings between the hinge knuckles in order to reduce friction.

ball-bearing pulverizer   *See* ball-and-race-type pulverizer.

ball bushing   [MECH ENG] A type of ball bearing that allows motion of the shaft in its axial direction.

ball catch   [DES ENG] A door fastener having a contained metal ball which is under pressure from a spring; the ball engages a striking plate and keeps the door from opening until force is applied.

ball float   [MECH ENG] A floating device, usually approximately spherical, which is used to operate a ball valve.

ball grinder   *See* ball mill.

ballhead   [MECH ENG] That part of the governor which contains flyweights whose force is balanced, at least in part, by the force of compression of a speeder spring.

ballistic coefficient   [MECH] The numerical measure of the ability of a missile to overcome air resistance; dependent upon the mass, diameter, and form factor.

ballistic conditions   [MECH] Conditions which affect the motion of a projectile in the bore and through the atmosphere, including muzzle velocity, weight of projectile, size and shape of projectile, rotation of the earth, density of the air, temperature or elasticity of the air, and the wind.

ballistic curve   [MECH] The curve described by the path of a bullet, a bomb, or other projectile as determined by the ballistic conditions, by the propulsive force, and by gravity.

ballistic deflection   [MECH] The deflection of a missile due to its ballistic characteristics.

ballistic density   [MECH] A representation of the atmospheric density encountered by a projectile in flight, expressed as a percentage of the density according to the standard artillery atmosphere.

ballistic efficiency   [MECH] 1. The ability of a projectile to overcome the resistance of the air; depends chiefly on the weight, diameter, and shape of the projectile. 2. The external efficiency of a rocket or other jet engine of a missile.

ballistic entry   [MECH] Movement of a ballistic body from without to within a planetary atmosphere.

ballistic limit   [MECH] The minimum velocity at which a particular armor-piercing projectile is expected to consistently and completely penetrate armor plate of given thickness and physical properties at a specified angle of obliquity.

ballistic measurement   [MECH] Any measurement in which an impulse is applied to a device such as the bob of a ballistic pendulum, or the moving part of a ballistic galvanometer, and the subsequent motion of the device is used to determine the magnitude of the impulse, and, from this magnitude, the quantity to be measured.

ballistics   [MECH] Branch of applied mechanics which deals with the motion and behavior characteristics of missiles, that is, projectiles, bombs, rockets, guided missiles, and so forth, and of accompanying phenomena.

ballistics of penetration   [MECH] That part of terminal ballistics which treats of the motion of a projectile as it forces its way into targets of solid or semisolid substances, such as earth, concrete, or steel.

ballistic table   [MECH] Compilation of ballistic data from which trajectory elements such as angle of fall, range to summit, time of flight, and ordinate at any time, can be obtained.

ballistic temperature [MECH] That temperature (in °F) which, when regarded as a surface temperature and used in conjunction with the lapse rate of the standard artillery atmosphere, would produce the same effect on a projectile as the actual temperature distribution encountered by the projectile in flight.

ballistic trajectory [MECH] The trajectory followed by a body being acted upon only by gravitational forces and resistance of the medium through which it passes.

ballistic uniformity [MECH] The capability of a propellant, when fired under identical conditions from round to round, to impart uniform muzzle velocity and produce similar interior ballistic results.

ballistic wave [MECH] An audible disturbance caused by compression of air ahead of a missile in flight.

ballistic wind [MECH] That constant wind which would produce the same effect upon the trajectory of a projectile as the actual wind encountered in flight.

ball joint *See* ball-and-socket joint.

ball mill [MECH ENG] A pulverizer that consists of a horizontal rotating cylinder, up to three diameters in length, containing a charge of tumbling or cascading steel balls, pebbles, or rods. Also known as ball grinder.

ballonet [AERO ENG] One of the air cells in a blimp, fastened to the bottom or sides of the envelope, which are used to maintain the required pressure in the envelope without adding or valving gas as the ship ascends or descends. Also spelled ballonnet.

balloon [AERO ENG] A nonporous, flexible spherical bag, inflated with a gas such as helium that is lighter than air, so that it will rise and float in the atmosphere; a large-capacity balloon can be used to lift a payload suspended from it.

balloon cover [AERO ENG] A cover which fits over a large, inflated balloon to facilitate handling in high or gusty winds. Also known as balloon shroud.

balloon shroud *See* balloon cover.

balloon-type rocket [AERO ENG] A liquid-fuel rocket, such as the Atlas, that requires the pressure of its propellants (or other gases) within it to give it structural integrity.

balloting [MECH] A tossing or bounding movement of a projectile, within the limits of the bore diameter, while moving through the bore under the influence of the propellant gases.

ballute [AERO ENG] A cross between a balloon and a parachute, used to brake the free fall of sounding rockets.

ball valve [MECH ENG] A valve in which the fluid flow is regulated by a ball moving relative to a spherical socket as a result of fluid pressure and the weight of the ball.

Banbury mixer [MECH ENG] Heavy-duty batch mixer, with two counterrotating rotors, for doughy material; used mainly in the plastics and rubber industries.

band [DES ENG] A strip or cord crossing the back of a book to which the sections are sewn.

band brake [MECH ENG] A brake in which the frictional force is applied by increasing the tension in a flexible band to tighten it around the drum.

band clamp [DES ENG] A two-piece metal clamp, secured by bolts at both ends; used to hold riser pipes.

band clutch [MECH ENG] A friction clutch in which a steel band, lined with fabric, contracts onto the clutch rim.

banding [DES ENG] A strip of fabric which is used for bands.

band saw [MECH ENG] A power-operated woodworking saw consisting basically of a flexible band of steel having teeth on one edge, running over two vertical pulleys, and operated under tension.

band wheel [MECH ENG] In a drilling operation, a large wheel that transmits power from the engine to the walking beam.

bang-bang control [CONT SYS] A type of automatic control system in which the applied control signals assume either their maximum or minimum values.

bank [AERO ENG] The lateral inward inclination of an airplane when it rounds a curve.

bank and turn indicator [AERO ENG] A device used to advise the pilot that the aircraft is turning at a certain rate, and that the wings are properly banked to preclude slipping or sliding of the aircraft as it continues in flight. Also known as bank indicator.

bank indicator See bank and turn indicator.

bar [MECH] A unit of pressure equal to 10^5 pascals, or 10^5 newtons per square meter, or 10^6 dynes per square centimeter.

barb bolt [DES ENG] A bolt having jagged edges to prevent its being withdrawn from the object into which it is driven. Also known as rag bolt.

bar clamp [DES ENG] A clamping device consisting of a long bar with adjustable clamping jaws; used in carpentry.

barker See bark spud.

bark spud [DES ENG] A tool which peels off bark. Also known as barker.

bar linkage [MECH ENG] A set of bars joined together at pivots by means of pins or equivalent devices; used to transmit power and information.

Barlow's equation [MECH] A formula, $t = DP/2S$, used in computing the strength of cylinders subject to internal pressures, where t is the thickness of the cylinder in inches, D the outside diameter in inches, P the pressure in pounds per square inch, and S the allowable tensile strength in pounds per square inch.

barodynamics [MECH] The mechanics of heavy structures which may collapse under their own weight.

barometric condenser [MECH ENG] A contact condenser that uses a long, vertical pipe into which the condensate and cooling liquid flow to accomplish their removal by the pressure created at the lower end of the pipe.

barometric draft regulator [MECH ENG] A damper usually installed in the breeching between a boiler and chimney; permits air to enter the breeching automatically as required, to maintain a constant overfire draft in the combustion chamber.

barometric fuel control [AERO ENG] A device that maintains the correct flow of fuel to an engine by adjusting to atmospheric pressure at different altitudes, as well as to impact pressure.

barometric gradient See pressure gradient.

baromil [MECH] The unit of length used in graduating a mercury barometer in the centimeter-gram-second system.

barrel [DES ENG] 1. A container having a circular lateral cross section that is largest in the middle, and ends that are flat; often made of staves held together by hoops. 2. A piece of small pipe inserted in the end of a cartridge to carry the squib to the powder. 3. That portion of a pipe having a constant bore and wall thickness. [MECH] Abbreviated bbl. 1. The unit of liquid volume equal to 31.5 gallons (approximately 119 liters). 2. The unit of liquid volume for petroleum equal to 42 gallons (approximately 158 liters). 3. The unit of dry volume equal to 105 quarts (approximately 116 liters). 4. A unit of weight that varies in size according to the commodity being weighed.

barrel bolt [DES ENG] A door bolt which moves in a cylindrical casing; not driven by a key. Also known as tower bolt.

barrel fitting [DES ENG] A short length of threaded connecting pipe.

barrelhead [DES ENG] The flat end of a barrel.

bar screen [MECH ENG] A sieve with parallel steel bars for separating small from large pieces of crushed rock.

bar strainer [DES ENG] A screening device consisting of a bar or a number or parallel bars; used to prevent objects from entering a drain.

bar turret lathe [MECH ENG] A turret lathe in which the bar stock is slid through the headstock and collet on line with the turning axis of the lathe and held firmly by the closed collet.

barycentric energy [MECH] The energy of a system in its center-of-mass frame.

barye [MECH] The pressure unit of the centimeter-gram-second system of physical units; equal to 1 dyne/cm^2 (0.001 millibar). Also known as microbar.

base circle [DES ENG] The circle on a gear such that each tooth-profile curve is an involute of it.

base drag [FL MECH] Drag owing to a base pressure lower than the ambient pressure; it is a part of the pressure drag.

base elbow [DES ENG] A cast-iron pipe elbow having a baseplate or flange which is cast on it and by which it is supported.

base plate [DES ENG] The part of a theodolite which carries the lower ends of the three foot screws and attaches the theodolite to the tripod for surveying.

base pressure [FL MECH] The pressure exerted on the base or extreme aft end of a body, as of a cylindrical or boat-tailed body or of a blunt-trailing-edge wing in fluid flow. [MECH] A pressure used as a reference base, for example, atmospheric pressure.

base tee [DES ENG] A pipe tee with a connected baseplate for supporting it.

basin [DES ENG] An open-top vessel with relatively low sloping sides for holding liquids.

basket [DES ENG] A lightweight container with perforations. [MECH ENG] A type of single-tube core barrel made from thin-wall tubing with the lower end notched into points, which is intended to pick up a sample of granular or plastic rock material by bending in on striking the bottom of the borehole or solid layer; may be used to recover an article dropped into a borehole. Also known as basket barrel; basket tube; sawtooth barrel.

basket barrel *See* basket.

basket tube *See* basket.

bastard-cut file [DES ENG] A file that has coarser teeth than a rough-cut file.

bastard thread [DES ENG] A screw thread that does not match any standard threads.

bat bolt [DES ENG] A bolt whose butt or tang is bashed or jagged.

batcher [MECH ENG] A machine in which the ingredients of concrete are measured and combined into batches before being discharged to the concrete mixer.

Batchinsky relation [FL MECH] The relation stating that the fluidity of a liquid is proportional to the difference between the specific volume and a characteristic specific volume, approximately equal to the specific volume appearing in the van der Waals equation.

batch mixer [MECH ENG] A machine which mixes concrete or mortar in batches, as opposed to a continuous mixer.

batch-type furnace [MECH ENG] A furnace used for heat treatment of materials, with or without direct firing; loading and unloading operations are carried out through a single door or slot.

batten [AERO ENG] Metal, wood, or plastic panels laced to the envelope of a blimp in the nose cone to add rigidity to the nose and provide a good point of attachment for mooring.

bay [AERO ENG] A space formed by structural partitions on an aircraft.

bayonet coupling [DES ENG] A coupling in which two or more pins extend out from a plug and engage in grooves in the side of a socket.

bayonet socket [DES ENG] A socket, having J-shaped slots on opposite sides, into which a bayonet base or coupling is inserted against a spring and rotated until its pins are seated firmly in the slots.

bayonet-tube exchanger [MECH ENG] A dual-tube apparatus with heating (or cooling) fluid flowing into the inner tube and out of the annular space between the inner and outer tubes; can be inserted into tanks or other process vessels to heat or cool the liquid contents.

BDC *See* bottom dead center.

bead [DES ENG] A projecting rim or band.

beaded tube end [MECH ENG] The exposed portion of a rolled tube which is rounded back against the sheet in which the tube is rolled.

beading plane [DES ENG] A plane having a curved cutting edge for shaping beads in wood. Also known as bead plane.

beam-deflection amplifier [MECH ENG] A jet-interaction fluidic device in which the direction of a supply jet is varied by flow from one or more control jets which are oriented at approximately 90° to the supply jet.

beam rider [AERO ENG] A missile for which the guidance system consists of standard reference signals transmitted in a radar beam which enable the missile to sense its location relative to the beam, correct its course, and thereby stay on the beam.

beam riding [AERO ENG] The maneuver of a spacecraft or other vehicle as it follows a beam.

bearing [MECH ENG] A machine part that supports another part which rotates, slides, or oscillates in or on it.

bearing capacity [MECH] Load per unit area which can be safely supported by the ground.

bearing pressure [MECH] Load on a bearing surface divided by its area. Also known as bearing stress.

bearing strain [MECH] The deformation of bearing parts subjected to a load.

bearing strength [MECH] The maximum load that a column, wall, footing, or joint will sustain at failure, divided by the effective bearing area.

bearing stress *See* bearing pressure.

beater [MECH ENG] A machine that cuts or beats paper stock.

beater mill *See* hammer mill.

Beattie and Bridgman equation [THERMO] An equation that relates the pressure, volume, and temperature of a real gas to the gas constant.

bêche [MECH ENG] A pneumatic forge hammer having an air-operated ram and an air-compressing cylinder integral with the frame.

bed [MECH ENG] The part of a machine having precisely machined ways or bearing surfaces which support or align other machine parts.

bell mouth [DES ENG] A flared mouth on a pipe opening or other orifice.

bellows expansion joint [DES ENG] In a run of piping, a joint formed with a flexible metal bellows which compress or stretch to compensate for linear expansion or contraction of the run of piping.

bellows seal [MECH ENG] A boiler seal in the form of a bellows which prevents leakage of air or gas.

belt [MECH ENG] A flexible band used to connect pulleys or to convey materials by transmitting motion and power.

belt conveyor [MECH ENG] A heavy-duty conveyor consisting essentially of a head or drive pulley, a take-up pulley, a level or inclined endless belt made of canvas, rubber, or metal, and carrying and return idlers.

belt drive [MECH ENG] The transmission of power between shafts by means of a belt connecting pulleys on the shafts.

belt feeder [MECH ENG] A short belt conveyor used to transfer granulated or powdered solids from a storage or supply point to an end-use point; for example, from a bin hopper to a chemical reactor.

belt sander [MECH ENG] A portable sanding tool having a power-driven abrasive-coated continuous belt.

belt shifter [MECH ENG] A device with fingerlike projections used to shift a belt from one pulley to another or to replace a belt which has slipped off a pulley.

belt slip [MECH ENG] The difference in speed between the driving drum and belt conveyor.

belt tightener [MECH ENG] In a belt drive, a device that takes up the slack in a belt that has become stretched and permanently lengthened.

bench lathe [MECH ENG] A small engine or toolroom lathe suitable for attachment to a workbench; bed length usually does not exceed 6 feet (1.8 meters) and workpieces are generally small.

bench plane [DES ENG] A plane used primarily in benchwork on flat surfaces, such as a block plane or jack plane.

bench sander [MECH ENG] A stationary power sander, usually mounted on a table or stand, which is equipped with a rotating abrasive disk or belt.

bend [DES ENG] **1.** The characteristic of an object, such as a machine part, that is curved. **2.** A section of pipe that is curved. **3.** A knot formed by a rope fastened to an object or another rope.

bend allowance [DES ENG] Length of the arc of the neutral axis between the tangent points of a bend in any material.

bender *See* bending machine.

bending brake [MECH ENG] A press brake for making sharply angular linear bends in sheet metal.

bending machine [MECH ENG] A machine for bending a metal or wooden part by pressure. Also known as bender.

bending moment [MECH] Algebraic sum of all moments located between a cross section and one end of a structural member; a bending moment that bends the beam convex downward is positive, and one that bends it convex upward is negative.

bending moment diagram [MECH] A diagram showing the bending moment at every point along the length of a beam plotted as an ordinate.

bending stress [MECH] An internal tensile or compressive longitudinal stress developed in a beam in response to curvature induced by an external load.

Bendix-Weiss universal joint [MECH ENG] A universal joint that provides for constant angular velocity of the driven shaft by transmitting the torque through a set of four balls lying in the plane that contains the bisector of, and is perpendicular to, the plane of the angle between the shafts.

bend radius [DES ENG] The radius corresponding to the curvature of a bent specimen or part, as measured at the inside surface of the bend.

bend wheel [MECH ENG] A wheel used to interrupt and change the normal path of travel of the conveying or driving medium; most generally used to effect a change in direction of conveyor travel from inclined to horizontal or a similar change.

bent-tube boiler [MECH ENG] A water-tube steam boiler in which the tubes terminate in upper and lower steam-and-water drums. Also known as drum-type boiler.

Bernoulli effect [FL MECH] As a consequence of the Bernoulli theorem, the pressure of a stream of fluid is reduced as its speed of flow is increased.

Bernoulli equation *See* Bernoulli theorem.

Bernoulli-Euler law [MECH] A law stating that the curvature of a beam is proportional to the bending moment.

Bernoulli law *See* Bernoulli theorem.

Bernoulli theorem [FL MECH] An expression of the conservation of energy in the steady flow of an incompressible, inviscid fluid; it states that the quantity $(p/\rho) + gz + (v^2/2)$ is constant along any streamline, where p is the fluid pressure, v is the fluid velocity, ρ is the mass density of the fluid, g is the acceleration due to gravity, and z is the vertical height. Also known as Bernoulli equation; Bernoulli law.

Betti's method [MECH] A method of finding the solution of the equations of equilibrium of an elastic body whose surface displacements are specified; it uses the fact that the dilatation is a harmonic function to reduce the problem to the Dirichlet problem.

Betti reciprocal theorem [MECH] A theorem in the mathematical theory of elasticity which states that if an elastic body is subjected to two systems of surface and body forces, then the work that would be done by the first system acting through the displacements resulting from the second system equals the work that would be done by the second system acting through the displacements resulting from the first system.

Betz momentum theory [MECH ENG] A theory of windmill performance that considers the deceleration in the air traversing the windmill disk.

bevel [DES ENG] **1.** The angle between one line or surface and another line or surface, or the horizontal, when this angle is not a right angle. **2.** A sloping surface or line.

bevel gear [MECH ENG] One of a pair of gears used to connect two shafts whose axes intersect.

beveling *See* chamfering.

Beveloid gear [DES ENG] An involute gear, of the Vinco Corporation, with tapered outside diameter, root diameter, and tooth thickness; used to reduce backlash in the drives of precision instruments.

bezel [DES ENG] **1.** A grooved rim used to hold a transparent glass or plastic window or lens for a meter, tuning dial, or some other indicating device. **2.** A sloping face on a cutting tool.

bhp *See* boiler horsepower; brake horsepower.

biaxial stress [MECH] The condition in which there are three mutually perpendicular principal stresses; two act in the same plane and one is zero.

bibb cock *See* bibcock.

bibcock [DES ENG] A faucet or stopcock whose nozzle is bent downward. Also spelled bibb cock.

bicable tramway [MECH ENG] A tramway consisting of two stationary cables on which the wheeled carriages travel, and an endless rope, which propels the carriages.

bicron *See* stigma.

biface tool [DES ENG] A tool, as an ax, made from a coil flattened on both sides to form a V-shaped cutting edge.

bifacial [DES ENG] Of a tool, having both sides alike.

bifilar micrometer *See* filar micrometer.

bilateral tolerance [DES ENG] The amount that the size of a machine part is allowed to vary above or below a basic dimension; for example, 3.650 ± 0.003 centimeters indicates a tolerance of ± 0.003 centimeter.

bill [DES ENG] One blade of a pair of scissors.

Bingham number [FL MECH] A dimensionless number used to study the flow of Bingham plastics.

Bingham plastic [FL MECH] A non-Newtonian fluid exhibiting a yield stress which must be exceeded before flow starts; thereafter the rate-of-shear versus shear stress curve is linear.

biosatellite [AERO ENG] An artificial satellite designed to contain and support humans, animals, or other living material in a reasonably normal manner for a period of time and to return safely to earth.

Biot-Fourier equation [THERMO] An equation for heat conduction which states that the rate of change of temperature at any point divided by the thermal diffusivity equals the Laplacian of the temperature.

Biot number [FL MECH] A dimensionless group, used in the study of mass transfer between a fluid and a solid, which gives the ratio of the mass-transfer rate at the interface to the mass-transfer rate in the interior of a solid wall of specified thickness.

biplane [AERO ENG] An aircraft with two wings fixed at different levels, especially one above and one below the fuselage.

Birmingham wire gage [DES ENG] A system of standard sizes of brass wire, telegraph wire, steel tubing, seamless tubing, sheet spring steel, strip steel, and steel plates, bands, and hoops. Abbreviated BWG.

bit [DES ENG] 1. A machine part for drilling or boring. 2. The cutting plate of a plane. 3. The blade of a cutting tool such as an ax. 4. A removable tooth of a saw. 5. Any cutting device which is attached to or part of a drill rod or drill string to bore or penetrate rocks.

bit blank [DES ENG] A steel bit in which diamonds or other cutting media may be inset by hand peening or attached by a mechanical process such as casting, sintering, or brazing. Also known as bit shank; blank; blank bit; shank.

bit drag [DES ENG] A rotary-drilling bit that has serrated teeth. Also known as drag bit.

bit shank *See* bit blank.

blackbody [THERMO] An ideal body which would absorb all incident radiation and reflect none. Also known as hohlraum; ideal radiator.

blackbody radiation [THERMO] The emission of radiant energy which would take place from a blackbody at a fixed temperature; it takes place at a rate expressed by the Stefan-Boltzmann law, with a spectral energy distribution described by Planck's equation.

blackbody temperature [THERMO] The temperature of a blackbody that emits the same amount of heat radiation per unit area as a given object; measured by a total radiation pyrometer. Also known as brightness temperature.

black-surface enclosure [THERMO] An enclosure for which the interior surfaces of the walls possess the radiation characteristics of a blackbody.

blacktop paver [MECH ENG] A construction vehicle that spreads a specified thickness of bituminous mixture over a prepared surface.

bladder press [MECH ENG] A machine which simultaneously molds and cures (vulcanizes) a pneumatic tire.

blade loading [AERO ENG] A rotor's thrust in a rotary-wing aircraft divided by the total area of the rotor blades.

Blake jaw crusher [MECH ENG] A crusher with one fixed jaw plate and one pivoted at the top so as to give the greatest movement on the smallest lump.

Blake number [FL MECH] A dimensionless number used in the study of beds of particles.

blank *See* bit blank.

blank bit *See* bit blank.

blankholder slide [MECH ENG] The outer slide of a double-action power press; it is usually operated by toggles or cams.

Blasius theorem [AERO ENG] A theorem that provides formulas for finding the force and moment on the airfoil profiler.

blast chamber [AERO ENG] A combustion chamber, especially in a gas-turbine, jet, or rocket engine.

blast deflector [AERO ENG] A device used to divert the exhaust of a rocket fired from a vertical position.

blast-off [AERO ENG] The takeoff of a rocket or missile.

BLC *See* boundary-layer control.

bleeder turbine [MECH ENG] A multistage turbine where steam is extracted (bled) at pressures intermediate between throttle and exhaust, for process or feedwater heating purposes.

bleeding cycle [MECH ENG] A steam cycle in which steam is drawn from the turbine at one or more stages and used to heat the feedwater. Also known as regenerative cycle.

blimp [AERO ENG] A name originally applied to nonrigid, pressure-type airships, usually of small size; now applied to airships with volumes of approximately 1,500,000 cubic feet (42,000 cubic meters).

blind controller system [CONT SYS] A process control arrangement that separates the in-plant measuring points (for example, pressure, temperature, and flow rate) and control points (for example, a valve actuator) from the recorder or indicator at the central control panel.

blind hole [DES ENG] A hole which does not pass completely through a workpiece.

blind landing [AERO ENG] Landing an aircraft solely by the use of instruments because of poor visibility.

blind nipple [MECH ENG] A short piece of piping or tubing having one end closed off; commonly used in boiler construction.

blink [MECH] A unit of time equal to 10^{-5} day or to 0.864 second.

block [DES ENG] **1.** A metal or wood case enclosing one or more pulleys; has a hook with which it can be attached to an object. **2.** *See* cylinder block.

block and fall *See* block and tackle.

block and tackle [MECH ENG] Combination of a rope or other flexible material and independently rotating frictionless pulleys. Also known as block and fall.

block brake [MECH ENG] A brake which consists of a block or shoe of wood bearing upon an iron or steel wheel.

block plane [DES ENG] A small type of hand plane, designed for cutting across the grain of the wood and for planing end grains.

block signal system [CONT SYS] An automatic railroad traffic control system in which the track is sectionalized into electrical circuits to detect the presence of trains, engines, or cars.

blowby [MECH ENG] Leaking of fluid between a cylinder and its piston during operation.

blowdown [MECH ENG] The difference between the pressure at which the safety valve opens and the closing pressure.

blowdown tunnel [AERO ENG] A wind tunnel in which stored compressed gas is allowed to expand through a test section to provide a stream of gas or air for model testing.

blowdown turbine [AERO ENG] A turbine attached to a reciprocating engine which receives exhaust gases separately from each cylinder, utilizing the kinetic energy of the gases.

blower [MECH ENG] A fan which operates where the resistance to gas flow is predominantly downstream of the fan.

blowing boundary-layer control [AERO ENG] A technique that is used in addition to purely geometric means to control boundary-layer flow; it consists of reenergizing the retarded flow in the boundary layer by supplying high-velocity flow through slots or jets on the surface of the body.

blowoff [AERO ENG] The action of applying an explosive force and separating a package section away from the remaining part of a rocket vehicle or reentry body, usually to retrieve an instrument or to obtain a record made during early flight.

blowoff valves [MECH ENG] Valves in boiler piping which facilitate removal of solid matter present in the boiler water.

bluff body [AERO ENG] A body having a broad, flattened front, as in some reentry vehicles.

blunt file [DES ENG] A file whose edges are parallel.

blunting [DES ENG] Slightly rounding a cutting edge to reduce the probability of edge chipping.

board drop hammer [MECH ENG] A type of drop hammer in which the ram is attached to wooden boards which slide between two rollers; after the ram falls freely on the forging, it is raised by friction between the rotating rollers. Also known as board hammer.

board hammer *See* board drop hammer.

boasting chisel [DES ENG] A broad chisel used in boasting stone.

boat spike [DES ENG] A long, square spike used in construction with heavy timbers. Also known as barge spike.

boattail [AERO ENG] Of an elongated body such as a rocket, the rear portion having decreasing cross-sectional area.

Bobillier's law [MECH] The law that, in general plane rigid motion, when a and b are the respective centers of curvature of points A and B, the angle between Aa and the tangent to the centrode of rotation (pole tangent) and the angle between Bb and a line from the centrode to the intersection of AB and ab (collineation axis) are equal and opposite.

Bodenstein number [FL MECH] A numberless group used in the study of diffusion in reactors.

body [AERO ENG] **1.** The main part or main central portion of an airplane, airship, rocket, or the like; a fuselage or hull. **2.** Any fabrication, structure, or other material form, especially one aerodynamically or ballistically designed; for example, an airfoil is a body designed to produce an aerodynamic reaction. [MECH ENG] The part of a drill which runs from the outer corners of the cutting lips to the shank or neck.

body angle [AERO ENG] The angle which the longitudinal axis of the airframe makes with some selected line.

body axis [AERO ENG] Any one of a system of mutually perpendicular reference axes fixed in an aircraft or a similar body and moving with it.

body cone [MECH] The cone in a rigid body that is swept out by the body's instantaneous axis during Poinsot motion. Also known as Polhode cone.

body force [MECH] An external force, such as gravity, which acts on all parts of a body.

bogey *See* bogie.

bogie Also spelled bogey; bogy. [AERO ENG] A type of landing-gear unit consisting of two sets of wheels in tandem with a central strut. [MECH ENG] The drive-wheel assembly and supporting frame comprising the four rear wheels of a six-wheel truck, mounted so that they can self-adjust to sharp curves and irregularities in the road.

bogy *See* bogie.

boiler [MECH ENG] A water heater for generating steam.

boiler air heater [MECH ENG] A component of a steam-generating unit that transfers heat from the products of combustion after they have passed through the steam-generating and superheating sections to combustion air, which recycles heat to the furnace.

boiler casing [MECH ENG] The gas-tight structure surrounding the component parts of a steam generator.

boiler circulation [MECH ENG] Circulation of water and steam in a boiler, which is required to prevent overheating of the heat-absorbing surfaces; may be provided naturally by gravitational forces, mechanically by pumps, or by a combination of both methods.

boiler code [MECH ENG] A code, established by professional societies and administrative units, which contains the basic rules for the safe design, construction, and materials for steam-generating units, such as the ASME code.

boiler controls [MECH ENG] Either manual or automatic devices which maintain desired boiler operating conditions with respect to variables such as feedwater flow, firing rate, and steam temperature.

boiler draft [MECH ENG] The difference between atmospheric pressure and some lower pressure existing in the furnace or gas passages of a steam-generating unit.

boiler economizer [MECH ENG] A component of a steam-generating unit that transfers heat from the products of combustion after they have passed through the steam-generating and superheating sections to the feedwater, which it receives from the boiler feed pump and delivers to the steam-generating section of the boiler.

boiler efficiency [MECH ENG] The ratio of heat absorbed in steam to the heat supplied in fuel, usually measured in percent.

boiler feedwater [MECH ENG] Water supplied to a steam-generating unit.

boiler feedwater regulation [MECH ENG] Addition of water to the steam-generating unit at a rate commensurate with the removal of steam from the unit.

boiler furnace [MECH ENG] An enclosed space provided for the combustion of fuel to generate steam in a boiler. Also known as steam-generating furnace.

boiler heat balance [MECH ENG] A means of accounting for the thermal energy entering a steam-generating system in terms of its ultimate useful heat absorption or thermal loss.

boiler horsepower [MECH ENG] A measurement of water evaporation rate; 1 boiler horsepower equals the evaporation per hour of 34½ pounds (15.7 kilograms) of water at 212°F (100°C) into steam at 212°F. Abbreviated bhp.

boiler hydrostatic test [MECH ENG] A procedure that employs water under pressure, in a new boiler before use or in old equipment after major alterations and repairs, to test the boiler's ability to withstand about 1½ times the design pressure.

boiler layup [MECH ENG] A significant length of time during which a boiler is inoperative in order to allow for repairs or preventive maintenance.

boiler-plate model [AERO ENG] A metal copy of a flight vehicle, the structure or components of which are heavier than the flight model.

boiler setting [MECH ENG] The supporting steel and gastight enclosure for a steam generator.

boiler storage [MECH ENG] A steam-generating unit that, when out of service, may be stored wet (filled with water) or dry (filled with protective gas).

boiler superheater [MECH ENG] A boiler component, consisting of tubular elements, in which heat is added to high-pressure steam to increase its temperature and enthalpy.

boiler trim [MECH ENG] Piping or tubing close to or attached to a boiler for connecting controls, gages, or other instrumentation.

boiler tube [MECH ENG] One of the tubes in a boiler that carry water (water-tube boiler) to be heated by the high-temperature gaseous products of combustion or that carry combustion products (fire-tube boiler) to heat the boiler water that surrounds them.

boiler walls [MECH ENG] The refractory walls of the boiler furnace, usually cooled by circulating water and capable of withstanding high temperatures and pressures.

boiler water [MECH ENG] Water in the steam-generating section of a boiler unit.

boil-off [THERMO] The vaporization of a liquid, such as liquid oxygen or liquid hydrogen, as its temperature reaches its boiling point under conditions of exposure, as in the tank of a rocket being readied for launch.

bolster plate [MECH ENG] A plate fixed on the bed of a power press to locate and support the die assembly.

bolt [DES ENG] A rod, usually of metal, with a square, round, or hexagonal head at one end and a screw thread on the other, used to fasten objects together.

bombardment aircraft *See* bomber.

bomb ballistics   [MECH]   The special branch of ballistics concerned with bombs dropped from aircraft.

bomb bay   [AERO ENG]   The compartment or bay in the fuselage of a bomber where the bombs are carried for release.

bomber   [AERO ENG]   An airplane specifically designed to carry and drop bombs. Also known as bombardment aircraft.

Bond and Wang theory   [MECH ENG]   A theory of crushing and grinding from which the energy, in horsepower-hours, required to crush a short ton of material is derived.

bonding strength   [MECH]   Structural effectiveness of adhesives, welds, solders, glues, or of the chemical bond formed between the metallic and ceramic components of a cermet, when subjected to stress loading, for example, shear, tension, or compression.

Bond number   [FL MECH]   A dimensionless number used in the study of atomization and the study of bubbles and drops, equal to $(\rho - \rho')L^2 g/\sigma$, where ρ is the density of a bubble or drop, ρ' is the density of the surrounding medium, L is a characteristic dimension, g is the acceleration of gravity, and σ is the surface tension of the bubble or drop.

Bond's law   [MECH ENG]   A statement that relates the work required for the crushing of solid materials (for example, rocks and ore) to the product size and surface area and the lengths of cracks formed. Also known as Bond's third theory.

Bond's third theory   *See* Bond's law.

boom cat   [MECH ENG]   A tractor supporting a boom and used in laying pipe.

boost   [AERO ENG]   **1.** An auxiliary means of propulsion such as by a booster. **2.** To supercharge. **3.** To launch or push along during a portion of a flight. **4.** *See* boost pressure.

booster   *See* booster engine; booster rocket; launch vehicle.

booster brake   [MECH ENG]   An auxiliary air chamber, operated from the intake manifold vacuum, and connected to the regular brake pedal, so that less pedal pressure is required for braking.

booster ejector   [MECH ENG]   A nozzle-shaped apparatus from which a high-velocity jet of steam is discharged to produce a continuous-flow vacuum for process equipment.

booster engine   [AERO ENG]   An engine, especially a booster rocket, that adds its thrust to the thrust of the sustainer engine. Also known as booster.

booster fan   [MECH ENG]   A fan used to increase either the total pressure or the volume of flow.

booster pump   [MECH ENG]   A machine used to increase pressure in a water or compressed-air pipe.

booster rocket   [AERO ENG]   Also known as booster. **1.** A rocket motor, either solid- or liquid-fueled, that assists the normal propulsive system or sustainer engine of a rocket or aeronautical vehicle in some phase of its flight. **2.** A rocket used to set a vehicle in motion before another engine takes over.

boost-glide vehicle   [AERO ENG]   An air vehicle capable of aerodynamic lift which is projected to an extreme altitude by reaction propulsion and then coasts down with little or no propulsion, gliding to increase its range when it reenters the sensible atmosphere.

boost pressure   [AERO ENG]   Manifold pressure greater than the ambient at atmospheric pressure, obtained by supercharging. Also known as boost.

bootstrap process [AERO ENG] A self-generating or self-sustaining process; specifically, the operation of liquid-propellant rocket engines in which, during main-stage operation, the gas generator is fed by the main propellants pumped by the turbopump, and the turbopump in turn is driven by hot gases from the gas generator system.

Borda mouthpiece [FL MECH] A reentrant tube in a hydraulic reservoir, whose contraction coefficient (the ratio of the cross section of the issuing jet of liquid to that of the opening) can be calculated more simply than for other discharge openings.

bore [DES ENG] Inside diameter of a pipe or tube. [MECH ENG] **1.** The diameter of a piston-cylinder mechanism as found in reciprocating engines, pumps, and compressors. **2.** To penetrate or pierce with a rotary tool. **3.** To machine a workpiece to increase the size of an existing hole in it.

borer [MECH ENG] An apparatus used to bore openings into the earth up to about 8 feet (2.4 meters) in diameter.

boring bar [MECH ENG] A rigid tool holder used to machine internal surfaces.

boring machine [MECH ENG] A machine tool designed to machine internal work such as cylinders, holes in castings, and dies; types are horizontal, vertical, jig, and single.

boring mill [MECH ENG] A boring machine tool used particularly for large workpieces; types are horizontal and vertical.

Bosch fuel injection pump [MECH ENG] A pump in the fuel injection system of an internal combustion engine, whose pump plunger and barrel are a very close lapped fit to minimize leakage.

Bosch metering system [MECH ENG] A system having a helical groove in the plunger which covers or uncovers openings in the barrel of the pump; most usually applied in diesel engine fuel-injection systems.

boss [DES ENG] Protuberance on a cast metal or plastic part to add strength, facilitate assembly, provide for fastenings, or so forth.

bottom dead center [MECH ENG] The position of the crank of a vertical reciprocating engine, compressor, or pump when the piston is at the end of its downstroke. Abbreviated BDC.

bottom-discharge bit *See* face-discharge bit.

bottoming drill [DES ENG] A flat-ended twist drill designed to convert a cone at the bottom of a drilled hole into a cylinder.

bottom tap [DES ENG] A tap with a chamfer 1–1½ threads in length.

bounce table [MECH ENG] A testing device which subjects devices and components to impacts such as might be encountered in accidental dropping.

boundary-layer control [FL MECH] Control over the development of a boundary layer by reduction of surface roughness and choice of surface contours. Abbreviated BLC.

boundary-layer flow [FL MECH] The flow of that portion of a viscous fluid which is in the neighborhood of a body in contact with the fluid and in motion relative to the fluid.

boundary-layer separation [FL MECH] That point where the boundary layer no longer continues to follow the contour of the boundary because the residual momentum of the fluid (left after overcoming viscous forces) may be insufficient to allow the flow to proceed into regions of increasing pressure. Also known as flow separation.

bound vector [MECH] A vector whose line of application and point of application are both prescribed, in addition to its direction.

Boussinesq approximation [FL MECH] The assumption (frequently used in the theory of convection) that the fluid is incompressible except insofar as the thermal expansion produces a buoyancy, represented by a term $g\alpha T$, where g is the acceleration of gravity, α is the coefficient of thermal expansion, and T is the perturbation temperature.

Boussinesq number [FL MECH] A dimensionless number used to study wave behavior in open channels.

bow [AERO ENG] The forward section of an aircraft.

Bowden cable [MECH ENG] A wire made of spring steel which is enclosed in a helical casing and used to transmit longitudinal motions over distances, particularly around corners.

bowl mill *See* bowl-mill pulverizer.

bowl-mill pulverizer [MECH ENG] A type of pulverizer which directly feeds a coal-fired furnace, in which springs press pivoted stationary rolls against a rotating bowl grinding ring, crushing the coal between them. Also known as a bowl mill.

bowl scraper [MECH ENG] A towed steel bowl hung within a fabricated steel frame, running on four or two wheels; transports soil, in addition to spreading and leveling it.

Bow's notation [MECH] A graphical method of representing coplanar forces and stresses, using alphabetical letters, in the solution of stresses or in determining the resultant of a system of concurrent forces.

bow wave [FL MECH] A shock wave occurring in front of a body, such as an airfoil, or apparently attached to the forward tip of the body.

box *See* boxing.

box header boiler [MECH ENG] A horizontal boiler with a front header and rear inclined rectangular header connected by tubes.

boxing [DES ENG] The threaded nut for the screw of a mounted auger drill. Also known as box.

Boyle's temperature [THERMO] For a given gas, the temperature at which the virial coefficient B in the equation of state $Pv = RT\,[1 + (B/v) + (C/v^2) + \ldots]$ vanishes.

brace [DES ENG] A cranklike device used for turning a bit.

brace and bit [DES ENG] A small hand tool to which is attached a metal- or wood-boring bit.

brachistochrone [MECH] The curve along which a smooth-sliding particle, under the influence of gravity alone, will fall from one point to another in the minimum time.

brake [MECH ENG] A machine element for applying friction to a moving surface to slow it (and often, the containing vehicle or device) down or bring it to rest.

brake band [MECH ENG] The contracting element of the band brake.

brake drum [MECH ENG] A rotating cylinder attached to a rotating part of machinery, which the brake band or brake shoe presses against.

brake horsepower [MECH ENG] The power developed by an engine as measured by the force applied to a friction brake or by an absorption dynamometer applied to the shaft or flywheel. Abbreviated bhp.

brake lining [MECH ENG] A covering, riveted or molded to the brake shoe or brake band, which presses against the rotating brake drum; made of either fabric or molded asbestos material.

brake mean-effective pressure [MECH ENG] Applied to reciprocating piston machinery, the average pressure on the piston during the power stroke, derived from the measurement of brake power output.

brake shoe [MECH ENG] The renewable friction element of a shoe brake. Also known as shoe.

brake thermal efficiency [MECH ENG] The ratio of brake power output to power input.

braking ellipses [AERO ENG] A series of ellipses, decreasing in size due to aerodynamic drag, followed by a spacecraft in entering a planetary atmosphere.

braking rocket *See* retrorocket.

branch gain *See* branch transmittance.

branch transmittance [CONT SYS] The amplification of current or voltage in a branch of an electrical network; used in the representation of such a network by a signal-flow graph. Also known as branch gain.

Brayton cycle [THERMO] A thermodynamic cycle consisting of two constant-pressure processes interspersed with two constant-entropy processes. Also known as complete-expansion diesel cycle; Joule cycle.

brazed shank tool [MECH ENG] A metal cutting tool made of a material different from the shank to which it is brazed.

breakaway phenomenon *See* breakoff phenomenon.

breaker cam [MECH ENG] A rotating, engine-driven device in the ignition system of an internal combustion engine which causes the breaker points to open, leading to a rapid fall in the primary current.

break frequency [CONT SYS] The frequency at which a graph of the logarithm of the amplitude of the frequency response versus the logarithm of the frequency has an abrupt change in slope. Also known as corner frequency; knee frequency.

breaking load [MECH] The stress which, when steadily applied to a structural member, is just sufficient to break or rupture it.

breaking strength [MECH] The ability of a material to resist breaking or rupture from a tension force.

breaking stress [MECH] The stress required to fracture a material whether by compression, tension, or shear.

breakoff phenomenon [AERO ENG] The feeling which sometimes occurs during high-altitude flight of being totally separated and detached from the earth and human society. Also known as breakaway phenomenon.

breast drill [DES ENG] A small, portable hand drill customarily used by handsetters to drill the holes in bit blanks in which diamonds are to be set; it includes a plate that is pressed against the worker's breast.

breather pipe [MECH ENG] A pipe that opens into a container for ventilation, as in a crankcase or oil tank.

breeching [MECH ENG] A duct through which the products of combustion are transported from the furnace to the stack; usually applied in steam boilers.

Brennan monorail car [MECH ENG] A type of car balanced on a single rail so that when the car starts to tip, a force automatically applied at the axle end is converted gyroscopically into a strong righting moment which forces the car back into a position of lateral equilibrium.

brennschluss [AERO ENG] 1. The cessation of burning in a rocket, resulting from consumption of the propellants, from deliberate shutoff, or from other cause. 2. The time at which this cessation occurs.

bridge crane [MECH ENG] A hoisting machine in which the hoisting apparatus is carried by a bridgelike structure spanning the area over which the crane operates.

bridge trolley [MECH ENG] Either of the wheeled attachments at the ends of the bridge of an overhead traveling crane, permitting the bridge to move backward and forward on elevated tracks.

bridge vibration [MECH] Mechanical vibration of a bridge superstructure due to natural and human-produced excitations.

bridgewall [MECH ENG] A wall in a furnace over which the products of combustion flow.

brightness temperature *See* blackbody temperature.

brine cooler [MECH ENG] The unit for cooling brine in a refrigeration system; the brine usually flows through tubes or pipes surrounded by evaporating refrigerant.

Brinkmann number [FL MECH] A dimensionless number used to study viscous flow.

British imperial pound [MECH] The British standard of mass, of which a standard is preserved by the government.

British thermal unit [THERMO] Abbreviated Btu. **1.** A unit of heat energy equal to the heat needed to raise the temperature of 1 pound of air-free water from 60° to 61°F at a constant pressure of 1 standard atmosphere; it is found experimentally to be equal to 1054.5 joules. Also known as sixty degrees Fahrenheit British thermal unit ($Btu_{60/61}$). **2.** A unit of heat energy that is equal to 1/180 of the heat needed to raise 1 pound of air-free water from 32°F (0°C) to 212°F (100°C) at a constant pressure of 1 standard atmosphere; it is found experimentally to be equal to 1055.79 joules. Also known as mean British thermal unit (Btu_{mean}). **3.** A unit of heat energy whose magnitude is such that 1 British thermal unit per pound equals 2326 joules per kilogram; it is equal to exactly 1055.05585262 joules. Also known as international table British thermal unit (Btu_{it}).

brittleness [MECH] That property of a material manifested by fracture without appreciable prior plastic deformation.

brittle temperature [THERMO] The temperature point below which a material, especially metal, is brittle; that is, the critical normal stress for fracture is reached before the critical shear stress for plastic deformation.

broach [MECH ENG] A multiple-tooth, barlike cutting tool; the teeth are shaped to give a desired surface or contour, and cutting results from each tooth projecting farther than the preceding one.

broaching [MECH ENG] The machine-shaping of metal or plastic by pushing or pulling a broach across a surface or through an existing hole in a workpiece.

broomy flow [FL MECH] A swirling flow of a fluid in a pipe after passing through a constricted section or after a sudden change of direction.

Btu *See* British thermal unit.

Btu$_{it}$ *See* British thermal unit.

bu *See* bushel.

bubble cavitation [FL MECH] **1.** Formation of vapor- or gas-filled cavities in liquids by mechanical forces. **2.** The formation of vapor-filled cavities in the interior of liquids in motion when the pressure is reduced without change in ambient temperature.

bucket carrier *See* bucket conveyor.

bucket conveyor [MECH ENG] A continuous bulk conveyor constructed of a series of buckets attached to one or two strands of chain or in some instances to a belt. Also called bucket carrier.

bucket dredge [MECH ENG] A floating mechanical excavator equipped with a bucket elevator.

bucket elevator [MECH ENG] A bucket conveyor operating on a steep incline or vertical path. Also known as elevating conveyor.

bucket excavator [MECH ENG] An elevating scraper, that is, one that does the work of a conventional scraper but has a bucket elevator mounted in front of the bowl.

bucket ladder *See* bucket-ladder dredge.

bucket-ladder dredge [MECH ENG] A dredge whose digging mechanism consists of a ladderlike truss on the periphery of which is attached an endless chain riding on sprocket wheels and carrying attached buckets. Also known as bucket ladder; bucket-line dredge; ladder-bucket dredge; ladder dredge.

bucket-ladder excavator *See* trench excavator.

bucket-line dredge *See* bucket-ladder dredge.

bucket loader [MECH ENG] A form of portable, self-feeding, inclined bucket elevator for loading bulk materials into cars, trucks, or other conveyors.

bucket-wheel excavator [MECH ENG] A continuous digging machine used extensively in large-scale stripping and mining. Abbreviated BWE. Also known as rotary excavator.

Buckingham's equations [MECH ENG] Equations which give the durability of gears and the dynamic loads to which they are subjected in terms of their dimensions, hardness, surface endurance, and composition.

buckling [MECH] Bending of a sheet, plate, or column supporting a compressive load.

buckling stress [MECH] Force exerted by the crippling load.

buckstay [MECH ENG] A structural support for a furnace wall.

buffeting [AERO ENG] 1. The beating of an aerodynamic structure or surfaces by unsteady flow, gusts, and so forth. 2. The irregular shaking or oscillation of a vehicle component owing to turbulent air or separated flow.

buffeting Mach number [AERO ENG] The free-stream Mach number of an aircraft when the local Mach number over the tops of the wings approaches unity.

buffing wheel [DES ENG] A flexible wheel with a surface of fine abrasive particles for buffing operations.

buhrstone mill [MECH ENG] A mill for grinding or pulverizing grain in which a flat siliceous rock (buhrstone), generally of cellular quartz, rotates against a stationary stone of the same material.

bulb angle [DES ENG] A steel angle iron enlarged to a bulbous thickening at one end.

bulk flow *See* convection.

bulk-handling machine [MECH ENG] Any of a diversified group of materials-handling machines designed for handling unpackaged, divided materials.

bulkhead [AERO ENG] A wall, partition, or such in a rocket, spacecraft, airplane fuselage, or similar structure, at right angles to the longitudinal axis of the structure and serving to strengthen, divide, or help give shape to the structure.

bulk modulus *See* bulk modulus of elasticity.

bulk modulus of elasticity [MECH] The ratio of the compressive or tensile force applied to a substance per unit surface area to the change in volume of the substance per unit volume. Also known as bulk modulus; compression modulus; hydrostatic modulus; modulus of compression; modulus of volume elasticity.

bulk strength [MECH] The strength per unit volume of a solid.

bulk transport [MECH ENG] Conveying, hoisting, or elevating systems for movement of solids such as grain, sand, gravel, coal, or wood chips.

bulldozer [MECH ENG] A wheeled or crawler tractor equipped with a reinforced, curved steel plate mounted in front, perpendicular to the ground, for pushing excavated materials.

bullet drop [MECH] The vertical drop of a bullet.

bull gear [DES ENG] A bull wheel with gear teeth.

bull-nose bit *See* wedge bit.

bull wheel [MECH ENG] **1.** The main wheel or gear of a machine, which is usually the largest and strongest. **2.** A cylinder which has a rope wound about it for lifting or hauling. **3.** A wheel attached to the base of a derrick boom which swings the derrick in a vertical plane.

Bulygen number [THERMO] A dimensionless number used in the study of heat transfer during evaporation.

bumpiness [AERO ENG] An atmospheric condition causing aircraft to experience sudden vertical jolts.

bumping *See* chugging.

bundling machine [MECH ENG] A device that automatically accumulates cans, cartons, or glass containers for semiautomatic or automatic loading or for shipping cartons by assembling the packages into units of predetermined count and pattern which are then machine-wrapped in paper, film paperboard, or corrugated board.

buoyancy [FL MECH] The resultant vertical force exerted on a body by a static fluid in which it is submerged or floating.

buoyancy parameter [FL MECH] The Grashof number divided by the square of the Reynolds number.

buoyant force [FL MECH] The force exerted vertically upward by a fluid on a body wholly or partly immersed in it; its magnitude is equal to the weight of the fluid displaced by the body.

burble [FL MECH] **1.** A separation or breakdown of the laminar flow past a body. **2.** The eddying or turbulent flow resulting from this occurrence.

burble angle *See* burble point.

burble point [FL MECH] A point reached in an increasing angle of attack at which burble begins. Also known as burble angle.

burner [MECH ENG] A unit of a steam boiler which mixes and directs the flow of fuel and air so as to ensure rapid ignition and complete combustion.

burning-rate constant [AERO ENG] A constant, related to initial grain temperature, used in calculating the burning rate of a rocket propellant grain.

burnout [AERO ENG] **1.** An act or instance of fuel or oxidant depletion or of depletion of both at once. **2.** The time at which this depletion occurs.

burnout velocity [AERO ENG] The velocity of a rocket at the time when depletion of the fuel or oxidant occurs. Also known as all-burnt velocity; burnt velocity.

Burnside boring machine [MECH ENG] A machine for boring in all types of ground with the feature of controlling water immediately if it is tapped.

burnt velocity *See* burnout velocity.

burst disk [AERO ENG] A diaphragm designed to burst at a predetermined pressure differential; sometimes used as a valve, for example, in a liquid-propellant line in a rocket. Also known as rupture disk.

bursting strength [MECH] A measure of the ability of a material to withstand pressure without rupture; it is the hydraulic pressure required to burst a vessel of given thickness.

burst pressure [MECH] The maximum inside pressure that a process vessel can safely withstand.

burst wave [FL MECH] Wave of compressed air caused by a bursting projectile or bomb; a detonation wave; it may produce extensive local damage.

burton [MECH ENG] A small hoisting tackle with two blocks, usually a single block and a double block, with a hook block in the running part of the rope.

bushel [MECH] Abbreviated bu. **1.** A unit of volume (dry measure) used in the United States, equal to 2150.42 cubic inches or approximately 35.239 liters. **2.** A unit of volume (liquid and dry measure) used in Britain, equal to 2219.36 cubic inches or 8 imperial gallons (approximately 36.369 liters).

bushing [MECH ENG] A removable piece of soft metal or graphite-filled sintered metal, usually in the form of a bearing, that lines a support for a shaft.

butt [DES ENG] The enlarged and squared-off end of a connecting rod or similar link in a machine.

butterfly nut *See* wing nut.

Butterworth head [MECH ENG] A mechanical hose head with revolving nozzles; used to wash down shipboard storage tanks.

button die [DES ENG] A mating member, usually replaceable, for a piercing punch. Also known as die bushing.

buttonhead [DES ENG] A screw, bolt, or rivet with a hemispherical head.

buttress thread [DES ENG] A screw thread whose forward face is perpendicular to the screw axis and whose back face is at an angle to the axis, so that the thread is both efficient in transmitting power and strong.

buzz [AERO ENG] Sustained oscillation of an aerodynamic control surface caused by intermittent flow separation on the surface, or by a motion of shock waves across the surface, or by a combination of flow separation and shock-wave motion on the surface. [CONT SYS] *See* dither. [FL MECH] In supersonic diffuser aerodynamics, a nonsteady shock motion and airflow associated with the shock system ahead of the inlet.

BWE *See* bucket-wheel excavator.

BWG *See* Birmingham wire gage.

C

c *See* calorie.

cabane [AERO ENG] The arrangement of struts used on early types of airplanes to brace the wings.

cabinet file [DES ENG] A coarse-toothed file with flat and convex faces used for woodworking.

cabinet hardware [DES ENG] Parts for the final trim of a cabinet, such as fastening hinges, drawer pulls, and knobs.

cabinet saw [DES ENG] A short saw, one edge used for ripping, the other for crosscutting.

cabinet scraper [DES ENG] A steel tool with a contoured edge used to remove irregularities on a wood surface.

cable [DES ENG] A stranded, ropelike assembly of wire or fiber.

cable conveyor [MECH ENG] A powered conveyor in which a trolley runs on a flexible, torque-transmitting cable that has helical threads.

cable-laid [DES ENG] Consisting of three ropes with a left-hand twist, each rope having three twisted strands.

cable railway [MECH ENG] An inclined track on which rail cars travel, with the cars fixed to an endless steel-wire rope at equal spaces; the rope is driven by a stationary engine.

cable-system drill *See* churn drill.

cableway [MECH ENG] A transporting system consisting of a cable extended between two or more points on which cars are propelled to transport bulk materials for construction operations.

cableway carriage [MECH ENG] A trolley that runs on main load cables stretched between two or more towers.

cage [MECH ENG] A frame for maintaining uniform separation between the balls or rollers in a bearing. Also known as separator.

cage mill [MECH ENG] Pulverizer used to disintegrate clay, press cake, asbestos, packing-house by-products, and various tough, gummy, high-moisture-content or low-melting-point materials.

cal *See* calorie.

calibrated airspeed [AERO ENG] The airspeed as read from a differential-pressure airspeed indicator which has been corrected for instrument and installation errors; equal to true airspeed for standard sea-level conditions.

caliper [DES ENG] An instrument with two legs or jaws that can be adjusted for measuring linear dimensions, thickness, or diameter.

caliper gage [DES ENG] An instrument, such as a micrometer, of fixed size for calipering.

Callendar's equation [THERMO] 1. An equation of state for steam whose temperature is well above the boiling point at the existing pressure, but less than the critical temperature: $(V - b) = (RT/p) - (a/T^n)$, where V is the volume, R is the gas constant, T is the temperature, p is the pressure, n equals 10/3, and a and b are constants. 2. A very accurate equation relating temperature and resistance of platinum, according to which the temperature is the sum of a linear function of the resistance of platinum and a small correction term, which is a quadratic function of temperature.

calorie [THERMO] Abbreviated cal; often designated c. 1. A unit of heat energy, equal to 4.1868 joules. Also known as International Table calorie (IT calorie). 2. A unit of energy, equal to the heat required to raise the temperature of 1 gram of water from 14.5° to 15.5°C at a constant pressure of 1 standard atmosphere; equal to 4.1855 ± 0.0005 joules. Also known as fifteen-degrees calorie; gram-calorie (g-cal); small calorie. 3. A unit of heat energy equal to 4.184 joules; used in thermochemistry. Also known as thermochemical calorie.

cam [MECH ENG] A plate or cylinder which communicates motion to a follower by means of its edge or a groove cut in its surface.

cam acceleration [MECH ENG] The acceleration of the cam follower.

camber [AERO ENG] The rise of the curve of an airfoil section, usually expressed as the ratio of the departure of the curve from a straight line joining the extremities of the curve to the length of this straight line. [DES ENG] Deviation from a straight line; the term is applied to a convex, edgewise sweep or curve, or to the increase in diameter at the center of rolled materials.

camber angle [MECH ENG] The inclination from the vertical of the steerable wheels of an automobile.

cam cutter [MECH ENG] A semiautomatic or automatic machine that produces the cam contour by swinging the work as it revolves; uses a master cam in contact with a roller.

cam dwell [DES ENG] That part of a cam surface between the opening and closing acceleration sections.

cam engine [MECH ENG] A piston engine in which a cam-and-roller mechanism seems to convert reciprocating motion into rotary motion.

cam follower [MECH ENG] The output link of a cam mechanism.

cam mechanism [MECH ENG] A mechanical linkage whose purpose is to produce, by means of a contoured cam surface, a prescribed motion of the output link.

cam nose [MECH ENG] The high point of a cam, which in a reciprocating engine holds valves open or closed.

cam pawl [MECH ENG] A pawl which prevents a wheel from turning in one direction by a wedging action, while permitting it to rotate in the other direction.

cam profile [DES ENG] The shape of the contoured cam surface by means of which motion is communicated to the follower. Also known as pitch line.

camshaft [MECH ENG] A rotating shaft to which a cam is attached.

can [DES ENG] A cylindrical metal vessel o container, usually with an open top or a removable cover.

canal [DES ENG] A groove on the underside of a corona.

canard [AERO ENG] 1. An aerodynamic vehicle in which horizontal surfaces used for trim and control are forward of the wing or main lifting surface. 2. The horizontal trim and control surfaces in such an arrangement.

canned motor [MECH ENG] A motor enclosed within a casing along with the driven element (that is, a pump) so that the motor bearings are lubricated by the same liquid that is being pumped.

canned pump [MECH ENG] A watertight pump that can operate under water.

cannular combustion chambers [AERO ENG] The separate combustion chambers in an aircraft gas turbine. Also known as can-type combustors.

canonical equations of motion *See* Hamilton's equations of motion.

canonical form [CONT SYS] A specific type of dynamical system representation in which the associated matrices possess specific row-column structures.

canonically conjugate variable [MECH] A generalized coordinate and its conjugate momentum.

canonical momentum *See* conjugate momentum.

canonical transformation [MECH] A transformation which occurs among the coordinates and momenta describing the state of a classical dynamical system and which leaves the form of Hamilton's equations of motion unchanged. Also known as contact transformation.

canopy [AERO ENG] 1. The umbrellalike part of a parachute which acts as its main supporting surface. 2. The overhead, transparent enclosure of an aircraft cockpit.

cant file [DES ENG] A fine-tapered file with a triangular cross section, used for sharpening saw teeth.

cant hook [DES ENG] A lever with a hooklike attachment at one end, used in lumbering.

cantilever spring [MECH ENG] A flat spring supported at one end and holding a load at or near the other end.

cantilever vibration [MECH] Transverse oscillatory motion of a body fixed at one end.

canting [MECH] Displacing the free end of a beam which is fixed at one end by subjecting it to a sideways force which is just short of that required to cause fracture.

can-type combustors *See* cannular combustion chambers.

cape chisel [DES ENG] A chisel that tapers to a flat, narrow cutting end; used to cut flat grooves.

cape foot [MECH] A unit of length equal to 1.033 feet or to 0.3148584 meter.

capillarity [FL MECH] The action by which the surface of a liquid where it contacts a solid is elevated or depressed, because of the relative attraction of the molecules of the liquid for each other and for those of the solid. Also known as capillary action.

capillary action *See* capillarity.

capillary attraction [FL MECH] The force of adhesion existing between a solid and a liquid in capillarity.

capillary depression [FL MECH] The depression of the meniscus of a liquid contained in a tube where the liquid does not wet the walls of the container, as in a mercury barometer; the meniscus has a convex shape, resulting in a depression.

capillary pressure [FL MECH] The difference in pressure across the interface of two immiscible fluid phases.

capillary ripple *See* capillary wave.

capillary rise [FL MECH] The rise of a liquid in a capillary tube times the radius of the tube.

capillary wave [FL MECH] 1. A wave occurring at the interface between two fluids, such as the interface between air and water on oceans and lakes, in which the principal restoring force is controlled by surface tension. 2. A water wave of less than 1.7 centimeters. Also known as capillary ripple; ripple.

cap screw [DES ENG] A screw which passes through a clear hole in the part to be joined, screws into a threaded hole in the other part, and has a head which holds the parts together.

capstan nut [DES ENG] A nut whose edge has several holes, in one of which a bar can be inserted for turning it.

capstan screw [DES ENG] A screw whose head has several radial holes, in one of which a bar can be inserted for turning it.

captive balloon [AERO ENG] A moored balloon, usually by steel cables.

captive fastener [DES ENG] A screw-type fastener that does not drop out after it has been unscrewed.

capture [AERO ENG] The process in which a missile is taken under control by the guidance system.

car *See* automobile.

Carathéodory's principle [THERMO] An expression of the second law of thermodynamics which says that in the neighborhood of any equilibrium state of a system, there are states which are not accessible by a reversible or irreversible adiabatic process. Also known as principle of inaccessibility.

carbide tool [DES ENG] A cutting tool made of tungsten, titanium, or tantalum carbides, having high heat and wear resistance.

carbon bit [DES ENG] A diamond bit in which the cutting medium is inset carbon.

carbon knock [MECH ENG] Premature ignition resulting in knocking or pinging in an internal combustion engine caused when the accumulation of carbon produces overheating in the cylinder.

carburetion [MECH ENG] The process of mixing fuel with air in a carburetor.

carburetor [MECH ENG] A device that makes and controls the proportions and quantity of fuel-air mixture fed to a spark-ignition internal combustion engine.

carburetor icing [MECH ENG] The formation of ice in an engine carburetor as a consequence of expansive cooling and evaporation of gasoline.

Cardan joint *See* Hooke's joint.

Cardan motion [MECH ENG] The straight-line path followed by a moving centrode in a four-bar centrode linkage.

Cardan shaft [MECH ENG] A shaft with a universal joint at its end to accommodate a varying shaft angle.

Cardan's suspension [DES ENG] An arrangement of rings in which a heavy body is mounted so that the body is fixed at one point; generally used in a gyroscope.

car dump [MECH ENG] Any one of several devices for unloading industrial or railroad cars by rotating or tilting the car.

cargo boom [MECH ENG] A long spar extending from the mast of a derrick to support or guide objects lifted or suspended.

cargo winch [MECH ENG] A motor-driven hoisting machine for cargo having a drum around which a chain or rope winds as the load is lifted.

Carnot cycle [THERMO] A hypothetical cycle consisting of four reversible processes in succession: an isothermal expansion and heat addition, an isentropic expansion, an isothermal compression and heat rejection process, and an isentropic compression.

Carnot efficiency [THERMO] The efficiency of a Carnot engine receiving heat at a temperature absolute T_1 and giving it up at a lower temperature absolute T_2; equal to $(T_1 - T_2)/T_1$.

Carnot engine [MECH ENG] An ideal, frictionless engine which operates in a Carnot cycle.

Carnot number [THERMO] A property of two heat sinks, equal to the Carnot efficiency of an engine operating between them.

Carnot's theorem [THERMO] 1. The theorem that all Carnot engines operating between two given temperatures have the same efficiency, and no cyclic heat engine operating between two given temperatures is more efficient than a Carnot engine. 2. The theorem that any system has two properties, the thermodynamic temperature T and the entropy S, such that the amount of heat exchanged in an infinitesimal reversible process is given by $dQ = TdS$; the thermodynamic temperature is a strictly increasing function of the empirical temperature measured on an arbitrary scale.

carpenter's level [DES ENG] A bar, usually of aluminum or wood, containing a spirit level.

carriage bolt [DES ENG] A round-head type of bolt with a square neck, used with a nut as a through bolt.

carriage stop [MECH ENG] A device added to the outer way of a lathe bed for accurately spacing grooves, turning multiple diameters and lengths, and cutting off pieces of specified thickness.

carrier [MECH ENG] Any machine for transporting materials or people.

carrier rocket [AERO ENG] A rocket vehicle used to carry something, as the carrier rocket of the first artificial earth satellite.

car shaker [MECH ENG] A device consisting of a heavy yoke on an open-top car's sides that actively vibrates and rapidly discharges a load, such as coal, gravel, or sand, when an unbalanced pulley attached to the yoke is rotated fast.

cartographic satellite [AERO ENG] An applications satellite that is used to prepare maps of the earth's surface and of the culture on it.

cartridge actuated initiator [AERO ENG] An item designed to provide gas pressure for activating various aircraft components such as canopy removers, thrusters, and catapults.

cartridge starter [MECH ENG] An explosive device which, when placed in an engine and detonated, moves a piston, thereby starting the engine.

cascade compensation [CONT SYS] Compensation in which the compensator is placed in series with the forward transfer function. Also known as series compensation; tandem compensation.

cascade control [CONT SYS] An automatic control system in which various control units are linked in sequence, each control unit regulating the operation of the next control unit in line.

cascade pulverizer [MECH ENG] A form of tumbling pulverizer that uses large lumps to do the pulverizing.

cascading [MECH ENG] An effect in ball-mill rotating devices when the upper level of crushing bodies breaks clear and falls to the top of the crop load.

casing [MECH ENG] A fire-resistant covering used to protect part or all of a steam generating unit.

casing nail [DES ENG] A nail about half a gage thinner than a common wire nail of the same length.

castellated bit [DES ENG] 1. A long-tooth, sawtooth bit. 2. A diamond-set coring bit with a few large diamonds or hard metal cutting points set in the face of each of several upstanding prongs separated from each other by deep waterways. Also known as padded bit.

castellated nut [DES ENG] A type of hexagonal nut with a cylindrical portion above through which slots are cut so that a cotter pin or safety wire can hold it in place.

Castigliano's principle *See* Castigliano's theorem.

Castigliano's theorem [MECH] The theorem that the component in a given direction of the deflection of the point of application of an external force on an elastic body is equal to the partial derivative of the work of deformation with respect to the component of the force in that direction. Also known as Castigliano's principle.

casting strain [MECH] Any strain that results from the cooling of a casting, causing casting stress.

casting stress [MECH] Any stress that develops in a casting due to geometry and casting shrinkage.

cast setting *See* mechanical setting.

cat-and-mouse engine [MECH ENG] A type of rotary engine, typified by the Tschudi engine, which is an analog of the reciprocating piston engine, except that the pistons travel in a circular motion. Also known as scissor engine.

catapult [AERO ENG] 1. A power-actuated machine or device for hurling an object at high speed, for example, a device which launches aircraft from a ship deck. 2. A device, usually explosive, for ejecting a person from an aircraft.

cataracting [MECH ENG] A motion of the crushed bodies in a ball mill in which some, leaving the top of the crop load, fall with impact to the toe of the load.

caterpillar [MECH ENG] A vehicle, such as a tractor or army tank, which runs on two endless belts, one on each side, consisting of flat treads and kept in motion by toothed driving wheels.

caterpillar chain [DES ENG] A short, endless chain on which dogs (grippers) or teeth are arranged to mesh with a conveyor.

Cauchy number [FL MECH] A dimensionless number used in the study of compressible flow, equal to the density of a fluid times the square of its velocity divided by its bulk modulus. Also known as Hooke number.

caulking iron [DES ENG] A tool for applying caulking to a seam.

causality [MECH] In classical mechanics, the principle that the specification of the dynamical variables of a system at a given time, and of the external forces acting on the system, completely determines the values of dynamical variables at later times. Also known as determinism.

causal system [CONT SYS] A system whose response to an input does not depend on values of the input at later times. Also known as nonanticipatory system; physical system.

cautious control [CONT SYS] A control law for a stochastic adaptive control system which hedges and uses lower gain when the estimates are uncertain.

cavitation [FL MECH] Formation of gas- or vapor-filled cavities within liquids by mechanical forces; broadly includes bubble formation when water is brought to a boil and effervescence of carbonated drinks; specifically, the formation of vapor-filled cavities in the interior or on the solid boundaries of vaporized liquids in motion where the pressure is reduced to a critical value without a change in ambient temperature.

cavitation number [FL MECH] The excess of the local static pressure head over the vapor pressure head divided by the velocity head.

cavity radiator [THERMO] A heated enclosure with a small opening which allows some radiation to escape or enter; the escaping radiation approximates that of a blackbody.

CAVU [AERO ENG] An operational term commonly used in aviation, which designates a condition wherein the ceiling is more than 10,000 feet (3048 meters) and the visibility is more than 10 miles (16 kilometers). Derived from ceiling and visibility unlimited.

ceiling and visibility unlimited *See* CAVU.

ceiling balloon [AERO ENG] A small balloon used to determine the height of the cloud base; the height is computed from the ascent velocity of the balloon and the time required for its disappearance into the cloud.

celo [MECH] A unit of acceleration equal to the acceleration of a body whose velocity changes uniformly by 1 foot per second (0.3048 meter per second) in 1 second.

Celsius degree [THERMO] Unit of temperature interval or difference equal to the kelvin.

Celsius temperature scale [THERMO] Temperature scale in which the temperature Θ_c in degrees Celsius (°C) is related to the temperature T_k in kelvins by the formula $\Theta_c = T_k - 273.15$; the freezing point of water at standard atmospheric pressure is very nearly 0°C and the corresponding boiling point is very nearly 100°C. Formerly known as centigrade temperature scale.

cement gun [MECH ENG] 1. A machine for mixing, wetting, and applying refractory mortars to hot furnace walls. Also known as cement injector. 2. A mechanical device for the application of cement or mortar to the walls or roofs of mine openings or building walls.

cement injector *See* cement gun.

cement mill [MECH ENG] A mill for grinding rock to a powder for cement.

cement pump [MECH ENG] A piston device used to move concrete through pipes.

cement valve [MECH ENG] A ball-, flapper-, or clack-type valve placed at the bottom of a string of casing, through which cement is pumped, so that when pumping ceases, the valve closes and prevents return of cement into the casing.

centare *See* centiare.

center gage [DES ENG] A gage used to check angles; for example, the angles of cutting tool points or screw threads, or the angular position of cutting tools.

centering machine [MECH ENG] A machine for drilling and countersinking work to be turned on a lathe.

centerless grinder [MECH ENG] A cylindrical metal-grinding machine that carries the work on a support or blade between two abrasive wheels.

center of attraction [MECH] A point toward which a force on a body or particle (such as gravitational or electrostatic force) is always directed; the magnitude of the force depends only on the distance of the body or particle from this point.

center of buoyancy [MECH] The point through which acts the resultant force exerted on a body by a static fluid in which it is submerged or floating; located at the centroid of displaced volume.

center of force [MECH] The point toward or from which a central force acts.

center of gravity [MECH] A fixed point in a material body through which the resultant force of gravitational attraction acts.

center of inertia *See* center of mass.

center of lift [AERO ENG] The mean of all the centers of pressure on an airfoil.

center of mass [MECH] That point of a material body or system of bodies which moves as though the system's total mass existed at the point and all external forces were applied at the point. Also known as center of inertia; centroid.

center of mass coordinate system [MECH] A reference frame which moves with the velocity of the center of mass, so that the center of mass is at rest in this system, and the total momentum of the system is zero. Also known as center of momentum coordinate system.

center of oscillation [MECH] Point in a physical pendulum, on the line through the point of suspension and the center of mass, which moves as if all the mass of the pendulum were concentrated there.

center of percussion [MECH] If a rigid body, free to move in a plane, is struck a blow at a point O, and the line of force is perpendicular to the line from O to the center of mass, then the initial motion of the body is a rotation about the center of percussion relative to O; it can be shown to coincide with the center of oscillation relative to O.

center of pressure [AERO ENG] The point in the chord of an airfoil section which is at the intersection of the chord (prolonged if necessary) and the line of action of the combined air forces (resultant air force).

center of pressure coefficient [AERO ENG] The ratio of the distance of a center of pressure from the leading edge of an airfoil to its chord length.

center of suspension [MECH] The intersection of the axis of rotation of a pendulum with a plane perpendicular to the axis that passes through the center of mass.

center of twist [MECH] A point on a line parallel to the axis of a beam through which any transverse force must be applied to avoid twisting of the section. Also known as shear center.

center plug [DES ENG] A small diamond-set circular plug, designed to be inserted into the annular opening in a core bit, thus converting it to a noncoring bit.

center punch [DES ENG] A tool similar to a prick punch but having the point ground to an angle of about 90°; used to enlarge prick-punch marks or holes.

center square [DES ENG] A straight edge with a sliding square; used to locate the center of a circle.

centiare [MECH] Unit of area equal to 1 square meter. Also spelled centare.

centibar [MECH] A unit of pressure equal to 0.01 bar or to 1000 pascals.

centigrade heat unit [THERMO] A unit of heat energy, equal to 0.01 of the quantity of heat needed to raise 1 pound of air-free water from 0 to 100°C at a constant pressure of 1 standard atmosphere; equal to 1900.44 joules. Symbolized CHU; (more correctly) CHU_{mean}.

centigrade temperature scale *See* Celsius temperature scale.

centigram [MECH] Unit of mass equal to 0.01 gram or 10^{-5} kilogram. Abbreviated cg.

centihg *See* centimeter of mercury.

centiliter [MECH] A unit of volume equal to 0.01 liter or to 10^{-5} cubic meter.

centimeter [MECH] A unit of length equal to 0.01 meter. Abbreviated cm.

centimeter of mercury [MECH] A unit of pressure equal to the pressure that would support a column of mercury 1 centimeter high, having a density of 13.5951 grams per cubic centimeter, when the acceleration of gravity is equal to its standard value (980.665 centimeters per second per second); it is equal to 1333.22387415 pascals;

it differs from the decatorr by less than 1 part in 7,000,000. Abbreviated cmHg. Also known as centihg.

centipoise   [FL MECH] A unit of viscosity which is equal to 0.01 poise. Abbreviated cp.

centistoke   [FL MECH] A cgs unit of kinematic viscosity in customary use, equal to the kinematic viscosity of a fluid having a dynamic viscosity of 1 centipoise and a density of 1 gram per cubic centimeter. Abbreviated cs.

central control   [AERO ENG] The place, facility, or activity at which the whole action incident to a test launch and flight is coordinated and controlled, from the make-ready at the launch site and on the range, to the end of the rocket flight down-range.

central force   [MECH] A force whose line of action is always directed toward a fixed point; the force may attract or repel.

central gear   [MECH ENG] The gear on the central axis of a planetary gear train, about which a pinion rotates. Also known as sun gear.

centrifugal   [MECH] Acting or moving in a direction away from the axis of rotation or the center of a circle along which a body is moving.

centrifugal atomizer   [MECH ENG] Device that atomizes liquids with a spinning disk; liquid is fed onto the center of the disk, and the whirling motion (3000 to 50,000 revolutions per minute) forces the liquid outward in thin sheets to cause atomization.

centrifugal barrier   [MECH] A steep rise, located around the center of force, in the effective potential governing the radial motion of a particle of nonvanishing angular momentum in a central force field, which results from the centrifugal force and prevents the particle from reaching the center of force, or causes its Schrödinger wave function to vanish there in a quantum-mechanical system.

centrifugal brake   [MECH ENG] A safety device on a hoist drum that applies the brake if the drum speed is greater than a set limit.

centrifugal clarification   [MECH ENG] The removal of solids from a liquid by centrifugal action which decreases the settling time of the particles from hours to minutes.

centrifugal classification   [MECH ENG] A type of centrifugal clarification purposely designed to settle out only the large particles (rather than all particles) in a liquid by reducing the centrifuging time.

centrifugal classifier   [MECH ENG] A machine that separates particles into size groups by centrifugal force.

centrifugal clutch   [MECH ENG] A clutch operated by centrifugal force from the speed of rotation of a shaft, as when heavy expanding friction shoes act on the internal surface of a rim clutch, or a flyball-type mechanism is used to activate clutching surfaces on cones and disks.

centrifugal collector   [MECH ENG] Device used to separate particulate matter of 0.1–1000 micrometers from an airstream; some types are simple cyclones, high-efficiency cyclones, and impellers.

centrifugal compressor   [MECH ENG] A machine in which a gas or vapor is compressed by radial acceleration in an impeller with a surrounding casing, and can be arranged multistage for high ratios of compression.

centrifugal discharge elevator   [MECH ENG] A high-speed bucket elevator from which free-flowing materials are discharged by centrifugal force at the top of the loop.

centrifugal fan   [MECH ENG] A machine for moving a gas, such as air, by accelerating it radially outward in an impeller to a surrounding casing, generally of scroll shape.

centrifugal filtration [MECH ENG] The removal of a liquid from a slurry by introducing the slurry into a rapidly rotating basket, where the solids are retained on a porous screen and the liquid is forced out of the cake by the centrifugal action.

centrifugal force [MECH] 1. An outward pseudo-force, in a reference frame that is rotating with respect to an inertial reference frame, which is equal and opposite to the centripetal force that must act on a particle stationary in the rotating frame. 2. The reaction force to a centripetal force.

centrifugal governor [MECH ENG] A governor whose flyweights respond to centrifugal force to sense speed.

centrifugal moment [MECH] The product of the magnitude of centrifugal force acting on a body and the distance to the center of rotation.

centrifugal pump [MECH ENG] A machine for moving a liquid, such as water, by accelerating it radially outward in an impeller to a surrounding volute casing.

centrifugal separation [MECH ENG] The separation of two immiscible liquids in a centrifuge within a much shorter period of time than could be accomplished solely by gravity.

centrifugal switch [MECH ENG] A switch opened or closed by centrifugal force; used on some induction motors to open the starting winding when the motor has almost reached synchronous speed.

centrifugal tachometer [MECH ENG] An instrument which measures the instantaneous angular speed of a shaft by measuring the centrifugal force on a mass rotating with it.

centrifuge [MECH ENG] 1. A rotating device for separating liquids of different specific gravities or for separating suspended colloidal particles, such as clay particles in an aqueous suspension, according to particle-size fractions by centrifugal force. 2. A large motor-driven apparatus with a long arm, at the end of which human and animal subjects or equipment can be revolved and rotated at various speeds to simulate the prolonged accelerations encountered in rockets and spacecraft.

centripetal [MECH] Acting or moving in a direction toward the axis of rotation or the center of a circle along which a body is moving.

centripetal acceleration [MECH] The radial component of the acceleration of a particle or object moving around a circle, which can be shown to be directed toward the center of the circle.

centripetal force [MECH] The radial force required to keep a particle or object moving in a circular path, which can be shown to be directed toward the center of the circle.

centrobaric [MECH] 1. Pertaining to the center of gravity, or to some method of locating it. 2. Possessing a center of gravity.

centrode [MECH] The path traced by the instantaneous center of a plane figure when it undergoes plane motion.

centroid *See* center of mass.

centroid of asymptotes [CONT SYS] The intersection of asymptotes in a root-locus diagram.

ceramic tool [DES ENG] A cutting tool made from metallic oxides.

certainty equivalence control [CONT SYS] An optimal control law for a stochastic adaptive control system which is obtained by solving the control problem in the case of known parameters and substituting the known parameters with their estimates.

cesium-ion engine [AERO ENG] An ion engine that uses a stream of cesium ions to produce a thrust for space travel.

cg *See* centigram.

chain [DES ENG] 1. A flexible series of metal links or rings fitted into one another; used for supporting, restraining, dragging, or lifting objects or transmitting power. 2. A mesh of rods or plates connected together, used to convey objects or transmit power.

chain belt [DES ENG] Belt of flat links to transmit power.

chain block [MECH ENG] A tackle which uses an endless chain rather than a rope, often operated from an overhead track to lift heavy weights especially in workshops. Also known as chain fall; chain hoist.

chain conveyor [MECH ENG] A machine for moving materials that carries the product on one or two endless linked chains with crossbars; allows smaller parts to be added as the work passes.

chain drive [MECH ENG] A flexible device for power transmission, hoisting, or conveying, consisting of an endless chain whose links mesh with toothed wheels fastened to the driving and driven shafts.

chain fall *See* chain block.

chain gear [MECH ENG] A gear that transmits motion from one wheel to another by means of a chain.

chain grate stoker [MECH ENG] A wide, endless chain used to feed, carry, and burn a noncoking coal in a furnace, control the air for combustion, and discharge the ash.

chain hoist *See* chain block.

chain saw [MECH ENG] A gasoline-powered saw for felling and bucking timber, operated by one person; has cutting teeth inserted in a sprocket chain that moves rapidly around the edge of an oval-shaped blade.

chain tongs [DES ENG] A tool for turning pipe, using a chain to encircle and grasp the pipe.

chain vise [DES ENG] A vise in which the work is encircled and held tightly by a chain.

chaldron [MECH] 1. A unit of volume in common use in the United Kingdom, equal to 36 bushels, or 288 gallons, or approximately 1.30927 cubic meters. 2. A unit of volume, formerly used for measuring solid substances in the United States, equal to 36 bushels, or approximately 1.26861 cubic meters.

chamber capacity *See* chamber volume.

chamber pressure [AERO ENG] The pressure of gases within the combustion chamber of a rocket engine.

chamber volume [AERO ENG] The volume of the rocket combustion chamber, including the convergent portion of the nozzle up to the throat. Also known as chamber capacity.

chamfer angle [DES ENG] The angle that a beveled surface makes with one of the original surfaces.

chamfering [MECH ENG] Machining operations to produce a beveled edge. Also known as beveling.

chamfer plane [DES ENG] A plane for chamfering edges of woodwork.

change gear [MECH ENG] A gear used to change the speed of a driven shaft while the speed of the driving remains constant.

channeling machine [MECH ENG] An electrically powered machine that operates by a chipping action of three to five chisels while traveling back and forth on a track;

used for primary separation from the rock ledge in marble, limestone, and soft sandstone quarries. Also known as channeler.

channel iron [DES ENG] A metal strip or beam with a U-shape.

channel wing [AERO ENG] A wing that is trough-shaped so as to surround partially a propeller to get increased lift at low speeds from the slipstream.

Chaplygin-Kármán-Tsien relation [FL MECH] The relation that in the case of isentropic flow of an ideal gas with negligible viscosity and thermal conductivity, the sum of the pressure and a constant times the reciprocal of the density of the fluid is constant along a streamline; a useful, although physically impossible, approximation.

Chapman-Jouguet plane [MECH] A hypothetical, infinite plane, behind the initial shock front, in which it is variously assumed that reaction (and energy release) has effectively been completed, that reaction product gases have reached thermodynamic equilibrium, and that reaction gases, streaming backward out of the detonation, have reached such a condition that a forward-moving sound wave located at this precise plane would remain a fixed distance behind the initial shock.

characteristic chamber length [AERO ENG] The length of a straight, cylindrical tube having the same volume as that of the chamber of a rocket engine if the chamber had no converging section.

characteristic exhaust velocity [AERO ENG] Of a rocket engine, a descriptive parameter, related to effective exhaust velocity and thrust coefficient. Also known as characteristic velocity.

characteristic length [MECH] A convenient reference length (usually constant) of a given configuration, such as overall length of an aircraft, the maximum diameter or radius of a body of revolution, or a chord or span of a lifting surface.

characteristic velocity *See* characteristic exhaust velocity.

charge [MECH ENG] 1. In refrigeration, the quantity of refrigerant contained in a system. 2. To introduce the refrigerant into a refrigeration system.

chase [DES ENG] A series of cuts, each having a path that follows the path of the cut before it; an example is a screw thread.

chase mortise [DES ENG] A mortise with a sloping edge from bottom to surface so that a tenon can be inserted when the outside clearance is small.

chase pilot [AERO ENG] A pilot who flies an escort airplane and advises another pilot who is making a check, training, or research flight in another craft.

chaser [AERO ENG] The vehicle that maneuvers in order to effect a rendezvous with an orbiting object.

chase ring [MECH ENG] In hobbing, the ring which restrains the blank from spreading during hob sinking.

chasing tool [DES ENG] A hammer or chisel used to decorate metal surfaces.

chassis punch [DES ENG] A hand tool used to make round or square holes in sheet metal.

chattering [CONT SYS] A mode of operation of a relay-type control system in which the relay switches back and forth infinitely fast.

check flight [AERO ENG] 1. A flight made to check or test the performance of an aircraft, rocket, or spacecraft, or a piece of its equipment, or to obtain measurements or other data on performance. 2. A familiarization flight in an aircraft, or a flight in which the pilot or the aircrew are tested for proficiency.

cheese head [DES ENG] A raised cylindrical head on a screw or bolt.

chemical pressurization [AERO ENG] The pressurization of propellant tanks in a rocket by means of high-pressure gases developed by the combustion of a fuel and oxidizer or by the decomposition of a substance.

cherry picker [AERO ENG] A crane used to remove the aerospace capsule containing astronauts from the top of the rocket in the event of a malfunction. [MECH ENG] Any of several small traveling cranes, especially one used to hoist a passenger on the end of a boom.

Chézy formula [FL MECH] For the velocity V of open-channel flow which is steady and uniform, $V = \sqrt{8g/f} \cdot \sqrt{mS}$, where f is the Darcy-Weisbach friction coefficient, m the hydraulic radius, S the energy dissipation per unit length, and g the acceleration of gravity.

Chicago boom [MECH ENG] A hoisting device that is supported on the structure being erected.

Chile mill [MECH ENG] A crushing mill having vertical rollers running in a circular enclosure with a stone or iron base or die. Also known as edge runner.

chimney core [MECH ENG] The inner section of a double-walled chimney which is separated from the outer section by an air space.

chimney effect [FL MECH] The tendency of air or gas in a vertical passage to rise when it is heated because its density is lower than that of the surrounding air or gas.

chip breaker [DES ENG] An irregularity or channel cut into the face of a lathe tool behind the cutting edge to cause removed stock to break into small chips or curls.

chip cap [DES ENG] A plate or cap on the upper part of the cutting iron of a carpenter's plane designed to give the tool rigidity and also to break up the wood shavings.

chisel [DES ENG] A tool for working the surface of various materials, consisting of a metal bar with a sharp edge at one end and often driven by a mallet.

chisel bit *See* chopping bit.

chisel-edge angle [DES ENG] The angle included between the chisel edge and the cutting edge, as seen from the end of the drill.

chisel-tooth saw [DES ENG] A circular saw with chisel-shaped cutting edges.

Chladni's figures [MECH] Figures produced by sprinkling sand or similar material on a horizontal plate and then vibrating the plate while holding it rigid at its center or along its periphery; indicate the nodal lines of vibration.

choke [MECH ENG] 1. To increase the fuel feed to an internal combustion engine through the action of a choke valve. 2. *See* choke valve.

choked flow [FL MECH] Flow in a duct or passage such that the flow upstream of a certain critical section cannot be increased by a reduction of downstream pressure.

choked neck [DES ENG] Container neck which has a narrowed or constricted opening.

choke valve [MECH ENG] A valve which supplies the higher suction necessary to give the excess fuel feed required for starting a cold internal combustion engine. Also known as choke.

choking [FL MECH] The condition prevailing in compressible fluid flow when the upper limit of mass flow is reached, or when the speed of sound is reached in a duct.

choking Mach number [FL MECH] The Mach number at some reference point in a duct or passage (for example, at the inlet) at which the flow in the passage becomes choked.

chopping bit [MECH ENG] A steel bit with a chisel-shaped cutting edge, attached to a string of drill rods to break up, by impact, boulders, hardpan, and a lost core in a drill hole. Also known as chisel bit.

chop-type feeder [MECH ENG] Device for semicontinuous feed of solid materials to a process unit, with intermittent opening and closing of a hopper gate (bottom closure) by a control arm actuated by an eccentric cam.

chord [AERO ENG] 1. A straight line intersecting or touching an airfoil profile at two points. 2. Specifically, that part of such a line between two points of intersection.

chordal thickness [DES ENG] The tangential thickness of a tooth on a circular gear, as measured along a chord of the pitch circle.

chord length [AERO ENG] The length of the chord of an airfoil section between the extremities of the section.

CHU *See* centigrade heat unit.

CHU_mean *See* centigrade heat unit.

chuck [DES ENG] A device for holding a component of an instrument rigid, usually by means of adjustable jaws or set screws, such as the workpiece in a metalworking or woodworking machine, or the stylus or needle of a phonograph pickup.

chucking [MECH ENG] The grasping of an outsize workpiece in a chuck or jawed device in a lathe.

chucking machine [MECH ENG] A lathe or grinder in which the outsize workpiece is grasped in a chuck or jawed device.

chuffing *See* chugging.

chugging [AERO ENG] Also known as bumping; chuffing. 1. A form of combustion instability in a rocket engine, characterized by a pulsing operation at a fairly low frequency, sometimes defined as occurring between particular frequency limits. 2. The noise that is made in this kind of combustion.

churn drill [MECH ENG] Portable drilling equipment, with drilling performed by a heavy string of tools tipped with a blunt-edge chisel bit suspended from a flexible cable, to which a reciprocating motion is imparted by its suspension from an oscillating beam or sheave, causing the bit to be raised and dropped. Also known as American system drill; cable-system drill.

churn shot drill [MECH ENG] A boring rig with both churn and shot drillings.

circle error probable *See* circle of equal probability.

circle of equal probability [AERO ENG] A measure of the accuracy with which a rocket or missile can be guided; the radius of the circle at a specific distance in which 50% of the reliable shots land. Also known as circle of probable error; circular error probable.

circle of probable error *See* circle of equal probability.

circle shear [MECH ENG] A shearing machine that cuts circular disks from a metal sheet rolling between the cutting wheels.

circular cutter [MECH ENG] A rotating blade with a square or knife edge used to slit or shear metal.

circular flow method [FL MECH] A method to determine viscosities of Newtonian fluids by measuring the torque from viscous drag of sample material between a closely spaced rotating plate–stationary cone assembly.

circular form tool [DES ENG] A round or disk-shaped tool with the cutting edge on the periphery.

circular inch [MECH] The area of a circle 1 inch (25.4 millimeters) in diameter.

circular mil [MECH] A unit equal to the area of a circle whose diameter is 1 mil (0.001 inch); used chiefly in specifying cross-sectional areas of round conductors. Abbreviated cir mil.

circular motion [MECH] 1. Motion of a particle in a circular path. 2. Motion of a rigid body in which all its particles move in circles about a common axis, fixed with respect to the body, with a common angular velocity.

circular pitch [DES ENG] The linear measure in inches along the pitch circle of a gear between corresponding points of adjacent teeth.

circular plane [DES ENG] A plane that can be adjusted for convex or concave surfaces.

circular saw [MECH ENG] Any of several power tools for cutting wood or metal, having a thin steel disk with a toothed edge that rotates on a spindle.

circular velocity [MECH] At any specific distance from the primary, the orbital velocity required to maintain a constant-radius orbit.

circulation [FL MECH] The flow or motion of fluid in or through a given area or volume.

cir mil *See* circular mil.

clamp [DES ENG] A tool for binding or pressing two or more parts together, by holding them firmly in their relative positions.

clamping coupling [MECH ENG] A coupling with a split cylindrical element which clamps the shaft ends together by direct compression, through bolts or rings, and by the wedge action of conical sections; not considered a permanent part of the shaft.

clamp screw [DES ENG] A screw that holds a part by forcing it against another part.

clamshell bucket [MECH ENG] A two-sided bucket used in a type of excavator to dig in a vertical direction; the bucket is dropped while its leaves are open and digs as they close. Also known as clamshell grab.

clamshell grab *See* clamshell bucket.

clamshell snapper [MECH ENG] A marine sediment sampler consisting of snapper jaws and a footlike projection which, upon striking the bottom, causes a spring mechanism to close the jaws, thus trapping a sediment sample.

Clapeyron-Clausius equation *See* Clausius-Clapeyron equation.

Clapeyron equation *See* Clausius-Clapeyron equation.

Clapeyron's theorem [MECH] The theorem that the strain energy of a deformed body is equal to one-half the sum over three perpendicular directions of the displacement component times the corresponding force component, including deforming loads and body forces, but not the six constraining forces required to hold the body in equilibrium.

clapper box [MECH ENG] A hinged device that permits a reciprocating cutting tool (as in a planer or shaper) to clear the work on the return stroke.

clarifying centrifuge [MECH ENG] A device that clears liquid of foreign matter by centrifugation.

clasp [DES ENG] A releasable catch which holds two or more objects together.

clasp lock [DES ENG] A spring lock with a self-locking feature.

clasp nut [DES ENG] A split nut that clasps a screw when closed around it.

classical mechanics [MECH] Mechanics based on Newton's laws of motion.

classifier [MECH ENG] Any apparatus for separating mixtures of materials into their constituents according to size and density.

clausius [THERMO] A unit of entropy equal to the increase in entropy associated with the absorption of 1000 international table calories of heat at a temperature of 1 K, or to 4186.8 joules per kelvin.

Clausius-Clapeyron equation [THERMO] An equation governing phase transitions of a substance, $dp/dT = \Delta H/(T\Delta V)$, in which p is the pressure, T is the temperature at which the phase transition occurs, ΔH is the change in heat content (enthalpy), and ΔV is the change in volume during the transition. Also known as Clapeyron-Clausius equation; Clapeyron equation.

Clausius equation [THERMO] An equation of state in reference to gases which applies a correction to the van der Waals equation: $\{P + (n^2a/[T(V + c)^2])\} (V - nb) = nRT$, where P is the pressure, T the temperature, V the volume of the gas, n the number of moles in the gas, R the gas constant, a depends only on temperature, b is a constant, and c is a function of a and b.

Clausius inequality [THERMO] The principle that for any system executing a cyclical process, the integral over the cycle of the infinitesimal amount of heat transferred to the system divided by its temperature is equal to or less than zero. Also known as Clausius theorem; inequality of Clausius.

Clausius law [THERMO] The law that an ideal gas's specific heat at constant volume does not depend on the temperature.

Clausius number [THERMO] A dimensionless number used in the study of heat conduction in forced fluid flow, equal to $V^3L\rho/k\Delta T$, where V is the fluid velocity, ρ is its density, L is a characteristic dimension, k is the thermal conductivity, and ΔT is the temperature difference.

Clausius' statement [THERMO] A formulation of the second law of thermodynamics, stating it is not possible that, at the end of a cycle of changes, heat has been transferred from a colder to a hotter body without producing some other effect.

Clausius theorem *See* Clausius inequality.

claw [DES ENG] A fork for removing nails or spikes.

claw clutch [MECH ENG] A clutch consisting of claws that interlock when pushed together.

claw coupling [MECH ENG] A loose coupling having projections or claws cast on each face which engage in corresponding notches in the opposite faces; used in situations in which shafts require instant connection.

claw hammer [DES ENG] A woodworking hammer with a flat working surface and a claw to pull nails.

clay digger [MECH ENG] A power-driven, hand-held spade for digging hard soil or soft rock.

cleanup [AERO ENG] Improving the external shape and smoothness of an aircraft to reduce its drag.

clearance [MECH ENG] 1. In a piston-and-cylinder mechanism, the space at the end of the cylinder when the piston is at dead-center position toward the end of the cylinder. 2. The ratio of the volume of this space to the piston displacement during a stroke.

clearance angle [MECH ENG] The angle between a plane containing the end surface of a cutting tool and a plane passing through the cutting edge in the direction of cutting motion.

clearance volume [MECH ENG] The volume remaining between piston and cylinder when the piston is at top dead center.

cleat [DES ENG] A fitting having two horizontally projecting horns around which a rope may be made fast.

clevis [DES ENG] A U-shaped metal fitting with holes in the open ends to receive a bolt or pin; used for attaching or suspending parts.

clevis pin [DES ENG] A fastener with a head at one end, used to join the ends of a clevis.

climate control *See* air conditioning.

climb [AERO ENG] The gain in altitude of an aircraft.

climbing crane [MECH ENG] A crane used on top of a high-rise construction that ascends with the building as work progresses.

climbing irons [DES ENG] Spikes attached to a steel framework worn on shoes to climb wooden utility poles and trees.

clinker building [DES ENG] A method of building ships and boilers in which the edge of the wooden planks or steel plates used for the outside covering overlap the edge of the plank or plate next to it; clinched nails fasten the planks together, and rivets fasten the steel plates.

clip [DES ENG] A device that fastens by gripping, clasping, or hooking one part to another.

clock control system [CONT SYS] A system in which a timing device is used to generate the control function. Also known as time-controlled system.

close-coupled pump [MECH ENG] Pump with built-in electric motor (sometimes a steam turbine), with the motor drive and pump impeller on the same shaft.

closed-belt conveyor [MECH ENG] Solids-conveying device with zipperlike teeth that mesh to form a closed tube wrapped snugly around the conveyed material; used with fragile materials.

closed cycle [THERMO] A thermodynamic cycle in which the thermodynamic fluid does not enter or leave the system, but is used over and over again.

closed-cycle turbine [MECH ENG] A gas turbine in which essentially all the working medium is continuously recycled, and heat is transferred through the walls of a closed heater to the cycle.

closed ecological system [AERO ENG] A system used in spacecraft that provides for the maintenance of life in an isolated living chamber through complete reutilization of the material available, in particular, by means of a cycle wherein exhaled carbon dioxide, urine, and other waste matter are converted chemically or by photosynthesis into oxygen, water, and food.

closed fireroom system [MECH ENG] A fireroom system in which combustion air is supplied via forced draft resulting from positive air pressure in the fireroom.

closed loop [CONT SYS] A family of automatic control units linked together with a process to form an endless chain; the effects of control action are constantly measured so that if the controlled quantity departs from the norm, the control units act to bring it back.

closed-loop control system *See* feedback control system.

closed nozzle [MECH ENG] A fuel nozzle having a built-in valve interposed between the fuel supply and combustion chamber.

closed pair [MECH] A pair of bodies that are subject to constraints which prevent any relative motion between them.

closed system [THERMO] A system which is isolated so that it cannot exchange matter or energy with its surroundings and can therefore attain a state of thermodynamic equilibrium.

closing line [MECH] The vector required to complete a polygon consisting of a set of vectors whose sum is zero (such as the forces acting on a body in equilibrium).

closing pressure [MECH ENG] The amount of static inlet pressure in a safety relief valve when the valve disk has a zero lift above the seat.

closing rate [AERO ENG] The speed at which two aircraft or missiles come closer together.

cloth wheel [DES ENG] A polishing wheel made of sections of cloth glued or sewn together.

clout nail [DES ENG] A nail with a large, thin, flat head used in building.

clusec [MECH ENG] A unit of power used to measure the power of evacuation of a vacuum pump, equal to the power associated with a leak rate of 1 centiliter per second at a pressure of 1 millitorr, or to approximately 1.33322×10^{-6} watt.

clutch [MECH ENG] A machine element for the connection and disconnection of shafts in equipment drives, especially while running.

cm *See* centimeter.

cmHg *See* centimeter of mercury.

coach screw [DES ENG] A large, square-headed, wooden screw used to join heavy timbers. Also known as lag bolt; lag screw.

Coanda effect [FL MECH] The tendency of a gas or liquid coming out of a jet to travel close to the wall contour even if the wall's direction of curvature is away from the jet's axis; a factor in the operation of a fluidic element.

coasting flight [AERO ENG] The flight of a rocket between burnout or thrust cutoff of one stage and ignition of another, or between burnout and summit altitude or maximum horizontal range.

coaxial [MECH] Sharing the same axes. [MECH ENG] Mounted on independent concentric shafts.

cockpit [AERO ENG] A space in an aircraft or spacecraft where the pilot sits.

coefficient of compressibility [MECH] The decrease in volume per unit volume of a substance resulting from a unit increase in pressure; it is the reciprocal of the bulk modulus.

coefficient of contraction [FL MECH] The ratio of the minimum cross-sectional area of a jet of liquid discharging from an orifice to the area of the orifice. Also known as contraction coefficient.

coefficient of cubical expansion [THERMO] The increment in volume of a unit volume of solid, liquid, or gas for a rise of temperature of 1° at constant pressure. Also known as coefficient of expansion; coefficient of thermal expansion; expansion coefficient; expansivity.

coefficient of discharge *See* discharge coefficient.

coefficient of eddy diffusion *See* eddy diffusivity.

coefficient of elasticity *See* modulus of elasticity.

coefficient of expansion *See* coefficient of cubical expansion.

coefficient of friction [MECH] The ratio of the frictional force between two bodies in contact, parallel to the surface of contact, to the force, normal to the surface of contact, with which the bodies press against each other. Also known as friction coefficient.

coefficient of friction of rest *See* coefficient of static friction.

coefficient of kinematic viscosity *See* kinematic viscosity.

coefficient of kinetic friction [MECH] The ratio of the frictional force, parallel to the surface of contact, that opposes the motion of a body which is sliding or rolling over another, to the force, normal to the surface of contact, with which the bodies press against each other.

coefficient of linear expansion [THERMO] The increment of length of a solid in a unit of length for a rise in temperature of 1° at constant pressure.

coefficient of performance [THERMO] In a refrigeration cycle, the ratio of the heat energy extracted by the heat engine at the low temperature to the work supplied to operate the cycle; when used as a heating device, it is the ratio of the heat delivered in the high-temperature coils to the work supplied.

coefficient of permeability *See* permeability coefficient.

coefficient of resistance [FL MECH] The ratio of the loss of head of fluid, issuing from an orifice or passing over a weir, to the remaining head.

coefficient of restitution [MECH] The constant e, which is the ratio of the relative velocity of two elastic spheres after direct impact to that before impact; e can vary from 0 to 1, with 1 equivalent to an elastic collision and 0 equivalent to a perfectly elastic collision.

coefficient of rigidity *See* modulus of elasticity in shear.

coefficient of rolling friction [MECH] The ratio of the frictional force, parallel to the surface of contact, opposing the motion of a body rolling over another, to the force, normal to the surface of contact, with which the bodies press against each other.

coefficient of sliding friction [MECH] The ratio of the frictional force, parallel to the surface of contact, opposing the motion of a body sliding over another, to the force, normal to the surface of contact, with which the bodies press against each other.

coefficient of static friction [MECH] The ratio of the maximum possible frictional force, parallel to the surface of contact, which acts to prevent two bodies in contact, and at rest with respect to each other, from sliding or rolling over each other, to the force, normal to the surface of contact, with which the bodies press against each other. Also known as coefficient of friction of rest.

coefficient of strain [MECH] For a substance undergoing a one-dimensional strain, the ratio of the distance along the strain axis between two points in the body, to the distance between the same points when the body is undeformed.

coefficient of thermal expansion *See* coefficient of cubical expansion.

coffin corner [AERO ENG] The range of Mach numbers between the buffeting Mach number and the stalling Mach number within which an aircraft must be operated.

cog [DES ENG] A tooth on the edge of a wheel.

cog belt [MECH ENG] A flexible device used for timing and for slip-free power transmission.

cogeneration [MECH ENG] The simultaneous on-site generation of electric energy and process steam or heat from the same plant.

cogwheel [DES ENG] A wheel with teeth around its edge.

cohesive strength [MECH] **1.** Strength corresponding to cohesive forces between atoms. **2.** Hypothetically, the stress causing tensile fracture without plastic deformation.

coil spring [DES ENG] A helical or spiral spring, such as one of the helical springs used over the front wheels in an automotive suspension.

coke breeze [MECH ENG] Undersized coke screenings passing through a screen opening of approximately ⅝ inch (16 millimeters).

coke knocker [MECH ENG] A mechanical device used to break loose coke within a drum or tower.

Colburn analogy [FL MECH] Dimensionless Reynolds equation for fluid-flow resistance modified to be analogous to the Colburn *j* factor heat-transfer equation.

Colburn j factor equation [THERMO] Dimensionless heat-transfer equation to calculate the natural convection movement of heat from vertical surfaces or horizontal cylinders to fluids (gases or liquids) flowing past these surfaces.

cold-air machine [MECH ENG] A refrigeration system in which air serves as the refrigerant in a cycle of adiabatic compression, cooling to ambient temperature, and adiabatic expansion to refrigeration temperature; the air is customarily reused in a closed superatmospheric pressure system. Also known as dense-air system.

cold chisel [DES ENG] A chisel specifically designed to cut or chip cold metal; made of specially tempered tool steel machined into various cutting edges. Also known as cold cutter.

cold cutter *See* cold chisel.

cold flow [MECH] Creep in polymer plastics.

cold-flow test [AERO ENG] A test of a liquid rocket without firing it to check or verify the integrity of a propulsion subsystem, and to provide for the conditioning and flow of propellants (including tank pressurization, propellant loading, and propellant feeding).

cold plate [MECH ENG] An aluminum or other plate containing internal tubing through which a liquid coolant is forced, to absorb heat transferred to the plate by transistors and other components mounted on it. Also known as liquid-cooled dissipator.

cold saw [MECH ENG] **1.** Any saw for cutting cold metal, as opposed to a hot saw. **2.** A disk made of soft steel or iron which rotates at a speed such that a point on its edge has a tangential velocity of about 15,000 feet per minute (75 meters per second), and which grinds metal by friction.

cold stress [MECH] Forces tending to deform steel, cement, and other materials, resulting from low temperatures.

cold trap [MECH ENG] A tube whose walls are cooled with liquid nitrogen or some other liquid to condense vapors passing through it; used with diffusion pumps and to keep vapors from entering a McLeod gage.

Colebrook equation [FL MECH] An empirical equation for the flow of liquids in ducts, relating the friction factor to the Reynolds number and the relative roughness of the duct.

coleopter [AERO ENG] An aircraft having an annular (barrel-shaped) wing, the engine and body being mounted within the circle of the wing.

collapse properties [MECH] Strength and dimensional attributes of piping, tubing, or process vessels, related to the ability to resist collapse from exterior pressure or internal vacuum.

collapsing pressure [MECH] The external pressure which causes a thin-walled body or structure to collapse.

collar [DES ENG] A ring placed around an object to restrict its motion, hold it in place, or cover an opening.

collar bearing [MECH ENG] A bearing that resists the axial force of a collar on a rotating shaft.

collet [DES ENG] A split, coned sleeve to hold small, circular tools or work in the nose of a lathe or other type of machine.

collision parameter [AERO ENG] In orbit computation, the distance between a center of attraction of a central force field and the extension of the velocity vector of a moving object at a great distance from the center.

colloid mill [MECH ENG] A grinding mill for the making of very fine dispersions of liquids or solids by breaking down particles in an emulsion or paste.

column crane [MECH ENG] A jib crane whose boom pivots about a post attached to a building column.

column drill [MECH ENG] A tunnel rock drill supported by a vertical steel column.

combination chuck [DES ENG] A chuck used in a lathe whose jaws either move independently or simultaneously.

combination collar [DES ENG] A collar that has left-hand threads at one end and right-hand threads at the other.

combination cycle *See* mixed cycle.

combination pliers [DES ENG] Pliers that can be used either for holding objects or for cutting and bending wire.

combination saw [MECH ENG] A saw made in various tooth arrangement combinations suitable for ripping and crosscut mitering.

combination square [DES ENG] A square head and steel rule that when used together have both a 45 and 90° face to allow the testing of the accuracy of two surfaces intended to have these angles.

combination wrench [DES ENG] A wrench that is an open-end wrench at one end and a socket wrench at the other.

combined flexure [MECH] The flexure of a beam under a combination of transverse and longitudinal loads.

combined stresses [MECH] Bending or twisting stresses in a structural member combined with direct tension or compression.

combustion chamber [AERO ENG] That part of the rocket engine in which the combustion of propellants takes place at high pressure. Also known as firing chamber. [MECH ENG] The space at the head end of an internal combustion engine cylinder where most of the combustion takes place.

combustion chamber volume [MECH ENG] The volume of the combustion chamber when the piston is at top dead center.

combustion engine [MECH ENG] An engine that operates by the energy of combustion of a fuel.

combustion engineering [MECH ENG] The design of combustion furnaces for a given performance and thermal efficiency, involving study of the heat liberated in the combustion process, the amount of heat absorbed by heat elements, and heat-transfer rates.

combustion instability [AERO ENG] Unsteadiness or abnormality in the combustion of fuel, as may occur in a rocket engine.

combustion knock *See* engine knock.

combustion turbine *See* gas turbine.

combustor [MECH ENG] The combustion chamber together with burners, igniters, and injection devices in a gas turbine or jet engine.

come-along [DES ENG] 1. A device for gripping and effectively shortening a length of cable, wire rope, or chain by means of two jaws which close when one pulls on a ring. 2. *See* puller.

command [CONT SYS] An independent signal in a feedback control system, from which the dependent signals are controlled in a predetermined manner.

command destruct [CONT SYS] A command control system that destroys a flightborne test rocket or a guided missile, actuated by the safety officer whenever the vehicle's performance indicates a safety hazard.

command module [AERO ENG] The spacecraft module that carries the crew, the main communication and telemetry equipment, and the reentry capsule during cruising flight.

commercial diesel cycle *See* mixed cycle.

comminution [MECH ENG] Breaking up or grinding into small fragments. Also known as pulverization.

comminutor [MECH ENG] A machine that breaks up solids.

communications satellite [AERO ENG] An orbiting, artificial earth satellite that relays radio, television, and other signals between ground terminal stations thousands of miles apart. Also known as radio relay satellite; relay satellite.

compactor [MECH ENG] Machine designed to consolidate earth and paving materials by kneading, weight, vibration, or impact, to sustain loads greater than those sustained in an uncompacted state.

companion body [AERO ENG] A nose cone, last-stage rocket, or other body that orbits along with an earth satellite or follows a space probe.

companion flange [DES ENG] A pipe flange that can be bolted to a similar flange on another pipe.

comparator [CONT SYS] A device which detects the value of the quantity to be controlled by a feedback control system and compares it continuously with the desired value of that quantity.

compartment mill [MECH ENG] A multisection pulverizing device divided by perforated partitions, with preliminary grinding at one end in a short ball-mill operation, and finish grinding at the discharge end in a longer tube-mill operation.

compass card [DES ENG] The part of a compass on which the direction graduations are placed, it is usually in the form of a thin disk or annulus graduated in degrees, clockwise from 0° at the reference direction to 360°, and sometimes also in compass points.

compass card axis [DES ENG] The line joining 0° and 180° on a compass card.

compass saw [DES ENG] A handsaw which has a handle with several attachable thin, tapering blades of varying widths, making it suitable for a variety of work, such as cutting circles and curves.

compensated pendulum [DES ENG] A pendulum made of two materials with different coefficients of expansion so that the distance between the point of suspension and center of oscillation remains nearly constant when the temperature changes.

compensating network [CONT SYS] A network used in a low-energy-level method for suppression of excessive oscillations in a control system.

compensation [CONT SYS] Introduction of additional equipment into a control system in order to reshape its root locus so as to improve system performance. Also known as stabilization.

compensator [CONT SYS] A device introduced into a feedback control system to improve performance and achieve stability. Also known as filter.

complete-expansion diesel cycle *See* Brayton cycle.

complex potential [FL MECH] An analytic function in ideal aerodynamics whose real part is the velocity potential and whose imaginary part is the stream function.

complex velocity [FL MECH] In ideal aerodynamic flow, the derivative of the complex potential with respect to $z = x + iy$, where x and y are the chosen coordinates.

compliance [MECH] The displacement of a linear mechanical system under a unit force.

compliance constant [MECH] Any one of the coefficients of the relations in the generalized Hooke's law used to express strain components as linear functions of the stress components. Also known as elastic constant.

composition of forces [MECH] The determination of a force whose effect is the same as that of two or more given forces acting simultaneously; all forces are considered acting at the same point.

composition of velocities law [MECH] A law relating the velocities of an object in two references frames which are moving relative to each other with a specified velocity.

compound engine [MECH ENG] A multicylinder-type displacement engine, using steam, air, or hot gas, where expansion proceeds successively (sequentially).

compounding [MECH ENG] The series placing of cylinders in an engine (such as steam) for greater ratios of expansion and consequent improved engine economy.

compound lever [MECH ENG] A train of levers in which motion or force is transmitted from the arm of one lever to that of the next.

compound rest [MECH ENG] A principal component of a lathe consisting of a base and an upper part dovetailed together; the base is graduated in degrees and can be swiveled to any angle; the upper part includes the tool post and tool holder.

compound screw [DES ENG] A screw having different or opposite pitches on opposite ends of the shank.

compound wave [FL MECH] A plane wave of finite amplitude in which neither the sum of the velocity potential and the component of velocity in the direction of wave motion, nor the difference of these two quantities, is constant.

compressadensity function [MECH] A function used in the acoustic levitation technique to determine either the density or the adiabatic compressibility of a submicroliter droplet suspended in another liquid, if the other property is known.

compressed air [MECH] Air whose density is increased by subjecting it to a pressure greater than atmospheric pressure.

compressed-air power [MECH ENG] The power delivered by the pressure of compressed air as it expands, utilized in tools such as drills, in hoists, grinders, riveters, diggers, pile drivers, motors, locomotives, and mine ventilating systems.

compressibility [MECH] The property of a substance capable of being reduced in volume by application of pressure; quantitively, the reciprocal of the bulk modulus.

compressibility burble [FL MECH] A region of disturbed flow, produced by and rearward of a shock wave.

compressibility correction [FL MECH] The correction of the calibrated airspeed caused by compressibility error.

compressibility error [FL MECH] The error in the readings of a differential-pressure-type airspeed indicator due to compression of the air on the forward part of the pitot tube component moving at high speeds.

compressibility factor [THERMO] The product of the pressure and the volume of a gas, divided by the product of the temperature of the gas and the gas constant; this factor may be inserted in the ideal gas law to take into account the departure of true gases from ideal gas behavior. Also known as deviation factor; gas-deviation factor; supercompressibility factor.

compressible flow [FL MECH] Flow in which the fluid density varies.

compressible-flow principle [FL MECH] The principle that when flow velocity is large, it is necessary to consider that the fluid is compressible rather than to assume that it has a constant density.

compression [MECH] Reduction in the volume of a substance due to pressure; for example in building, the type of stress which causes shortening of the fibers of a wooden member. [MECH ENG] See compression ratio.

compression coupling [MECH ENG] 1. A means of connecting two perfectly aligned shafts in which a slotted tapered sleeve is placed over the junction and two flanges are drawn over the sleeve so that they automatically center the shafts and provide sufficient contact pressure to transmit medium loads. 2. A type of tubing fitting.

compression ignition [MECH ENG] Ignition produced by compression of the air in a cylinder of an internal combustion engine before fuel is admitted.

compression-ignition engine See diesel engine.

compression machine See compressor.

compression modulus See bulk modulus of elasticity.

compression pressure [MECH ENG] That pressure developed in a reciprocating piston engine at the end of the compression stroke without combustion of fuel.

compression ratio [MECH ENG] The ratio in internal combustion engines between the volume displaced by the piston plus the clearance space, to the volume of the clearance space. Also known as compression.

compression refrigeration [MECH ENG] The cooling of a gaseous refrigerant by first compressing it to liquid form (with resultant heat buildup), cooling the liquid by heat exchange, then releasing pressure to allow the liquid to vaporize (with resultant absorption of latent heat of vaporization and a refrigerative effect).

compression release [MECH ENG] Release of compressed gas resulting from incomplete closure of intake or exhaust valves.

compression strength [MECH] Property of a material to resist rupture under compression.

compression stroke [MECH ENG] The phase of a positive displacement engine or compressor in which the motion of the piston compresses the fluid trapped in the cylinder.

compression wave [FL MECH] A wave in a fluid in which a compression is propagated.

compressive strength [MECH] The maximum compressive stress a material can withstand without failure.

compressive stress [MECH] A stress which causes an elastic body to shorten in the direction of the applied force.

compressor [MECH ENG] A machine used for increasing the pressure of a gas or vapor. Also known as compression machine.

compressor blade [MECH ENG] The vane components of a centrifugal or axial-flow, air or gas compressor.

compressor station [MECH ENG] A permanent facility which increases the pressure on gas to move it in transmission lines or into storage.

compressor valve [MECH ENG] A valve in a compressor, usually automatic, which operates by pressure difference (less than 5 pounds per square inch or 35 kilopascals) on the two sides of a movable, single-loaded member and which has no mechanical linkage with the moving parts of the compressor mechanism.

computer control [CONT SYS] Process control in which the process variables are fed into a computer and the output of the computer is used to control the process.

computer-controlled system [CONT SYS] A feedback control system in which a computer operates on both the input signal and the feedback signal to effect control.

concave bit [DES ENG] A type of tungsten carbide drill bit having a concave cutting edge; used for percussive boring.

concentrated load [MECH] A force that is negligible because of a small contact area; a beam supported on a girder represents a concentrated load on the girder.

concentric groove *See* locked groove.

concentric locating [DES ENG] The process of making the axis of a tooling device coincide with the axis of the workpiece.

concentric orifice plate [DES ENG] A fluid-meter orifice plate whose edges have a circular shape and whose center coincides with the center of the pipe.

concrete mixer [MECH ENG] A machine with a rotating drum in which the components of concrete are mixed.

concrete pump [MECH ENG] A device which drives concrete to the placing position through a pipeline of 6-inch (15-centimeter) diameter or more, using a special type of reciprocating pump.

concrete vibrator [MECH ENG] Vibrating device used to achieve proper consolidation of concrete; the three types are internal, surface, and form vibrators.

condensate strainer [MECH ENG] A screen used to remove solid particles from the condensate prior to its being pumped back to the boiler.

condensate well [MECH ENG] A chamber into which condensed vapor falls for convenient accumulation prior to removal.

condensation [MECH] An increase in density.

condensation shock wave [FL MECH] A sheet of discontinuity associated with a sudden condensation and fog formation in a field of flow; it occurs, for example, on a wing where a rapid drop in pressure causes the temperature to drop considerably below the dew point.

condenser [MECH ENG] A heat-transfer device that reduces a thermodynamic fluid from its vapor phase to its liquid phase, suchas in a vapor-compression refrigeration plant or in a condensing steam power plant.

condenser tubes [MECH ENG] Metal tubes used in a heat-transfer device, with condenser vapor as the heat source and flowing liquid such as water as the receiver.

condensing engine [MECH ENG] A steam engine in which the steam exhausts from the cylinder to a vacuum space, where the steam is liquefied.

condensing flow [FL MECH] The flow and simultaneous condensation (partial or complete) of vapor through a cooled pipe or other closed conduit or container.

conditionally periodic motion [MECH] Motion of a system in which each of the coordinates undergoes simple periodic motion, but the associated frequencies are not all rational fractions of each other so that the complete motion is not simply periodic.

conductance [FL MECH] For a component of a vacuum system, the amount of a gas that flows through divided by the pressure difference across the component. [THERMO] *See* thermal conductance.

cone bearing [MECH ENG] A cone-shaped journal bearing running in a correspondingly tapered sleeve.

cone brake [MECH ENG] A type of friction brake whose rubbing parts are cone-shaped.

cone classifier [MECH ENG] Inverted-cone device for the separation of heavy particulates (such as sand, ore, or other mineral matter) from a liquid stream; feed enters the top of the cone, heavy particles settle to the bottom where they can be withdrawn, and liquid overflows the top edge, carrying the smaller particles or those of lower gravity over the rim; used in the mining and chemical industries.

cone clutch [MECH ENG] A clutch which uses the wedging action of mating conical surfaces to transmit friction torque.

cone crusher [MECH ENG] A machine that reduces the size of materials such as rock by crushing in the tapered space between a truncated revolving cone and an outer chamber.

conehead rivet [DES ENG] A rivet with a head shaped like a truncated cone.

cone key [DES ENG] A taper saddle key placed on a shaft to adapt it to a pulley with a too-large hole.

cone mandrel [DES ENG] A mandrel in which the diameter can be changed by moving conical sleeves.

cone nozzle [DES ENG] A cone-shaped nozzle that disperses fluid in an atomized mist.

cone of visibility [AERO ENG] Generally, the right conical space which has its apex at some ground target and within which an aircraft must be located if the pilot is to be able to discern the target while flying at a specified altitude.

cone pulley *See* step pulley.

cone rock bit [MECH ENG] A rotary drill with two hardened knurled cones which cut the rock as they roll. Also known as roller bit.

configuration [AERO ENG] A particular type of specific aircraft, rocket, or such, which differs from others of the same model by the arrangement of its components or by the addition or omission of auxiliary equipment; for example, long-range configuration or cargo configuration.   [MECH] The positions of all the particles in a system.

congruent melting point [THERMO] A point on a temperature composition plot of a nonstoichiometric compound at which the one solid phase and one liquid phase are adjacent.

conical ball mill [MECH ENG] A cone-shaped tumbling pulverizer in which the steel balls are classified, with the larger balls at the feed end where larger lumps are crushed, and the smaller balls at the discharge end where the material is finer.

conical bearing [MECH ENG] An antifriction bearing employing tapered rollers.

conical pendulum [MECH] A weight suspended from a cord or light rod and made to rotate in a horizontal circle about a vertical axis with a constant angular velocity.

conical refiner [MECH ENG] In paper manufacture, a cone-shaped continuous refiner having two sets of bars mounted on the rotating plug and fixed shell for beating unmodified cellulose fibers.

conjugate momentum [MECH] If q_j $(j = 1, 2, \ldots)$ are generalized coordinates of a classical dynamical system, and L is its Lagrangian, the momentum conjugate to q_j is $p_j = \partial L/\partial q_j$. Also known as canonical momentum; generalized momentum.

connecting rod [MECH ENG] Any straight link that transmits motion or power from one linkage to another within a mechanism, especially linear to rotary motion, as in a reciprocating engine or compressor.

conservation of momentum [MECH] The principle that, when a system of masses is subject only to internal forces that masses of the system exert on one another, the total vector momentum of the system is constant; no violation of this principle has been found. Also known as momentum conservation.

conservation of vorticity [FL MECH] 1. The principle that the vertical component of the absolute vorticity of each particle in an inviscid, autobarotropic fluid flowing horizontally remains constant. 2. The hypothesis that the vorticity of fluid particles remains constant during the turbulent mixing of the fluid.

conservative force field [MECH] A field of force in which the work done on a particle in moving it from one point to another depends only on the particle's initial and final positions.

conservative property [THERMO] A property of a system whose value remains constant during a series of events.

consolute temperature [THERMO] The upper temperature of immiscibility for a two-component liquid system. Also known as upper consolute temperature; upper critical solution temperature.

constant-force spring [MECH ENG] A spring which has a constant restoring force, regardless of displacement.

constant-level balloon [AERO ENG] A balloon designed to float at a constant pressure level. Also known as constant-pressure balloon.

constant of gravitation *See* gravitational constant.

constant of motion [MECH] A dynamical variable of a system which remains constant in time.

constant-pressure balloon *See* constant-level balloon.

constant-pressure combustion [MECH ENG] Combustion occurring without a pressure change.

constant-speed drive [MECH ENG] A mechanism transmitting motion from one shaft to another that does not allow the velocity ratio of the shafts to be varied, or allows it to be varied only in steps.

constant-speed propeller [AERO ENG] A variable-pitch propeller having a governor which automatically changes the pitch to maintain constant engine speed.

constant-velocity universal joint [MECH ENG] A universal joint that transmits constant angular velocity from the driving to the driven shaft, such as the Bendix-Weiss universal joint.

constrained mechanism [MECH ENG] A mechanism in which all members move only in prescribed paths.

constraint [MECH] A restriction on the natural degrees of freedom of a system; the number of constraints is the difference between the number of natural degrees of freedom and the number of actual degrees of freedom.

constrictor [AERO ENG] The exit portion of the combustion chamber in some designs of ramjets, where there is a narrowing of the tube at the exhaust.

construction [DES ENG] The number of strands in a wire rope and the number of wires in a strand; expressed as two numbers separated by a multiplication sign.

construction equipment [MECH ENG] Heavy power machines which perform specific construction or demolition functions.

construction weight [AERO ENG] The weight of a rocket exclusive of propellant, load, and crew if any. Also known as structural weight.

construction wrench [DES ENG] An open-end wrench with a long handle; the handle is used to align matching rivet or bolt holes.

contact [FL MECH] The surface between two immiscible fluids contained in a reservoir.

contact angle *See* angle of contact.

contact condenser [MECH ENG] A device in which a vapor, such as steam, is brought into direct contact with a cooling liquid, such as water, and is condensed by giving up its latent heat to the liquid. Also known as direct-contact condenser.

contact gear ratio *See* contact ratio.

contact-initiated discharge machining [MECH ENG] An electromachining process in which the discharge is initiated by allowing the tool and workpiece to come into contact, after which the tool is withdrawn and an arc forms.

contactor control system [CONT SYS] A feedback control system in which the control signal is a discontinuous function of the sensed error and may therefore assume one of a limited number of discrete values.

contact ratio [DES ENG] The ratio of the length of the path of contact of two gears to the base pitch, equal to approximately the average number of pairs of teeth in contact. Also known as contact gear ratio.

contact transformation *See* canonical transformation.

continued fraction [CONT SYS] The sum of a number and a fraction whose denominator is the sum of a number and a fraction, and so forth; it may have either a finite or an infinite number of terms.

continuous brake [MECH ENG] A train brake that operates on all cars but is controlled from a single point.

continuous bucket elevator [MECH ENG] A bucket elevator on an endless chain or belt.

continuous bucket excavator [MECH ENG] A bucket excavator with a continuous bucket elevator mounted in front of the bowl.

continuous control [CONT SYS] Automatic control in which the controlled quantity is measured continuously and corrections are a continuous function of the deviation.

continuous-flow conveyor [MECH ENG] A totally enclosed, continuous-belt conveyor pulled transversely through a mass of granular, powdered or small-lump material fed from an overhead hopper.

continuous mixer [MECH ENG] A mixer in which materials are introduced, mixed, and discharged in a continuous flow.

continuous system [CONT SYS] A system whose inputs and outputs are capable of changing at any instant of time. Also known as continuous-time signal system.

continuous-type furnace [MECH ENG] A furnace used for heat treatment of materials, with or without direct firing; pieces are loaded through one door, progress continuously through the furnace, and are discharged from another door.

contour machining [MECH ENG] Machining of an irregular surface.

contour turning [MECH ENG] Making a three-dimensional reproduction of the shape of a template by controlling the cutting tool with a follower that moves over the surface of a template.

contraction [MECH] The action or process of becoming smaller or pressed together, as a gas on cooling.

contraction coefficient *See* coefficient of contraction.

contraction loss [FL MECH] In fluid flow, the loss in mechanical energy in a stream flowing through a closed duct or pipe when there is a sudden contraction of the cross-sectional area of the passage.

contrarotating propellers [MECH ENG] A pair of propellers on concentric shafts, turning in opposite directions.

control [CONT SYS] A means or device to direct and regulate a process or sequence of events.

control accuracy [CONT SYS] The degree of correspondence between the ultimately controlled variable and the ideal value in a feedback control system.

control column [AERO ENG] A cockpit control lever pivoted or sliding in front of the pilot; controls operation of the elevator and aileron.

control element [CONT SYS] The portion of a feedback control system that acts on the process or machine being controlled.

control feel [AERO ENG] The impression of the stability and control of an aircraft that a pilot receives through the cockpit controls, either from the aerodynamic forces acting on the control surfaces or from forces simulating these aerodynamic forces.

controllability [AERO ENG] The quality of an aircraft or guided weapon which determines the ease of producing changes in flight direction or in altitude by operation of its controls. [CONT SYS] Property of a system for which, given any initial state and any desired state, there exists a time interval and an input signal which brings the system from the initial state to the desired state during the time interval.

controllable-pitch propeller [MECH ENG] An aircraft or ship propeller in which the pitch of the blades can be changed while the propeller is in motion; five types used for aircraft are two-position, variable-pitch, constant-speed, feathering, and reversible-pitch. Abbreviated CP propeller.

controlled-leakage system [AERO ENG] A system that provides for the maintenance of life in an aircraft or spacecraft cabin by a controlled escape of carbon dioxide and other waste from the cabin, with replenishment provided by stored oxygen and food.

controlled variable [CONT SYS] In process automatic-control work, that quantity or condition of a controlled system that is directly measured or controlled.

controller *See* automatic controller.

controller-structure interaction [CONT SYS] Feedback of an active control algorithm in the process of model reduction; this occurs through observation spillover and control spillover.

control-moment gyro [AERO ENG] An internal momentum storage device that applies torques to the attitude-control system through large rotating gyros.

control plane [AERO ENG] An aircraft from which the movements of another craft are controlled remotely.

control rocket [AERO ENG] A vernier engine, retrorocket, or other such rocket used to change the attitude of, guide, or make small changes in the speed of a rocket, spacecraft, or the like.

control signal [CONT SYS] The signal applied to the device that makes corrective changes in a controlled process or machine.

control spillover [CONT SYS] The excitation by an active control system of modes of motion that have been omitted from the control algorithm in the process of model reduction.

control surface [AERO ENG] 1. Any movable airfoil used to guide or control an aircraft, guided missile, or the like in the air, including the rudder, elevators, ailerons, spoiler flaps, and trim tabs. 2. In restricted usage, one of the main control surfaces, such as the rudder, an elevator, or an aileron.

control system feedback [CONT SYS] A signal obtained by comparing the output of a control system with the input, which is used to diminish the difference between them.

control vane [AERO ENG] A movable vane used for control, especially a movable air vane or jet vane on a rocket used to control flight altitude.

control variable [CONT SYS] One of the input variables of a control system, such as motor torque or the opening of a valve, which can be varied directly by the operator to maximize some measure of performance of the system.

convection [FL MECH] Diffusion in which the fluid as a whole is moving in the direction of diffusion. Also known as bulk flow.

convection coefficient *See* film coefficient.

convection modulus [FL MECH] An intrinsic property of a fluid which is important in determining the Nusselt number, equal to the acceleration of gravity times the volume coefficient of thermal expansion divided by the product of the kinematic viscosity and the thermal diffusivity.

convergent-divergent nozzle [DES ENG] A nozzle in which supersonic velocities are attained; has a divergent portion downstream of the contracting section. Also known as supersonic nozzle.

convertiplane [AERO ENG] A hybrid form of heavier-than-air craft capable, because of one or more horizontal rotors or units acting as rotors, of taking off, hovering, and landing in a fashion similar to a helicopter; and once aloft and moving forward, capable, by means of a mechanical conversion, of flying purely as a fixed-wing aircraft, especially in higher speed ranges.

conveyor [MECH ENG] Any materials-handling machine designed to move individual articles such as solids or free-flowing bulk materials over a horizontal, inclined, declined, or vertical path of travel with continuous motion.

cooler nail [DES ENG] A thin, cement-coated wire nail.

cooling coil [MECH ENG] A coiled arrangement of pipe or tubing for the transfer of heat between two fluids.

cooling correction [THERMO] A correction that must be employed in calorimetry to allow for heat transfer between a body and its surroundings. Also known as radiation correction.

cooling curve [THERMO] A curve obtained by plotting time against temperature for a solid-liquid mixture cooling under constant conditions.

cooling degree day [MECH ENG] A unit for estimating the energy needed for cooling a building; one unit is given for each degree Fahrenheit that the daily mean temperature exceeds 75°F (24°C).

cooling fin [MECH ENG] The extended element of a heat-transfer device that effectively increases the surface area.

cooling load [MECH ENG] The total amount of heat energy that must be removed from a system by a cooling mechanism in a unit time, equal to the rate at which heat is generated by people, machinery, and processes, plus the net flow of heat into the system not associated with the cooling machinery.

cooling method [THERMO] A method of determining the specific heat of a liquid in which the times taken by the liquid and an equal volume of water in an identical vessel to cool through the same range of temperature are compared.

cooling power [MECH ENG] A parameter devised to measure the air's cooling effect upon a human body; it is determined by the amount of heat required by a device to maintain the device at a constant temperature (usually 34°C); the entire system should be made to correspond, as closely as possible, to the external heat exchange mechanism of the human body.

cooling range [MECH ENG] The difference in temperature between the hot water entering and the cold water leaving a cooling tower.

cooling stress [MECH] Stress resulting from uneven contraction during cooling of metals and ceramics due to uneven temperature distribution.

coordinating holes [DES ENG] Holes in two parts of an assembly which form a single continuous hole when the parts are joined.

coordinator *See* second-level controller.

cope chisel [DES ENG] A chisel used to cut grooves in metal.

coping [MECH ENG] Shaping stone or other nonmetallic substance with a grinding wheel.

coping saw [DES ENG] A type of handsaw that has a narrow blade, usually about ⅛ inch (3 millimeters) wide, held taut by a U-shaped frame equipped with a handle; used for shaping and cutout work.

coplanar forces [MECH] Forces that act in a single plane; thus the forces are parallel to the plane and their points of application are in the plane.

cord tire [DES ENG] A pneumatic tire made with cords running parallel to the tread.

core barrel [DES ENG] A hollow cylinder attached to a specially designed bit; used to obtain a continuous section of the rocks penetrated in drilling.

core bit [DES ENG] The hollow, cylindrical cutting part of a core drill.

core catcher *See* split-ring core lifter.

core drill [MECH ENG] A mechanism designed to rotate and to cause an annular-shaped rock-cutting bit to penetrate rock formations, produce cylindrical cores of the formations penetrated, and lift such cores to the surface, where they may be collected and examined.

core gripper *See* split-ring core lifter.

core lifter *See* split-ring core lifter.

Coriolis acceleration [MECH] **1.** An acceleration which, when added to the acceleration of an object relative to a rotating coordinate system and to its centripetal acceleration, gives the acceleration of the object relative to a fixed coordinate system. **2.** A vector which is equal in magnitude and opposite in direction to that of the first definition.

Coriolis deflection *See* Coriolis effect.

Coriolis effect [MECH] Also known as Coriolis deflection. **1.** The deflection relative to the earth's surface of any object moving above the earth, caused by the Coriolis force; an object moving horizontally is deflected to the right in the Northern Hemisphere, to the left in the Southern. **2.** The effect of the Coriolis force in any rotating system.

Coriolis force [MECH] A velocity-dependent pseudoforce in a reference frame which is rotating with respect to an inertial reference frame; it is equal and opposite to the product of the mass of the particle on which the force acts and its Coriolis acceleration.

Corliss valve [MECH ENG] An oscillating type of valve gear with a trip mechanism for the admission and exhaust of steam to and from an engine cylinder.

corner chisel [DES ENG] A chisel with two cutting edges at right angles.

corner frequency *See* break frequency.

cornering tool [DES ENG] A cutting tool with a curved edge, used to round off sharp corners.

cornice brake [MECH ENG] A machine used to bend sheet metal into different forms.

correction time [CONT SYS] The time required for the controlled variable to reach and stay within a predetermined band about the control point following any change of the independent variable or operating condition in a control system. Also known as settling time.

corrective action [CONT SYS] The act of varying the manipulated process variable by the controlling means in order to modify overall process operating conditions.

corrugated bar [DES ENG] Steel bar with transverse ridges; used in reinforced concrete.

corrugated fastener [DES ENG] A thin corrugated strip of steel that can be hammered into a wood joint to fasten it.

corrugating [DES ENG] Forming straight, parallel, alternate ridges and grooves in sheet metal, cardboard, or other material.

cosmonaut [AERO ENG] A Soviet astronaut.

Cosmos satellites [AERO ENG] A series of earth satellites launched by the Soviet Union starting in 1962 to conduct geophysical experiments.

cotter [DES ENG] A tapered piece that can be driven in a tapered hole to hold together an assembly of machine or structural parts.

cottered joint [MECH ENG] A joint in which a cotter, usually a flat bar tapered on one side to ensure a tight fit, transmits power by shear on an area at right angles to its length.

cotter pin [DES ENG] A split pin, inserted into a hole, to hold a nut or cotter securely to a bolt or shaft, or to hold a pair of hinge plates together.

Couette flow [FL MECH] Low-speed, steady motion of a viscous fluid between two infinite plates moving parallel to each other.

Coulomb friction [MECH] Friction occurring between dry surfaces.

count [AERO ENG] 1. To proceed from one point to another in a countdown or plus count, normally by calling a number to signify the point reached. 2. To proceed in a countdown, for example, T minus 90 and counting.

countdown [AERO ENG] 1. The process in the engineering definition, used in leading up to the launch of a large or complicated rocket vehicle, or in leading up to a captive test, a readiness firing, a mock firing, or other firing test. 2. The act of counting inversely during this process.

counterbalance See counterweight.

counterblow hammer [MECH ENG] A forging hammer in which the ram and anvil are driven toward each other by compressed air or steam.

counterbore [DES ENG] A flat-bottom enlargement of the mouth of a cylindrical bore to enlarge a borehole and give it a flat bottom.

countercurrent flow [MECH ENG] A sensible heat-transfer system in which the two fluids flow in opposite directions.

counterpoise See counterweight.

countershaft [MECH ENG] A secondary shaft that is driven by a main shaft and from which power is supplied to a machine part.

countersink [DES ENG] The tapered and relieved cutting portion in a twist drill, situated between the pilot drill and the body.

countersinking [MECH ENG] Drilling operation to form a flaring depression around the rim of a hole.

counterweight [MECH ENG] 1. A device which counterbalances the original load in elevators and skip and mine hoists, going up when the load goes down, so that the engine must only drive against the unbalanced load and overcome friction. 2. Any weight placed on a mechanism which is out of balance so as to maintain static equilibrium. Also known as counterbalance; counterpoise.

couple [MECH] A system of two parallel forces of equal magnitude and opposite sense.

coupled engine [MECH ENG] A locomotive engine having the driving wheels connected by a rod.

coupled oscillators [MECH] A set of particles subject to elastic restoring forces and also to elastic interactions with each other.

coupled wave [FL MECH] A surface wave which is being continuously generated by another wave having the same phase velocity. Also known as C wave.

coupling [MECH ENG] The mechanical fastening that connects shafts together for power transmission. Also known as shaft coupling.

course programmer [CONT SYS] An item which initiates and processes signals in a manner to establish a vehicle in which it is installed along one or more projected courses.

Cowell method [AERO ENG] A method of orbit computation using direct step-by-step integration in rectangular coordinates of the total acceleration of the orbiting body.

cowling [AERO ENG] The streamlined metal cover of an aircraft engine.

cp *See* centipoise.

cramp [DES ENG] A metal plate with bent ends used to hold blocks together.

crampon [DES ENG] A device for holding heavy objects such as rock or lumber to be lifted by a crane or hoist; shaped like scissors, with points bent inward for grasping the load. Also spelled crampoon.

crampoon *See* crampon.

crane [MECH ENG] A hoisting machine with a power-operated inclined or horizontal boom and lifting tackle for moving loads vertically and horizontally.

crane hoist [MECH ENG] A mobile construction machine built principally for lifting loads by means of cables and consisting of an undercarriage on which the unit moves, a cab or house which envelops the main frame and contains the power units and controls, and a movable boom over which the cables run.

crane hook [DES ENG] A hoisting fixture designed to engage a ring or link of a lifting chain, or the pin of a shackle or cable socket.

crane truck [MECH ENG] A crane with a jiblike boom mounted on a truck. Also known as yard crane.

crank [MECH ENG] A link in a mechanical linkage or mechanism that can turn about a center of rotation.

crank angle [MECH ENG] 1. The angle between a crank and some reference direction. 2. Specifically, the angle between the crank of a slider crank mechanism and a line from crankshaft to the piston.

crank arm [MECH ENG] The arm of a crankshaft attached to a connecting rod and piston.

crank axle [MECH ENG] 1. An axle containing a crank. 2. An axle bent at both ends so that it can accommodate a large body with large wheels.

crankcase [MECH ENG] The housing for the crankshaft of an engine, where, in the case of an automobile, oil from hot engine parts is collected and cooled before returning to the engine by a pump.

crankpin [DES ENG] A cylindrical projection on a crank which holds the connecting rod.

crank press [MECH ENG] A punch press that applies power to the slide by means of a crank.

crankshaft [MECH ENG] The shaft about which a crank rotates.

crank throw [MECH ENG] 1. The web or arm of a crank. 2. The displacement of a crankpin from the crankshaft.

crank web [MECH ENG] The arm of a crank connecting the crankshaft to crankpin, or connecting two adjacent crankpins.

crater [MECH ENG] A depression in the face of a cutting tool worn down by chip contact.

crawler [MECH ENG] 1. One of a pair of an endless chain of plates driven by sprockets and used instead of wheels by certain power shovels, tractors, bulldozers, drilling machines, and such, as a means of propulsion. 2. Any machine mounted on such tracks.

crawler crane [MECH ENG] A self-propelled crane mounted on two endless tracks that revolve around wheels.

crawler tractor [MECH ENG] A tractor that propels itself on two endless tracks revolving around wheels.

crawler wheel [MECH ENG] A wheel that drives a continuous metal belt, as on a crawler tractor.

creep [MECH] A time-dependent strain of solids caused by stress.

creep buckling [MECH] Buckling that may occur when a compressive load is maintained on a member over a long period, leading to creep which eventually reduces the member's bending stiffness.

creeping flow [FL MECH] Fluid flow in which the velocity of flow is very small.

creep limit [MECH] The maximum stress a given material can withstand in a given time without exceeding a specified quantity of creep.

creep recovery [MECH] Strain developed in a period of time after release of load in a creep test.

creep rupture strength [MECH] The stress which, at a given temperature, will cause a material to rupture in a given time.

creep strength [MECH] The stress which, at a given temperature, will result in a creep rate of 1 percent deformation in 100,000 hours.

cremone bolt [DES ENG] A fastening for double doors or casement windows; employs vertical rods that move up and down to engage the top and bottom of the frame.

crest [DES ENG] The top of a screw thread.

crest clearance [DES ENG] The clearance, in a radial direction, between the crest of the thread of a screw and the root of the thread with which the screw mates.

crinal [MECH] A unit of force equal to 0.1 newton.

crith [MECH] A unit of mass, used for gases, equal to the mass of 1 liter of hydrogen at standard pressure and temperature; it is found experimentally to equal 8.9885×10^{-5} kilogram.

critical altitude [AERO ENG] The maximum altitude at which a supercharger can maintain a pressure in the intake manifold of an engine equal to that existing during normal operation at rated power and speed at sea level without the supercharger.

critical angle of attack [AERO ENG] The angle of attack of an airfoil at which the flow of air about the airfoil changes abruptly so that lift is sharply reduced and drag is sharply increased. Also known as stalling angle of attack.

critical compression ratio [MECH ENG] The lowest compression ratio which allows compression ignition of a specific fuel.

critical exponent [THERMO] A parameter n that characterizes the temperature dependence of a thermodynamic property of a substance near its critical point; the temperature dependence has the form $|T - T_c|^n$, where T is the temperature and T_c is the critical temperature.

critical flow [FL MECH] The rate of flow of a fluid equivalent to the speed of sound in that fluid.

critical Mach number [AERO ENG] The free-stream Mach number at which a local Mach number of 1.0 is attained at any point on the body under consideration.

critical pressure [FL MECH] For a nozzle whose cross section at each point is such that a fluid in isentropic flow just fills it, the pressure at the section of minimum area of the nozzle; if the nozzle is cut off at this point with no diverging section, decrease in the discharge pressure below the critical pressure (at constant admission pressure) does not result in increased flow. [THERMO] The pressure of the liquid-vapor critical point.

critical pressure ratio [FL MECH] The ratio of the critical pressure of a nozzle to the admission pressure of the nozzle (equals 0.53 for gases).

critical Reynolds number [FL MECH] The Reynolds number at which there is a transition from laminar to turbulent flow.

critical speed [FL MECH] *See* critical velocity. [MECH ENG] The angular speed at which a rotating shaft becomes dynamically unstable with large lateral amplitudes, due to resonance with the natural frequencies of lateral vibration of the shaft.

critical velocity [AERO ENG] In rocketry, the speed of sound at the conditions prevailing at the nozzle throat. Also known as throat velocity. [MECH] **1.** The speed of flow equal to the local speed of sound. Also known as critical speed. **2.** The speed of fluid flow through a given conduit above which it becomes turbulent.

critical zone [FL MECH] In fluid flow, the area on a graph of the Reynolds number versus friction factor indicating unstable flow (Reynolds number 2000 to 4000) between laminar flow and the transition to turbulent flow.

Crocco's equation [FL MECH] A relationship, expressed as $v \times \omega = -T$ grad S, between vorticity and entropy gradient for the steady flow of an inviscid compressible fluid; v is the fluid velocity vector, ω (= curl v) is the vorticity vector, T is the fluid temperature, and S is the entropy per unit mass of the fluid.

crochet file [DES ENG] A thin, flat, round-edged file that tapers to a point.

crocodile shears *See* lever shears.

cross axle [MECH ENG] **1.** A shaft operated by levers at its ends. **2.** An axle with cranks set at 90°.

cross-belt drive [DES ENG] A belt drive having parallel shafts rotating in opposite directions.

crossbolt [DES ENG] A lock bolt with two parts which can be moved in opposite directions.

cross box [MECH ENG] A boxlike structure for the connection of circulating tubes to the longitudinal drum of a header-type boiler.

crosscurrent [FL MECH] A current that flows across or opposite to another current.

crosscut file [DES ENG] A file with a rounded edge on one side and a thin edge on the other; used to sharpen straight-sided saw teeth with round gullets.

crosscut saw [DES ENG] A type of saw for cutting across the grain of the wood; designed with about eight teeth per inch.

cross drum boiler [MECH ENG] A sectional header or box header type of boiler in which the axis of the horizontal drum is perpendicular to the axis of the main bank of tubes.

crossed belt [MECH ENG] A pulley belt arranged so that the sides cross, thereby making the pulleys rotate in opposite directions.

crossed-field accelerator [AERO ENG] A plasma engine for space travel in which plasma serves as a conductor to carry current across a magnetic field, so that a resultant force is exerted on the plasma.

cross-flow [AERO ENG] A flow going across another flow, as a spanwise flow over a wing.

cross-flow plane [AERO ENG] A plane at right angles to the free-stream velocity.

crosshaul [MECH ENG] A device for loading objects onto vehicles, consisting of a chain that is hooked on opposite sides of a vehicle, looped under the object, and connected to a power source and that rolls the object onto the vehicle.

crosshead [MECH ENG] A block sliding between guides and containing a wrist pin for the conversion of reciprocating to rotary motion, as in an engine or compressor.

crossover spiral *See* lead-over groove.

cross slide [MECH ENG] A part of a machine tool that allows the tool carriage to move at right angles to the main direction of travel.

cross turret [MECH ENG] A turret that moves horizontally and at right angles to the lathe guides.

crowbar [DES ENG] An iron or steel bar that is usually bent and has a wedge-shaped working end; used as a lever and for prying.

crown saw [DES ENG] A saw consisting of a hollow cylinder with teeth around its edge; used for cutting round holes. Also known as hole saw.

crown sheet [MECH ENG] The structural element which forms the top of a furnace in a fire-tube boiler.

crown wheel [DES ENG] A gear that is light and crown-shaped.

cruciform wing [AERO ENG] An aircraft wing in the shape of a cross.

cruise missile [AERO ENG] A pilotless airplane that can be launched from a submarine, surface ship, ground vehicle, or another airplane; range can be up to 1500 miles (2400 kilometers), flying at a constant altitude that can be as low as 60 meters.

crusher [MECH ENG] A machine for crushing rock and other bulk materials.

crushing strain [MECH] Compression which causes the failure of a material.

crushing strength [MECH] The compressive stress required to cause a solid to fail by fracture; in essence, it is the resistance of the solid to vertical pressure placed upon it.

cryology [MECH ENG] The study of low-temperature (approximately 200°R, or −160°C) refrigeration.

cryosorption pump [MECH ENG] A high-vacuum pump that employs a sorbent such as activated charcoal or synthetic zeolite cooled by nitrogen or some other refrigerant; used to reduce pressure from atmospheric pressure to a few millitorr.

crystal holder [DES ENG] A housing designed to provide proper support, mechanical protection, and connections for a quartz crystal plate.

cs *See* centistoke.

cu *See* cubic.

cubic [MECH] Denoting a unit of volume, so that if x is a unit of length, a cubic x is the volume of a cube whose sides have length $1x$; for example, a cubic meter, or a meter cubed, is the volume of a cube whose sides have a length of 1 meter. Abbreviated cu.

cubical dilation [MECH] The isotropic part of the strain tensor describing the deformation of an elastic solid, equal to the fractional increase in volume.

cubic foot per second *See* cusec.

cubic measure [MECH] A unit or set of units to measure volume.

cumec [MECH] A unit of volume flow rate equal to 1 cubic meter per second.

cumulative compound motor [MECH ENG] A motor with operating characteristics between those of the constant-speed (shunt-wound) and the variable-speed (series-wound) types.

cup [DES ENG] A cylindrical part with only one end open.

Curie principle [THERMO] The principle that a macroscopic cause never has more elements of symmetry than the effect it produces; for example, a scalar cause cannot produce a vectorial effect.

curling dies [MECH ENG] A set of tools that shape the ends of a piece of work into a form with a circular cross section.

curling machine [MECH ENG] A machine with curling dies; used to curl the ends of cans.

current function *See* Lagrange stream function.

cursor [DES ENG] A clear or amber-colored filter that can be placed over a radar screen and rotated until an etched diameter line on the filter passes through a target echo; the bearing from radar to target can then be read accurately on a stationary 360° scale surrounding the filter.

Curtis stage [MECH ENG] A velocity-staged impulse turbine using reversing buckets between stages.

Curtis turbine [MECH ENG] A velocity-staged, impulse-type steam engine.

curve resistance [MECH] The force opposing the motion of a railway train along a track due to track curvature.

curvilinear motion [MECH] Motion along a curved path.

cusec [MECH] A unit of volume flow rate, used primarily to describe pumps, equal to a uniform flow of 1 cubic foot in 1 second. Also known as cubic foot per second (cfs).

cut nail [DES ENG] A flat, tapered nail sheared from steel plate; it has greater holding power than a wire nail and is generally used for fastening flooring.

cutoff [MECH ENG] **1.** The shutting off of the working fluid to an engine cylinder. **2.** The time required for this process.

cutoff point [MECH ENG] **1.** The point at which there is a transition from spiral flow in the housing of a centrifugal fan to straight-line flow in the connected duct. **2.** The point on the stroke of a steam engine where admission of steam is stopped.

cutoff tool [MECH ENG] A tool used on bar-type lathes to separate the finished piece from the bar stock.

cutoff valve [MECH ENG] A valve used to stop the flow of steam to the cylinder of a steam engine.

cutoff wheel [MECH ENG] A thin wheel impregnated with an abrasive used for severing or cutting slots in a material or part.

cutter *See* cutting tool.

cutter bar [MECH ENG] The bar that supports the cutting tool in a lathe or other machine.

cutterhead [MECH ENG] A device on a machine tool for holding a cutting tool.

cutter sweep [MECH ENG] The section that is cut off or eradicated by the milling cutter or grinding wheel in entering or leaving the flute.

cutting angle [MECH ENG] The angle that the cutting face of a tool makes with the work surface back of the tool.

cutting drilling [MECH ENG] A rotary drilling method in which drilling occurs through the action of the drill steel rotating while pressed against the rock.

cutting edge [DES ENG] 1. The point or edge of a diamond or other material set in a drill bit. Also known as cutting point. 2. The edge of a lathe tool in contact with the work during a machining operation.

cutting-off machine [MECH ENG] A machine for cutting off metal bars and shapes; includes the lathe type using single-point cutoff tools, and several types of saws.

cutting pliers [DES ENG] Pliers with cutting blades on the jaws.

cutting point *See* cutting edge.

cutting speed [MECH ENG] The speed of relative motion between the tool and workpiece in the main direction of cutting. Also known as feed rate; peripheral speed.

cutting tool [MECH ENG] The part of a machine tool which comes into contact with and removes material from the workpiece by the use of a cutting medium. Also known as cutter.

C wave *See* coupled wave.

cycle [FL MECH] A system of phases through which the working substance passes in an engine, compressor, pump, turbine, power plant, or refrigeration system.

cyclic coordinate [MECH] A generalized coordinate on which the Lagrangian of a system does not depend explicitly.

cyclic train [MECH ENG] A set of gears, such as an epicyclic gear system, in which one or more of the gear axes rotates around a fixed axis.

cycling [CONT SYST] A periodic change of the controlled variable from one value to another in an automatic control system. Also known as oscillation.

cycloidal gear teeth [DES ENG] Gear teeth whose profile is formed by the trace of a point on a circle rolling without slippage on the outside or inside of the pitch circle of a gear; now used only for clockwork and timer gears.

cycloidal pendulum [MECH] A modification of a simple pendulum in which a weight is suspended from a cord which is slung between two pieces of metal shaped in the form of cycloids; as the bob swings, the cord wraps and unwraps on the cycloids; the pendulum has a period that is independent of the amplitude of the swing.

cycloidal wave [FL MECH] A very steep, symmetrical wave in the form of a cycloid whose crest forms an angle of 120°.

cyclone [MECH ENG] Any cone-shaped air-cleaning apparatus operated by centrifugal separation that is used in particle collecting and fine grinding operations.

cyclone classifier *See* cyclone separator.

cyclone separator [MECH ENG] A funnel-shaped device for removing particles from air or other fluids by centrifugal means; used to remove dust from air or other fluids, steam from water, and water from steam, and in certain applications to separate particles into two or more size classes. Also known as cyclone classifier.

cyclostrophic flow   [FL MECH] A form of gradient flow in which the centripetal acceleration exactly balances the horizontal pressure force.

cylinder   *See* engine cylinder.

cylinder actuator   [MECH ENG] A device that converts hydraulic power into useful mechanical work by means of a tight-fitting piston moving in a closed cylinder.

cylinder block   [DES ENG] The metal casting comprising the piston chambers of a multicylinder internal combustion engine. Also known as block; engine block.

cylinder bore   [DES ENG] The internal diameter of the tube in which the piston of an engine or pump moves.

cylinder head   [MECH ENG] The cap that serves to close the end of the piston chamber of a reciprocating engine, pump, or compressor.

cylinder liner   [MECH ENG] A separate cylindrical sleeve inserted in an engine block which serves as the cylinder.

cylindrical cam   [MECH ENG] A cam mechanism in which the cam follower undergoes translational motion parallel to the camshaft as a roller attached to it rolls in a groove in a circular cylinder concentric with the camshaft.

cylindrical cutter   [DES ENG] Any cutting tool with a cylindrical shape, such as a milling cutter.

cylindrical grinder   [MECH ENG] A machine for doing work on the peripheries or shoulders of workpieces composed of concentric cylindrical or conical shapes, in which a rotating grinding wheel cuts a workpiece rotated from a power headstock and carried past the face of the wheel.

D

D *See* Deborah number.

Da₁ *See* Darcy number 1.

Da₂ *See* Darcy number 2.

dado head [MECH ENG] A machine consisting of two circular saws with one or more chippers in between; used for cutting flat-bottomed grooves in wood.

dado plane [DES ENG] A narrow plane for cutting flat grooves in woodwork.

Dahlin's algorithm [CONT SYS] A digital control algorithm in which the requirement of minimum response time used in the deadbeat algorithm is relaxed to reduce ringing in the system response.

d'Alembert's paradox [FL MECH] The paradox that no forces act on a body moving at constant velocity in a straight line through a large mass of incompressible, inviscid fluid which was initially at rest, or in uniform motion.

d'Alembert's principle [MECH] The principle that the resultant of the external forces and the kinetic reaction acting on a body equals zero.

Dall tube [MECH ENG] Fluid-flow measurement device, similar to a venturi tube, inserted as a section of a fluid-carrying pipe; flow rate is measured by pressure drop across a restricted throat.

Dalton's temperature scale [THERMO] A scale for measuring temperature such that the absolute temperature T is given in terms of the temperature on the Dalton scale τ by $T = 273.15(373.15/273.15)^{\tau/100}$.

damaging stress [MECH] The minimum unit stress for a given material and use that will cause damage to the member and make it unfit for its expected length of service.

Damköhler number V *See* Reynold's number.

damper [MECH ENG] A valve or movable plate for regulating the flow of air or the draft in a stove, furnace, or fireplace.

damping capacity [MECH] A material's capability in absorbing vibrations.

damping coefficient *See* resistance.

damping constant *See* resistance.

dandy roll [MECH ENG] A roll in a Fourdrinier paper-making machine; used to compact the sheet and sometimes to imprint a watermark.

Darcy number 1 [FL MECH] A dimensionless group, equal to four times the Fanning friction factor. Symbolized Da_1. Also known as Darcy-Weisbach coefficient; resistance coefficient 2.

Darcy number 2 [FL MECH] A dimensionless group used in the study of the flow of fluids in porous media, equal to the fluid velocity times the flow path divided by the permeability of the medium. Symbolized Da_2.

Darcy's law [FL MECH] The law that the rate at which a fluid flows through a permeable substance per unit area is equal to the permeability, which is a property only of the substance through which the fluid is flowing, times the pressure drop per unit length of flow, divided by the viscosity of the fluid.

Darcy-Weisbach coefficient *See* Darcy number 1.

Darcy-Weisbach equation [FL MECH] An equation for the loss of head due to friction h_f during turbulent flow of a fluid through a duct of any shape; in the case of a circular pipe, $h_f = f(L/d)(V^2/2g)$, where L and d are the length and diameter of the pipe, V is the fluid velocity, g the acceleration of gravity, and f a dimensionless number called Darcy number 1.

dark satellite [AERO ENG] Satellite that gives no information to a friendly ground environment, either because it is controlled or because the radiating equipment is inoperative.

dart configuration [AERO ENG] An aerodynamic configuration in which the control surfaces are at the tail of the vehicle.

dashpot [MECH ENG] A device used to dampen and control a motion, in which an attached piston is loosely fitted to move slowly in a cylinder containing oil.

DaV *See* Reynolds number.

Davis correction [FL MECH] Empirical relation of flow-line diameters used to correct data calculated from the Atherton equation (friction loss in annular passages).

Davis wing [AERO ENG] A narrow-chord wing that has comparatively low drag and a stable center of pressure and develops lift at relatively small angles of attack.

dead axle [MECH ENG] An axle that carries a wheel but does not drive it.

deadbeat [MECH] Coming to rest without vibration or oscillation, as when the pointer of a meter moves to a new position without overshooting. Also known as deadbeat response.

deadbeat algorithm [CONT SYS] A digital control algorithm which attempts to follow set-point changes in minimum time, assuming that the controlled process can be modeled approximately as a first-order plus dead-time system.

deadbeat response *See* deadbeat.

dead bolt [DES ENG] A lock bolt that is moved directly by the turning of a knob or key, not by spring action.

dead center [MECH ENG] 1. A position of a crank in which the turning force applied to it by the connecting rod is zero; occurs when the crank and rod are in a straight line. 2. A support for the work on a lathe which does not turn with the work.

dead load *See* static load.

deadlocking latch bolt *See* auxiliary dead latch.

deadman's brake [MECH ENG] An emergency device that automatically is activated to stop a vehicle when the driver removes his foot from the pedal.

deadman's handle [MECH ENG] A handle on a machine designed so that the operator must continuously press on it in order to keep the machine running.

dead stick [AERO ENG] The propeller of an airplane that is not rotating because the engine has stopped.

dead-stroke [MECH ENG] Having a recoilless or nearly recoilless stroke.

dead-stroke hammer [MECH ENG] A power hammer provided with a spring on the hammer head to reduce recoil.

dead time [CONT SYS] The time interval between a change in the input signal to a process control system and the response to the signal.

dead-time compensation [CONT SYS] The modification of a controller to allow for time delays between the input to a control system and the response to the signal.

deaerator [MECH ENG] A device in which oxygen and carbon dioxide are removed from boiler feedwater.

deal [DES ENG] 1. A face on which numbers are registered by means of a pointer. 2. A disk usually with a series of markings around its border, which can be turned to regulate the operation of a machine or electrical device.

Dean number [FL MECH] A dimensionless number giving the ratio of the viscous force acting on a fluid flowing in a curved pipe to the centrifugal force; equal to the Reynolds number times the square root of the ratio of the radius of the pipe to its radius of curvature. Symbolized N_D.

Deborah number [MECH] A dimensionless number used in rheology, equal to the relaxation time for some process divided by the time it is observed. Symbolized D.

debriefing [AERO ENG] The relating of factual information by a flight crew at the termination of a flight, consisting of flight weather encountered, the condition of the aircraft, or facilities along the airways or at the airports.

decaliter [MECH] A unit of volume, equal to 10 liters, or to 0.01 cubic meter.

decameter [MECH] A unit of length in the metric system equal to 10 meters.

decastere [MECH] A unit of volume, equal to 10 cubic meters.

deceleration [MECH] The rate of decrease of speed of a motion.

deceleration parachute *See* drogue.

deceleron [AERO ENG] A lateral control surface of an airplane that is divided so as to combine the functions of an airbrake and an aileron.

deciare [MECH] A unit of area, equal to 0.1 are or 10 square meters.

decibar [MECH] A metric unit of pressure equal to one-tenth bar.

decigram [MECH] A unit of mass, equal to 0.1 gram.

deciliter [MECH] A unit of volume, equal to 0.1 liter, or 10^{-4} cubic meter.

decimeter [MECH] A metric unit of length equal to one-tenth meter.

dedendum [DES ENG] The difference between the radius of the pitch circle of a gear and the radius of its root circle.

dedendum circle [DES ENG] A circle tangent to the bottom of the spaces between teeth on a gear wheel.

Deep Space Network [AERO ENG] A spacecraft network operated by NASA which tracks, commands, and receives telemetry for all types of spacecraft sent to explore deep space, the moon, and solar system planets. Abbreviated DSN.

deep-space probe [AERO ENG] A spacecraft designed for exploring space beyond the gravitational and magnetic fields of the earth.

deep-well pump [MECH ENG] A multistage centrifugal pump for lifting water from deep, small-diameter wells; a surface electric motor operates the shaft. Also known as vertical turbine pump.

deflecting torque [MECH] An instrument's moment, resulting from the quantity measured, that acts to cause the pointer's deflection.

deflection bit [DES ENG] A long, cone-shaped, noncoring bit used to drill past a deflection wedge in a borehole.

deflection curve [MECH] The curve, generally downward, described by a shot deviating from its true course.

deflection wedge [DES ENG] A wedge-shaped tool inserted into a borehole to direct the drill bit.

deformation [MECH] Any alteration of shape or dimensions of a body caused by stresses, thermal expansion or contraction, chemical or metallurgical transformations, or shrinkage and expansions due to moisture change.

deformation curve [MECH] A curve showing the relationship between the stress or load on a structure, structural member, or a specimen and the strain or deformation that results. Also known as stress-strain curve.

deformation ellipsoid *See* strain ellipsoid.

defrost [THERMO] To thaw out from a frozen state.

degradation [THERMO] The conversion of energy into forms that are increasingly difficult to convert into work, resulting from the general tendency of entropy to increase.

degree [FL MECH] One of the units in any of various scales of specific gravity, such as the Baumé scale. [THERMO] One of the units of temperature or temperature difference in any of various temperature scales, such as the Celsius, Fahrenheit, and Kelvin temperature scales (the Kelvin degree is now known as the kelvin).

degree-day [MECH ENG] A measure of the departure of the mean daily temperature from a given standard; one degree-day is recorded for each degree of departure above (or below) the standard during a single day; used to estimate energy requirements for building heating and, to a lesser extent, for cooling.

degree Engler [FL MECH] A measure of viscosity; the ratio of the time of flow of 200 milliliters of the liquid through a viscometer devised by Engler, to the time for the flow of the same volume of water.

degree Kelvin *See* kelvin.

degree of freedom [MECH] Of a gyro, the number of orthogonal axes about which the spin axis is free to rotate, the spin axis freedom not being counted; this is not a universal convention; for example, the free gyro is frequently referred to as a three-degree-of-freedom gyro, the spin axis being counted. [MECH ENG] Any one of the number of ways in which the space configuration of a mechanical system may change.

dehumidification [MECH ENG] The process of reducing the moisture in the air; serves to increase the cooling power of air.

dehumidifier [MECH ENG] Equipment designed to reduce the amount of water vapor in the ambient atmosphere.

deicer [AERO ENG] Any device to keep the wings and propeller of an airplane free of ice.

dekapoise [FL MECH] A unit of absolute viscosity, equal to 10 poises.

Delaunay orbit element [MECH] In the n-body problem, certain functions of variable elements of an ellipse with a fixed focus along which one of the bodies travels; these functions have rates of change satisfying simple equations.

de Laval nozzle [AERO ENG] A converging-diverging nozzle used in certain rockets. Also known as Laval nozzle.

delayed repeater satellite [AERO ENG] Satellite which stores information obtained from a ground terminal at one location, and upon interrogation by a terminal at a different location, transmits the stored message.

delay time [CONT SYS] The amount of time by which the arrival of a signal is retarded after transmission through physical equipment or systems.

delta wing [AERO ENG] A triangularly shaped wing of an aircraft.

demister [MECH ENG] A series of ducts in automobiles arranged so that hot, dry air directed from the heat source is forced against the interior of the windscreen or windshield to prevent condensation.

demon of Maxwell [THERMO] Hypothetical creature who controls a trapdoor over a microscopic hole in an adiabatic wall between two vessels filled with gas at the same temperature, so as to supposedly decrease the entropy of the gas as a whole and thus violate the second law of thermodynamics. Also known as Maxwell's demon.

dense-air refrigeration cycle *See* reverse Brayton cycle.

dense-air system *See* cold-air machine.

density [MECH] The mass of a given substance per unit volume.

density airspeed [AERO ENG] Calibrated airspeed corrected for pressure altitude and true air temperature.

density correction [AERO ENG] A correction made necessary because the airspeed indicator is calibrated only for standard air pressure; it is applied to equivalent airspeed to obtain true airspeed, or to calibrated airspeed to obtain density airspeed.

density error [AERO ENG] The error in the indications of a differential-pressure-type airspeed indicator due to nonstandard atmospheric density.

density specific impulse [AERO ENG] The product of the specific impulse of a propellant combination and the average specific gravity of the propellants.

dental coupling [MECH ENG] A type of flexible coupling used to join a steam turbine to a reduction-gear pinion shaft; consists of a short piece of shaft with gear teeth at each end, and mates with internal gears in a flange at the ends of the two shafts to be joined.

deorbit [AERO ENG] To recover a spacecraft from earth orbit by providing a new orbit which intersects the earth's atmosphere.

depth gage [DES ENG] An instrument or tool for measuring the depth of depression to a thousandth inch.

depth micrometer [DES ENG] A micrometer used to measure the depths of holes, slots, and distances of shoulders and projections.

depth of engagement [DES ENG] The depth of contact, in a radial direction, between mating threads.

depth of thread [DES ENG] The distance, in a radial direction, from the crest of a screw thread to the base.

derivative action [CONT SYS] Control action in which the speed at which a correction is made depends on how fast the system error is increasing. Also known as derivative compensation; rate action.

derivative compensation *See* derivative action.

derivative network [CONT SYS] A compensating network whose output is proportional to the sum of the input signal and its derivative. Also known as lead network.

derrick [MECH ENG] A hoisting machine consisting usually of a vertical mast, a slanted boom, and associated tackle; may be operated mechanically or by hand.

derrick crane *See* stiffleg derrick.

descending branch [MECH] That portion of a trajectory which is between the summit and the point where the trajectory terminates, either by impact or air burst, and along which the projectile falls, with altitude constantly decreasing. Also known as descent trajectory.

descending node [AERO ENG] That point at which an earth satellite crosses to the south side of the equatorial plane of its primary. Also known as southbound node.

descent [AERO ENG] Motion of a craft in which the path is inclined with respect to the horizontal.

descent trajectory *See* descending branch.

describing function [CONT SYS] A function used to represent a nonlinear transfer function by an approximately equivalent linear transfer function; it is the ratio of the phasor representing the fundamental component of the output of the nonlinearity, determined by Fourier analysis, to the phasor representing a sinusoidal input signal.

design gross weight [AERO ENG] The gross weight at takeoff that an aircraft, rocket, or such is expected to have, used in design calculations.

design load [DES ENG] The most stressful combination of weight or other forces a building, structure, or mechanical system or device is designed to sustain.

design pressure [DES ENG] The pressure used in the calculation of minimum thickness or design characteristics of a boiler or pressure vessel in recognized code formulas; static head may be added where appropriate for specific parts of the structure.

design standards [DES ENG] Generally accepted uniform procedures, dimensions, materials, or parts that directly affect the design of a product or facility.

design stress [DES ENG] A permissible maximum stress to which a machine part or structural member may be subjected, which is large enough to prevent failure in case the loads exceed expected values, or other uncertainties turn out unfavorably.

design thickness [DES ENG] The sum of required thickness and corrosion allowance utilized for individual parts of a boiler or pressure vessel.

desilter [MECH ENG] Wet, mechanical solids classifier (separator) in which silt particles settle as the carrier liquid is slowly stirred by horizontally revolving rakes; solids are plowed outward and removed at the periphery of the container bowl.

deslimer [MECH ENG] Apparatus, such as a bowl-type centrifuge, used to remove fine, wet particles (slime) from cement rocks and to size pigments and abrasives.

destruct [AERO ENG] The deliberate action of destroying a rocket vehicle after it has been launched, but before it has completed its course.

destruct line [AERO ENG] On a rocket test range, a boundary line on each side of the down-range course beyond which a rocket cannot fly without being destroyed under destruct procedures; or a line beyond which the impact point cannot pass.

detached shock wave [FL MECH] A shock wave not in contact with the body which originates it.

detent [MECH ENG] A catch or lever in a mechanism which initiates or locks movement of a part, especially in escapement mechanisms.

determinant [CONT SYS] The product of the partial return differences associated with the nodes of a signal-flow graph.

determinate structure [MECH] A structure in which the equations of statics alone are sufficient to determine the stresses and reactions.

determinism *See* causality.

detonating rate [MECH] The velocity at which the explosion wave passes through a cylindrical charge.

detonation [MECH ENG] Spontaneous combustion of the compressed charge after passage of the spark in an internal combustion engine; it is accompanied by knock.

detonation wave [FL MECH] A shock wave that accompanies detonation and has a shock front followed by a region of decreasing pressure in which the reaction occurs.

developed planform [AERO ENG] The plan of an airfoil as drawn with the chord lines at each section rotated about the airfoil axis into a plane parallel to the plane of projection and with the airfoil axis rotated or developed and projected into the plane of projection.

deviation factor *See* compressibility factor.

deviatonic stress [MECH] The portion of the total stress that differs from an isostatic hydrostatic pressure; it is equal to the difference between the total stress and the spherical stress.

devil's pitchfork [DES ENG] A tool with flexible prongs used in recovery of a bit, underreamer, cutters, or such lost during drilling.

dewaterer [MECH ENG] Wet-type mechanical classifier (solids separator) in which solids settle out of the carrier liquid and are concentrated for recovery.

diabatic [THERMO] Involving a thermodynamic change of state of a system in which there is a transfer of heat across the boundaries of the system.

diagonal pliers [DES ENG] Pliers with cutting jaws at an angle to the handles to permit cutting off wires close to terminals.

diagonal stay [MECH ENG] A diagonal member between the tube sheet and shell in a fire-tube boiler.

diagram factor [MECH ENG] The ratio of the actual mean effective pressure, as determined by an indicator card, to the map of the ideal cycle for a steam engine.

dial [DES ENG] A separate scale or other device for indicating the value to which a control is set.

dial cable [DES ENG] Braided cord or flexible wire cable used to make a pointer move over a dial when a separate control knob is rotated, or used to couple two shafts together mechanically.

dial cord [DES ENG] A braided cotton, silk, or glass fiber cord used as a dial cable.

dial feed [MECH ENG] A device that rotates workpieces into position successively so they can be acted on by a machine.

dial indicator [DES ENG] Meter or gage with a calibrated circular face and a pivoted pointer to give readings.

dial press [MECH ENG] A punch press with dial feed.

diameter group [MECH ENG] A dimensionless group, used in the study of flow machines such as turbines and pumps, equal to the fourth root of pressure number 2 divided by the square root of the delivery number.

diametral pitch [DES ENG] A gear tooth design factor expressed as the ratio of the number of teeth to the diameter of the pitch circle measured in inches.

diamond bit [DES ENG] A rotary drilling bit crowned with bort-type diamonds, used for rock boring. Also known as bort bit.

diamond chisel [DES ENG] A chisel having a V-shaped or diamond-shaped cutting edge.

diamond count [DES ENG] The number of diamonds set in a diamond crown bit.

diamond crown [DES ENG] The cutting bit used in diamond drilling; it consists of a steel shell set with black diamonds on the face and cutting edges.

diamond drill [DES ENG] A drilling machine with a hollow, diamond-set bit for boring rock and yielding continuous and columnar rock samples.

diamond matrix [DES ENG] The metal or alloy in which diamonds are set in a drill crown.

diamond orientation [DES ENG] The set of a diamond in a cutting tool so that the crystal face will be in contact with the material being cut.

diamond-particle bit [DES ENG] A diamond bit set with small fragments of diamonds.

diamond pattern [DES ENG] The arrangement of diamonds set in a diamond crown.

diamond point [DES ENG] A cutting tool with a diamond tip.

diamond-point bit *See* mud auger.

diamond reamer [DES ENG] A diamond-inset pipe behind, and larger than, the drill bit and core barrel that is used for enlarging boreholes.

diamonds [FL MECH] The pattern of shock waves often visible in a rocket exhaust which resembles a series of diamond shapes placed end to end.

diamond saw [DES ENG] A circular, band, or frame saw inset with diamonds or diamond dust for cutting sections of rock and other brittle substances.

diamond tool [DES ENG] 1. Any tool using a diamond-set bit to drill a borehole. 2. A diamond shaped to the contour of a single-pointed cutting tool, used for precision machining.

diamond wheel [DES ENG] A grinding wheel in which synthetic diamond dust is bonded as the abrasive to cut very hard materials such as sintered carbide or quartz.

diaphragm compressor [MECH ENG] Device for compression of small volumes of a gas by means of a reciprocally moving diaphragm, in place of pistons or rotors.

diaphragm pump [MECH ENG] A metering pump which uses a diaphragm to isolate the operating parts from pumped liquid in a mechanically actuated diaphragm pump, or from hydraulic fluid in a hydraulically actuated diaphragm pump.

diathermous envelope [THERMO] A surface enclosing a thermodynamic system in equilibrium that is not an adiabatic envelope; intuitively, this means that heat can flow through the surface.

dicing cutter [MECH ENG] A cutting mill for sheet material; sheet is first slit into horizontal strands by blades, then fed against a rotating knife for dicing.

die [DES ENG] A tool or mold used to impart shapes to, or to form impressions on, materials such as metals and ceramics.

dieing machine [MECH ENG] A vertical press with the slide activated by pull rods attached to the drive mechanism below the bed of the press.

diesel cycle [THERMO] An internal combustion engine cycle in which the heat of compression ignites the fuel.

diesel electric locomotive [MECH ENG] A locomotive with a diesel engine driving an electric generator which supplies electric power to traction motors for propelling the vehicle. Also known as diesel locomotive.

diesel electric power generation [MECH ENG] Electric power generation in which the generator is driven by a diesel engine.

diesel engine [MECH ENG] An internal combustion engine operating on a thermodynamic cycle in which the ratio of compression of the air charge is sufficiently high to ignite the fuel subsequently injected into the combustion chamber. Also known as compression-ignition engine.

diesel index [MECH ENG] Diesel fuel rating based on ignition qualities; high-quality fuel has a high index number.

diesel knock [MECH ENG] A combustion knock caused when the delayed period of ignition is long so that a large quantity of atomized fuel accumulates in the combustion chamber; when combustion occurs, the sudden high pressure resulting from the accumulated fuel causes diesel knock.

diesel locomotive　*See* diesel electric locomotive.

diesel rig　[MECH ENG] Any diesel engine apparatus or machinery.

die shoe　[MECH ENG] A block placed beneath the lower part of a die upon which the die holder is mounted; spreads the impact over the die bed, thereby reducing wear.

die slide　[MECH ENG] A device in which the lower die of a power press is mounted; it slides in and out of the press for easy access and safety in feeding the parts.

Dieterici equation of state　[THERMO] An empirical equation of state for gases, $pe^{a/vRT}(v-b) = RT$, where p is the pressure, T is the absolute temperature, v is the molar volume, R is the gas constant, and a and b are constants characteristic of the substance under consideration.

differential　[CONT SYS] The difference between levels for turn-on and turn-off operation in a control system.　[MECH ENG] Any arrangement of gears forming an epicyclic train in which the angular speed of one shaft is proportional to the sum or difference of the angular speeds of two other gears which lie on the same axis; allows one shaft to revolve faster than the other, the speed of the main driving member being equal to the algebraic mean of the speeds of the two shafts. Also known as differential gear.

differential brake　[MECH ENG] A brake in which operation depends on a difference between two motions.

differential calorimetry　[THERMO] Technique for measurement of and comparison (differential) of process heats (reaction, absorption, hydrolysis, and so on) for a specimen and a reference material.

differential effects　[MECH] The effects upon the elements of the trajectory due to variations from standard conditions.

differential game　[CONT SYS] A two-sided optimal control problem.

differential gap controller　[CONT SYS] A two-position (on-off) controller that actuates when the manipulated variable reaches the high or low value of its range (differential gap).

differential gear　*See* differential.

differential heat of solution　[THERMO] The partial derivative of the total heat of solution with respect to the molal concentration of one component of the solution, when the concentration of the other component or components, the pressure, and the temperature are held constant.

differential indexing　[MECH ENG] A method of subdividing a circle based on the difference between movements of the index plate and index crank of a dividing engine.

differential motion　[MECH ENG] A mechanism in which the follower has two driving elements; the net motion of the follower is the difference between the motions that would result from either driver acting alone.

differential pressure fuel valve　[MECH ENG] A needle or spindle normally closed, with seats at the back side of the valve orifice.

differential pulley　[MECH ENG] A tackle in which an endless cable passes through a movable lower pulley, which carries the load, and two fixed coaxial upper pulleys having different diameters; yields a high mechanical advantage.

differential screw　[MECH ENG] A type of compound screw which produces a motion equal to the difference in motion between the two component screws.

differential thermal analysis　[THERMO] A method of determining the temperature at which thermal reactions occur in a material undergoing continuous heating to elevated temperatures; also involves a determination of the nature and intensity of such reactions.

differential thermogravimetric analysis [THERMO] Thermal analysis in which the rate of material weight change upon heating versus temperature is plotted; used to simplify reading of weight-versus-temperature thermogram peaks that occur close together.

differential windlass [MECH ENG] A windlass in which the barrel has two sections, each having a different diameter; the rope winds around one section, passes through a pulley (which carries the load), then winds around the other section of the barrel.

diffluence [FL MECH] A region of fluid flow in which the fluid is diverging from the direction of flow.

diffusion [MECH ENG] The conversion of air velocity into static pressure in the diffuser casing of a centrifugal fan, resulting from increases in the radius of the air spin and in area.

diffusion number [FL MECH] A dimensionless number used in the study of mass transfer, equal to the diffusivity of a solute through a stationary solution contained in the solid, times a characteristic time, divided by the square of the distance from the midpoint of the solid to the surface. Symbolized β.

diffusion velocity [FL MECH] **1.** The relative mean molecular velocity of a selected gas undergoing diffusion in a gaseous atmosphere, commonly taken as a nitrogen (N_2) atmosphere; a molecular phenomenon that depends upon the gaseous concentration as well as upon the pressure and temperature gradients present. **2.** The velocity or speed with which a turbulent diffusion process proceeds as evidenced by the motion of individual eddies.

diffusivity [THERMO] The quantity of heat passing normally through a unit area per unit time divided by the product of specific heat, density, and temperature gradient. Also known as thermometric conductivity.

digital control [CONT SYS] The use of digital or discrete technology to maintain conditions in operating systems as close as possible to desired values despite changes in the operating environment.

dihedral [AERO ENG] The upward or downward inclination of an airplane's wing or other supporting surface in respect to the horizontal; in some contexts, the upward inclination only.

Dings magnetic separator [MECH ENG] A device which is suspended above a belt conveyor to pull out and separate magnetic material from burden as thick as 40 inches (1 meter) and at belt speeds up to 750 feet (229 meters) per minute.

dinking [MECH ENG] Using a sharp, hollow punch for cutting light-gage soft metals or nonmetallic materials.

dipper dredge [MECH ENG] A power shovel resembling a grab crane mounted on a flat-bottom boat for dredging under water. Also known as dipper shovel.

dipper stick [MECH ENG] A straight shaft connecting the digging bucket of an excavating machine or power shovel with the boom.

dipper trip [MECH ENG] A device which releases the door of a shovel bucket.

direct-acting pump [MECH ENG] A displacement reciprocating pump in which the steam or power piston is connected to the pump piston by means of a rod, without crank motion or flywheel.

direct air cycle [AERO ENG] A thermodynamic propulsion cycle involving a nuclear reactor and gas turbine or ramjet engine, in which air is the working fluid. Also known as direct cycle.

direct-connected [MECH ENG] The connection between a driver and a driven part, as a turbine and an electric generator, without intervening speed-changing devices, such as gears.

direct-contact condenser *See* contact condenser.

direct control function *See* regulatory control function.

direct-coupled [MECH ENG] Joined without intermediate connections.

direct coupling [MECH ENG] The direct connection of the shaft of a prime mover (such as a motor) to the shaft of a rotating mechanism (such as a pump or compressor).

direct cycle *See* direct air cycle.

direct digital control [CONT SYS] The use of a digital computer generally on a time-sharing or multiplexing basis, for process control in petroleum, chemical, and other industries.

direct drive [MECH ENG] A drive in which the driving part is directly connected to the driven part.

direct-drive vibration machine [MECH ENG] A vibration machine in which the vibration table is forced to undergo a displacement by a positive linkage driven by a direct attachment to eccentrics or camshafts.

direct-expansion coil [MECH ENG] A finned coil, used in air cooling, inside of which circulates a cold fluid or evaporating refrigerant. Abbreviated DX coil.

direct feedback system [CONT SYS] A system in which electrical feedback is used directly, as in a tachometer.

direct-geared [MECH ENG] Joined by a gear on the shaft of one machine meshing with a gear on the shaft of another machine.

directional gyro [AERO ENG] A flight instrument incorporating a gyro that holds its position in azimuth and thus can be used as a directional reference. Also known as direction indicator. [MECH] A two-degrees-of-freedom gyro with a provision for maintaining its spin axis approximately horizontal.

directional stability [AERO ENG] The property of an aircraft, rocket, or such, enabling it to restore itself from a yawing or side-slipping condition. Also known as weathercock stability.

direction indicator *See* directional gyro.

dirigible [AERO ENG] A lighter-than-air craft equipped with means of propelling and steering for controlled flight.

discharge [FL MECH] The flow rate of a fluid at a given instant expressed as volume per unit of time.

discharge channel [MECH ENG] The passage in a pressure-relief device through which the fluid is released to the outside of the device.

discharge coefficient [FL MECH] In a nozzle or other constriction, the ratio of the mass flow rate at the discharge end of the nozzle to that of an ideal nozzle which expands an identical working fluid from the same initial conditions to the same exit pressure. Also known as coefficient of discharge.

discharge head [MECH ENG] Vertical distance between the intake level of a water pump and the level at which it discharges water freely to the atmosphere.

discharge tube [MECH ENG] A tube through which steam and water are released into a boiler drum.

DISCOS *See* disturbance compensation system.

discrete system [CONT SYS] A control system in which signals at one or more points may change only at discrete values of time. Also known as discrete-time system.

discrete-time system *See* discrete system.

disintegrator [MECH ENG] An apparatus used for pulverizing or grinding substances, consisting of two steel cages which rotate in opposite directions.

disk attrition mill *See* disk mill.

disk brake [MECH ENG] A type of brake in which disks attached to a fixed frame are pressed against disks attached to a rotating axle or against the inner surfaces of a rotating housing.

disk cam [MECH ENG] A disk with a contoured edge which rotates about an axis perpendicular to the disk, communicating motion to the cam follower which remains in contact with the edge of the disk.

disk canvas wheel [DES ENG] A polishing wheel made of disks of canvas sewn together with heavy twine or copper wire, and reinforced by steel side plates and side rings with bolts or screws.

disk centrifuge [MECH ENG] A centrifuge with a large bowl having a set of disks that separate the liquid into thin layers to create shallow settling chambers.

disk clutch [MECH ENG] A clutch in which torque is transmitted by friction between friction disks with specially prepared friction material riveted to both sides and contact plates keyed to the inner surface of an external hub.

disk coupling [MECH ENG] A flexible coupling in which the connecting member is a flexible disk.

disk engine [MECH ENG] A rotating engine in which the piston is a disk.

disk grinder [MECH ENG] A grinding machine that employs abrasive disks.

disk grinding [MECH ENG] Grinding with the flat side of a rigid, bonded abrasive disk or segmental wheel.

disk leather wheel [DES ENG] A polishing wheel made of leather disks glued together.

disk loading [AERO ENG] A measure which expresses the design gross weight of a helicopter as a function of the swept areas of the lifting rotor.

disk mill [MECH ENG] Size-reduction apparatus in which grinding of feed solids takes place between two disks, either or both of which rotate. Also known as disk attrition mill.

disk sander [MECH ENG] A machine that uses a circular disk coated with abrasive to smooth or shape surfaces.

disk spring [MECH ENG] A mechanical spring that consists of a disk or washer supported by one force (distributed by a suitable chuck or holder) at the periphery and by an opposing force on the center or hub of the disk.

disk wheel [DES ENG] A wheel in which a solid metal disk, rather than separate spokes, joins the hub to the rim.

dispersion [AERO ENG] Deviation from a prescribed flight path; specifically, circular dispersion especially as applied to missiles.

dispersion mill [MECH ENG] Size-reduction apparatus that disrupts clusters or agglomerates of solids, rather than breaking down individual particles; used for paint pigments, food products, and cosmetics.

displacement [FL MECH] **1.** The weight of fluid which is displaced by a floating body, equal to the weight of the body and its contents; the displacement of a ship is generally measured in long tons (1 long ton = 2240 pounds). **2.** The volume of fluid which is displaced by a floating body. [MECH] **1.** The linear distance from the initial to the final position of an object moved from one place to another, regardless of the length of path followed. **2.** The distance of an oscillating particle from its equilibrium position.

displacement compressor [MECH ENG] A type of compressor that depends on displacement of a volume of air by a piston moving in a cylinder.

displacement engine *See* piston engine.

displacement pump [MECH ENG] A pump that develops its action through the alternate filling and emptying of an enclosed volume as in a piston-cylinder construction.

dissipation function *See* Rayleigh's dissipation function; viscous dissipation function.

dissipation trail [FL MECH] A clear rift left behind an aircraft as it flies in a thin cloud layer; the opposite of a condensation trail. Also known as distrail.

distance [MECH] The spatial separation of two points, measured by the length of a hypothetical line joining them.

distance/velocity lag [CONT SYS] The delay caused by the amount of time required to transport material or propagate a signal or condition from one point to another. Also known as transportation lag; transport lag.

distrail *See* dissipation trail.

distributed control system [CONT SYS] A collection of modules, each with its own specific function, interconnected tightly to carry out an integrated data acquisition and control application.

distributor gear [MECH ENG] A gear which meshes with the camshaft gear to rotate the distributor shaft.

district heating [MECH ENG] The supply of heat, either in the form of steam or hot water, from a central source to a group of buildings.

disturbance [CONT SYS] An undesired command signal in a control system.

disturbance compensation system [AERO ENG] A system applied to navigational satellites to remove the along-the-track component of drag and radiation forces. Abbreviated DISCOS.

ditcher *See* trench excavator.

ditching [AERO ENG] A forced landing on water, or the process of making such a landing.

dither [CONT SYS] A force having a controlled amplitude and frequency, applied continuously to a device driven by a servomotor so that the device is constantly in small-amplitude motion and cannot stick at its null position. Also known as buzz.

Dittus-Boelter equation [FL MECH] An equation used to calculate the surface coefficient of heat transfer for fluids in turbulent flow inside clean, round pipes.

dive [AERO ENG] A rapid descent by an aircraft or missile, nose downward, with or without power or thrust.

dive bomber [AERO ENG] An aircraft designed to release bombs during a steep dive.

dive brake [AERO ENG] An air brake designed for operation in a dive; flaps at the following edge of one wing that can be extended into the airstream to increase drag and hold the aircraft to its "never exceed" dive speed in a vertical dive; used on dive bombers and sailplanes.

divergence [FL MECH] The ratio of the area of any section of fluid emerging from a nozzle to the area of the throat of the nozzle.

divergence speed [AERO ENG] The speed of an aircraft above which no statically stable equilibrium condition exists and the deformation will increase to a point of structural failure.

divergent nozzle [DES ENG] A nozzle whose cross section becomes larger in the direction of flow.

diverging duct [DES ENG] Fluid-flow conduit whose internal cross-sectional area increases in the direction of flow.

diverging yaw [AERO ENG] In the flight of a projectile, an angle of yaw increasing from the initial yaw, so that the projectile is unstable.

divided pitch [DES ENG] In a screw with multiple threads, the distance between corresponding points on two adjacent threads measured parallel to the axis.

divider [DES ENG] A tool like a compass, used in metalworking to lay out circles or arcs and to space holes or other dimensions.

division plate [MECH ENG] A diaphragm which surrounds the piston rod of a crosshead-type engine and separates the crankcase from the lower portion of the cylinder.

docking [AERO ENG] The mechanical coupling of two or more man-made orbiting objects.

dodge chain [DES ENG] A chain with detachable bearing blocks between the links.

dog [DES ENG] 1. Any of various simple devices for holding, gripping, or fastening, such as a hook, rod, or spike with a ring, claw, or lug at the end. 2. An iron for supporting logs in a fireplace. 3. A drag for the wheel of a vehicle.

dog clutch [DES ENG] A clutch in which projections on one part fit into recesses on the other part.

dog iron [DES ENG] 1. A short iron bar with ends bent at right angles. 2. An iron pin that can be inserted in stone or timber in order to lift it.

dog screw [DES ENG] A screw with an eccentric head; used to mount a watch in its case.

domestic refrigerator [MECH ENG] A refrigeration system for household use which typically has a compression machine designed for continuous automatic operation and for conservation of the charges of refrigerant and oil, and is usually motor-driven and air-cooled. Also known as refrigerator.

domestic satellite [AERO ENG] A satellite in stationary orbit 22,300 miles (35,680 kilometers) above the equator for handling up to 12 separate color television programs, up to 14,000 private-line telephone calls, or an equivalent number of channels for other communication services within the United States. Abbreviated DOMSAT.

DOMSAT *See* domestic satellite.

donkey engine [MECH ENG] A small auxiliary engine which is usually portable or semiportable and powered by steam, compressed air, or other means, particularly one used to power a windlass to lift cargo on shipboard or to haul logs.

Donohue equation [THERMO] Equation used to determine the heat-transfer film coefficient for a fluid on the outside of a baffled shell-and-tube heat exchanger.

doodlebug [MECH ENG] 1. A small tractor. 2. A motor-driven railcar used for maintenance and repair work.

door check *See* door closer.

door closer [DES ENG] 1. A device that makes use of a spring for closing, and a compression chamber from which liquid or air escapes slowly, to close a door at a controlled speed. Also known as door check. 2. In elevators, a device or assembly of devices which closes an open car or hoistway door by the use of gravity or springs.

Dorr agitator [MECH ENG] A tank used for batch washing of precipitates which cannot be leached satisfactorily in a tank; equipped with a slowly rotating rake at the bottom, which moves settled solids to the center, and an air lift that lifts slurry to the launders. Also known as Dorr thickener.

Dorr classifier [MECH ENG] A horizontal flow classifier consisting of a rectangular tank with a sloping bottom, a rake mechanism for moving sands uphill along the bottom, an inlet for feed, and outlets for sand and slime.

Dorr thickener *See* Dorr agitator.

double-acting [MECH ENG] Acting in two directions, as with a reciprocating piston in a cylinder with a working chamber at each end.

double-acting compressor [MECH ENG] A reciprocating compressor in which both ends of the piston act in working chambers to compress the fluid.

double-acting pawl [MECH ENG] A double pawl which can drive in either direction.

double-action mechanical press [MECH ENG] A press having two slides which move one within the other in parallel movements.

double-block brake [MECH ENG] Two single-block brakes in symmetrical opposition, where the operating force on one lever is the reaction on the other.

double-cone bit [DES ENG] A type of roller bit having only two cone-shaped cutting members.

double-core barrel drill [DES ENG] A core drill consisting of an inner and an outer tube; the inner member can remain stationary while the outer one revolves.

double-crank press [MECH ENG] A mechanical press with a single wide slide operated by a crankshaft having two crank pins.

double-cut file [DES ENG] A file covered with two series of parallel ridges crossing at angles to each other.

double-cut planer [MECH ENG] A planer designed to cut in both the forward and reverse strokes of the table.

double-cut saw [DES ENG] A saw with teeth that cut during the forward and return strokes.

double-drum hoist [MECH ENG] A hoisting device consisting of two cable drums which rotate in opposite directions and can be operated separately or together.

double Hooke's joint [MECH ENG] A universal joint which eliminates the variation in angular displacement and angular velocity between driving and driven shafts, consisting of two Hooke's joints with an intermediate shaft.

double-housing planer [MECH ENG] A planer having two housings to support the cross rail, with two heads on the cross rail and one sidehead on each housing.

double-integrating gyro [MECH] A single-degree-of-freedom gyro having essentially no restraint of its spin axis about the output axis.

double jack [DES ENG] A heavy hammer, weighing about 10 pounds (4.5 kilograms), requiring the use of both hands.

double pendulum [MECH] Two masses, one suspended from a fixed point by a weightless string or rod of fixed length, and the other similarly suspended from the first; often the system is constrained to remain in a vertical plane.

double-roll crusher [MECH ENG] A machine which crushes materials between teeth on two roll surfaces; used mainly for coal.

doublet [FL MECH] A source and a sink separated by an infinitesimal distance, each having an infinitely large strength so that the product of this strength and the separation is finite.

doublet flow [FL MECH] The motion of a fluid in the vicinity of a doublet; can be superposed with uniform flow to yield flow around a cylinder or a sphere.

dovetail joint [DES ENG] A joint consisting of a flaring tenon in a fitting mortise.

dovetail saw [DES ENG] A short stiff saw with a thin blade and fine teeth; used for accurate woodwork.

dowel [DES ENG] 1. A headless, cylindrical pin which is sunk into corresponding holes in adjoining parts, to locate the parts relative to each other or to join them together. 2. A round wooden stick from which dowel pins are cut.

dowel screw [DES ENG] A dowel with threads at both ends.

downdraft carburetor [MECH ENG] A carburetor in which the fuel is fed into a downward current of air.

down lock [AERO ENG] An airplane mechanism that locks the landing gear in a down position after the gear is lowered.

downrange [AERO ENG] Any area along the flight course of a rocket or missile test range.

downwash [FL MECH] The downward deflection of air, relative to the direction of motion of an airfoil.

dr *See* dram.

drachm *See* dram.

draft Also spelled draught. [FL MECH] 1. An air current in a confined space, such as that in a cooling tower or chimney. 2. The difference between atmospheric pressure and some lower pressure in a confined space that causes air to flow, such as exists in the furnace or gas passages of a steam-generating unit or in a chimney.

draft differential [FL MECH] The difference in static pressure between two locations of gas flow.

draft loss [MECH ENG] A decrease in the static pressure of a gas in a furnace or boiler due to flow resistance.

draft tube [MECH ENG] The piping system for a reaction-type hydraulic turbine that allows the turbine to be set safely above tail water and yet utilize the full head of the site from head race to tail race.

drag [FL MECH] Resistance caused by friction in the direction opposite to that of the motion of the center of gravity of a moving body in a fluid.

drag-chain conveyor [MECH ENG] A conveyor in which the open links of a chain drag material along the bottom of a hard-faced concrete or cast iron trough. Also known as dragline conveyor.

drag chute *See* drag parachute.

drag classifier [MECH ENG] A continuous belt containing transverse rakes, used to separate coarse sand from fine; the belt moves up through an inclined trough, and fast-settling sands are dragged along by the rakes.

drag coefficient [FL MECH] A characteristic of a body in a flowing inviscous fluid, equal to the ratio of twice the force on the body in the direction of flow to the product of the density of the fluid, the square of the flow velocity, and the effective cross-sectional area of the body.

drag direction [AERO ENG] In stress analysis of a given airfoil, the direction of the relative wind.

dragline [MECH ENG] An excavator operated by pulling a bucket on ropes towards the jib from which it is suspended. Also known as dragline excavator.

dragline conveyor *See* drag-chain conveyor.

dragline excavator *See* dragline.

dragline scraper [MECH ENG] A machine with a flat, plowlike blade or partially open bucket pulled on rope for withdrawing piled material, such as stone or coal, from a stockyard to the loading platform; the empty bucket is subsequently returned to the pile of material by means of a return rope.

drag link [MECH ENG] A four-bar linkage in which both cranks traverse full circles; the fixed member must be the shortest link.

drag parachute [AERO ENG] Any of various types of parachutes that can be deployed from the rear of an aircraft, especially during landings, to decrease speed and also, under certain flight conditions, to control and stabilize the aircraft. Also known as drag chute.

drag rope [AERO ENG] A long, heavy rope carried in the basket of a balloon and permitted to hang over the side and drag on the ground in order to lighten the basket.

dragsaw [DES ENG] A saw that cuts on the pulling stroke; used in power saws for cutting felled trees.

drag truss [AERO ENG] A truss that is positioned horizontally between the wing spars; used to stiffen the wing structure and as a resistance for drag forces acting on the airplane wing.

drag-weight ratio [AERO ENG] The ratio of the drag of a missile to its total weight.

drag wire [AERO ENG] A part of the truss in an airplane wing and also in the wing support; used to sustain the backward reaction due to the wing's drag.

dram [MECH] 1. A unit of mass, used in the apothecaries' system of mass units, equal to ⅛ apothecaries' ounce or 60 grains or 3.8879346 grams. Also known as apothecaries' dram (dram ap); drachm (British). 2. A unit of mass, formerly used in the United Kingdom, equal to 1/16 ounce (avoirdupois) or approximately 1.77185 grams. Abbreviated dr.

draught *See* draft.

drawbar horsepower [MECH ENG] The horsepower available at the drawbar in the rear of a locomotive or tractor to pull the vehicles behind it.

drawbar pull [MECH ENG] The force with which a locomotive or tractor pulls vehicles on a drawbar behind it.

drawknife [DES ENG] A woodcutting tool with a long, narrow blade and two handles mounted at right angles to the blade.

dredge [MECH ENG] A floating excavator used for widening or deepening channels, building canals, constructing levees, raising material from stream or harbor bottoms to be used elsewhere as fill, or mining.

dress [MECH ENG] 1. To shape a tool. 2. To restore a tool to its original shape and sharpness.

Dressler kiln [MECH ENG] The first successful muffle-type tunnel kiln.

drift [MECH ENG] The water lost in a cooling tower as mist or droplets entrained by the circulating air, not including the evaporative loss.

drifter [MECH ENG] A rock drill, similar to but usually larger than a jack hammer, mounted for drilling holes up to 4½ inches (11.4 centimeters) in diameter.

driftpin [DES ENG] A round, tapered metal rod that is driven into matching rivet holes of two metal parts for stretching the parts and bringing them into alignment.

drill angle gage *See* drill grinding gage.

drill capacity [MECH ENG] The length of drill rod of specified size that the hoist on a diamond or rotary drill can lift or that the brake can hold on a single line.

drill carriage [MECH ENG] A platform or frame on which several rock drills are mounted and which moves along a track, for heavy drilling in large tunnels. Also known as jumbo.

drill chuck [DES ENG] A chuck for holding a drill or other cutting tool on a spindle.

drill collar [DES ENG] A ring which holds a drill bit and gives it radial location with respect to a bearing.

driller *See* drilling machine.

drill feed [MECH ENG] The mechanism by which the drill bit is fed into the borehole during drilling.

drill gage [DES ENG] A thin, flat steel plate that has accurate holes for many sizes of drills; each hole, identified as to drill size, enables the diameter of a drill to be checked.

drill grinding gage [DES ENG] A tool that checks the angle and length of a twist drill while grinding it. Also known as drill angle gage; drill point gage.

drilling machine [MECH ENG] A device, usually motor-driven, fitted with an end cutting tool that is rotated with sufficient power either to create a hole or to enlarge an existing hole in a solid material. Also known as driller.

drilling rate [MECH ENG] The number of lineal feet drilled per unit of time.

drill jig [MECH ENG] A device fastened to the work in repetition drilling to position and guide the drill.

drill point gage *See* drill grinding gage.

drill press [MECH ENG] A drilling machine in which a vertical drill moves into the work, which is stationary.

drill stem *See* drill string.

drill string [MECH ENG] The assemblage of drill rods, core barrel, and bit, or of drill rods, drill collars, and bit in a borehole, which is connected to and rotated by the drill collar of the borehole. Also known as drill stem.

drive [MECH ENG] The means by which a machine is given motion or power (as in steam drive, diesel-electric drive), or by which power is transferred from one part of a machine to another (as in gear drive, belt drive).

drive chuck [MECH ENG] A mechanism at the lower end of a diamond-drill drive rod on the swivel head by means of which the motion of the drive rod can be transmitted to the drill string.

drive fit [DES ENG] A fit in which the larger (male) part is pressed into a smaller (female) part; the assembly must be effected through the application of an external force.

driven gear [MECH ENG] The member of a pair of gears to which motion and power are transmitted by the other.

drive pulley [MECH ENG] The pulley that drives a conveyor belt.

drivescrew [DES ENG] A screw that is driven all the way in, or nearly all the way in, with a hammer.

drive shaft [MECH ENG] A shaft which transmits power from a motor or engine to the rest of a machine.

drive shoe [DES ENG] A sharp-edged steel sleeve attached to the bottom of a drivepipe or casing to act as a cutting edge and protector.

driving pinion [MECH ENG] The input gear in the differential of an automobile.

driving point function [CONT SYS] A special type of transfer function in which the input and output variables are voltages or currents measured between the same pair of terminals in an electrical network.

driving resistance [MECH] The force exerted by soil on a pile being driven into it.

driving wheel [MECH ENG] A wheel that supplies driving power.

drogue [AERO ENG] 1. A small parachute attached to a body for stabilization and deceleration. Also known as deceleration parachute. 2. A funnel-shaped device at the end of the hose of a tanker aircraft in flight, to receive the probe of another aircraft that will take on fuel.

drone [AERO ENG] A pilotless aircraft usually subordinated to the controlling influences of a remotely located command station, but occasionally preprogrammed.

drooped ailerons [AERO ENG] Ailerons that are of the hinged trailing-edge type and are so arranged that both the right and left one have a 10 to 15° positive downward deflection with the control column in a neutral position.

droop governor [MECH ENG] A governor whose equilibrium speed decreases as the load on the machinery controlled by the governor increases.

drop [FL MECH] The quantity of liquid that coalesces into a single globule; sizes vary according to physical conditions and the properties of the fluid itself.

drop bar [MECH ENG] A bar that guides sheets of paper into a printing or folding machine.

drop hammer *See* pile hammer.

drop press *See* punch press.

drop tank [AERO ENG] A fuel tank on an airplane that may be jettisoned.

drop weight [FL MECH] The weight of the largest drop that can hang from the end of a tube of given radius.

drop-weight method [FL MECH] A method of measuring surface tension by measuring the weight of a slowly increasing drop of the liquid hanging from the end of a tube, just before it is detached from the tube.

drum [DES ENG] 1. A hollow, cylindrical container. 2. A metal cylindrical shipping container for liquids having a capacity of 12–110 gallons (45–416 liters). [MECH ENG] A horizontal cylinder about which rope or wire rope is wound in a hoisting mechanism. Also known as hoisting drum.

drum brake [MECH ENG] A brake in which two curved shoes fitted with heat- and wear-resistant linings are forced against the surface of a rotating drum.

drum cam [MECH ENG] A device consisting of a drum with a contoured surface which communicates motion to a cam follower as the drum rotates around an axis.

drum dryer [MECH ENG] A machine for removing water from substances such as milk, in which a thin film of the product is moved over a turning steam-heated drum and a knife scrapes it from the drum after moisture has been removed.

drum feeder [MECH ENG] A rotating drum with vanes or buckets to lift and carry parts and drop them into various orienting or chute arrangements. Also known as tumbler feeder.

drum filter [MECH ENG] A cylindrical drum that rotates through thickened ore pulp, extracts liquid by a vacuum, and leaves solids, in the form of a cake, on a permeable membrane on the drum end. Also known as rotary filter; rotary vacuum filter.

drum-type boiler *See* bent-tube boiler.

dry abrasive cutting [MECH ENG] Frictional cutting using a rotary abrasive wheel without the use of a liquid coolant.

dry cooling tower [MECH ENG] A structure in which water is cooled by circulation through finned tubes, transferring heat to air passing over the fins; there is no loss of water by evaporation because the air does not directly contact the water.

dry friction [MECH] Resistance between two dry solid surfaces, that is, surfaces free from contaminating films or fluids.

dry measure [MECH] A measure of volume for commodities that are dry.

dry mill [MECH ENG] Grinding device used to powder or pulverize solid materials without an associated liquid.

dry pint *See* pint.

dry pipe [MECH ENG] A perforated metal pipe above the normal water level in the steam space of a boiler which prevents moisture or extraneous matter from entering steam outlet lines.

dry-pit pump [MECH ENG] A pump operated with the liquid conducted to and from the unit by piping.

dry sleeve [MECH ENG] A cylinder liner which is not in contact with the coolant.

dry steam drum [MECH ENG] **1.** Pressurized chamber into which steam flows from the steam space of a boiler drum. **2.** That portion of a two-stage furnace that extends forward of the main combustion chamber; fuel is dried and gasified therein, with combustion of gaseous products accomplished in the main chamber; the refractory walls of the Dutch oven are sometimes water-cooled.

dry storage [MECH ENG] Cold storage in which refrigeration is provided by chilled air.

Drzwiecki theory [MECH ENG] In theoretical investigations of windmill performance, a theory concerning the air forces produced on an element of the blade.

DSN *See* Deep Space Network.

dual-bed dehumidifier [MECH ENG] A sorbent dehumidifier with two beds, one bed dehumidifying while the other bed is reactivating, thus providing a continuous flow of air.

dual control [CONT SYS] An optimal control law for a stochastic adaptive control system that gives a balance between keeping the control errors and the estimation errors small.

dual-flow oil burner [MECH ENG] An oil burner with two sets of tangential slots in its atomizer for use at different capacity levels.

dual-fuel engine [MECH ENG] Internal combustion engine that can operate on either of two fuels, such as natural gas or gasoline.

dual-mode control [CONT SYS] A type of control law which consists of two distinct types of operation; in linear systems, these modes usually consist of a linear feedback mode and a bang-bang-type mode.

dual thrust [AERO ENG] A rocket thrust derived from two propellant grains and using the same propulsion section of a missile.

dual-thrust motor [AERO ENG] A solid-propellant rocket engine built to obtain dual thrust.

duckbill [MECH ENG] A shaking type of combination loader and conveyor whose loading end is generally shaped like a duck's bill.

duct [MECH ENG] A fluid flow passage which may range from a few inches in diameter to many feet in rectangular cross section, usually constructed of galvanized steel, aluminum, or copper, through which air flows in a ventilation system or to a compressor, supercharger, or other equipment at speeds ranging to thousands of feet per minute.

ducted fan [MECH ENG] A propeller or multibladed fan inside a coaxial duct or cowling. Also known as ducted propeller; shrouded propeller.

ducted-fan engine [AERO ENG] An aircraft engine incorporating a fan or propeller enclosed in a duct; especially, a jet engine in which an enclosed fan or propeller is used to ingest ambient air to augment the gases of combustion in the jetstream.

ducted rocket *See* rocket ramjet.

ductile fracture *See* fibrous fracture.

duct propulsion [AERO ENG] A means of propelling a vehicle by ducting a surrounding fluid through an engine, adding momentum by mechanical or thermal means, and ejecting the fluid to obtain a reactive force.

Dufour effect [THERMO] Energy flux due to a mass gradient occurring as a coupled effect of irreversible processes.

Dufour number [THERMO] A dimensionless number used in studying thermodiffusion, equal to the increase in enthalpy of a unit mass during isothermal mass transfer divided by the enthalpy of a unit mass of mixture. Symbol Du_2.

Duhem-Margules equation [THERMO] An equation showing the relationship between the two constituents of a liquid-vapor system and their partial vapor pressures:

$$\frac{d \ln p_A}{d \ln x_A} = \frac{d \ln p_B}{d \ln x_B}$$

where x_A and x_B are the mole fractions of the two constituents, and p_A and p_B are the partial vapor pressures.

Dulong-Petit law [THERMO] The law that the product of the specific heat per gram and the atomic weight of many solid elements at room temperature has almost the same value, about 6.3 calories (264 joules) per degree Celsius.

dumbwaiter [MECH ENG] An industrial elevator which carries small objects but is not permitted to carry people.

dump bucket [MECH ENG] A large bucket with movable discharge gates at the bottom; used to move soil or other construction materials by a crane or cable.

dump car [MECH ENG] Any of several types of narrow-gage rail cars with bodies which can easily be tipped to dump material.

duplex lock [DES ENG] A lock with two independent pin-tumbler cylinders on the same bolt.

duplex pump [MECH ENG] A reciprocating pump with two parallel pumping cylinders.

duplex tandem compressor [MECH ENG] A compressor having cylinders on two parallel frames connected through a common crankshaft.

Dupré equation [THERMO] The work W_{LS} done by adhesion at a gas-solid-liquid interface, expressed in terms of the surface tensions γ of the three phases, is $W_{LS} = \gamma_{GS} + \gamma_{GL} - \gamma_{LS}$.

duration [MECH] A basic concept of kinetics which is expressed quantitatively by time measured by a clock or comparable mechanism.

dwell [DES ENG] That part of a cam that allows the cam follower to remain at maximum lift for a period of time. [ENG] A pause in the application of pressure to a mold.

dwt *See* pennyweight.

DX coil *See* direct-expansion coil.

dynamical similarity [MECH] Two flow fields are dynamically similar if one can be transformed into the other by a change of length and velocity scales. All dimensionless numbers of the flows must be the same.

dynamical variable [MECH] One of the quantities used to describe a system in classical mechanics, such as the coordinates of a particle, the components of its velocity, the momentum, or functions of these quantities.

dynamic augment [MECH ENG] Force produced by unbalanced reciprocating parts in a steam locomotive.

dynamic balance [MECH] The condition which exists in a rotating body when the axis about which it is forced to rotate, or to which reference is made, is parallel with a principal axis of inertia; no products of inertia about the center of gravity of the body exist in relation to the selected rotational axis.

dynamic boundary condition [FL MECH] The condition that the pressure must be continuous across an internal boundary or free surface in a fluid.

dynamic braking [MECH] A technique of electric braking in which the retarding force is supplied by the same machine that originally was the driving motor.

dynamic compressor [MECH ENG] A compressor which uses rotating vanes or impellers to impart velocity and pressure to the fluid.

dynamic creep [MECH] Creep resulting from fluctuations in a load or temperature.

dynamic equilibrium Also known as kinetic equilibrium. [MECH] The condition of any mechanical system when the kinetic reaction is regarded as a force, so that the resultant force on the system is zero according to d'Alembert's principle.

dynamic factor [AERO ENG] A ratio formed from the load carried by any airplane part when the airplane is accelerating or subjected to abnormal conditions to the load carried in the conditions of normal flight.

dynamic fluidity [FL MECH] The reciprocal of the dynamic viscosity.

dynamic instability *See* inertial instability.

dynamic load [AERO ENG] With respect to aircraft, rockets, or spacecraft, a load due to an acceleration of craft, as imposed by gusts, by maneuvering, by landing, by firing rockets, and so on.

dynamic pressure [FL MECH] 1. The pressure that a moving fluid would have if it were brought to rest by isentropic flow against a pressure gradient. Also known as impact pressure; stagnation pressure; total pressure. 2. The difference between the quantity in the first definition and the static pressure.

dynamics [MECH] That branch of mechanics which deals with the motion of a system of material particles under the influence of forces, especially those which originate outside the system under consideration.

dynamic similarity [MECH ENG] A relation between two mechanical systems (often referred to as model and prototype) such that by proportional alterations of the units of length, mass, and time, measured quantities in the one system go identically (or with a constant multiple for each) into those in the other; in particular, this implies constant ratios of forces in the two systems.

dynamic stability [MECH] The characteristic of a body, such as an aircraft, rocket, or ship, that causes it, when disturbed from an original state of steady motion in an upright position, to damp the oscillations set up by restoring moments and gradually return to its original state. Also known as stability.

dynamic unbalance [MECH ENG] Failure of the rotation axis of a piece of rotating equipment to coincide with one of the principal axes of inertia due to forces in a single axial plane and on opposite sides of the rotation axis, or in different axial planes.

dyne [MECH] The unit of force in the centimeter-gram-second system of units, equal to the force which imparts an acceleration of 1 cm/s^2 to a 1 gram mass.

earthmover [MECH ENG] A machine used to excavate, transport, or push earth.

earth resources technology satellite [AERO ENG] One of a series of satellites designed primarily to measure the natural resources of the earth; functions include mapping, cataloging water resources, surveying crops and forests, tracing sources of water and air pollution, identifying soil and rock formations, and acquiring oceanographic data. Abbreviated ERTS.

earth satellite [AERO ENG] An artificial satellite placed into orbit about the earth.

eccentric bit [DES ENG] A modified chisel for drilling purposes having one end of the cutting edge extended further from the center of the bit than the other.

eccentric cam [DES ENG] A cylindrical cam with the shaft displaced from the geometric center.

eccentric gear [DES ENG] A gear whose axis deviates from the geometric center.

eccentricity [MECH] The distance of the geometric center of a revolving body from the axis of rotation.

eccentric rotor engine [MECH ENG] A rotary engine, such as the Wankel engine, wherein motion is imparted to a shaft by a rotor eccentric to the shaft.

Echo satellite [AERO ENG] An aluminized-surface, Mylar balloon about 100 feet (30 meters) in diameter, placed in orbit as a passive communications satellite for reflecting microwave signals from a transmitter to receivers beyond the horizon.

economizer [MECH ENG] A forced-flow, once-through, convection-heat-transfer tube bank in which feedwater is raised in temperature on its way to the evaporating section of a steam boiler, thus lowering flue gas temperature, improving boiler efficiency, and saving fuel.

eddy [FL MECH] A vortexlike motion of a fluid running contrary to the main current.

eddy conductivity [THERMO] The exchange coefficient for eddy heat conduction.

eddy-current brake [MECH ENG] A control device or dynamometer for regulating rotational speed, as of flywheels, in which energy is converted by eddy currents into heat.

eddy-current clutch [MECH ENG] A type of electromagnetic clutch in which torque is transmitted by means of eddy currents induced by a magnetic field set up by a coil carrying direct current in one rotating member.

eddy-current damper [AERO ENG] A device used to damp nutation and other unwanted vibration in spacecraft, based on the principle that eddy currents induced in conducting material by motion relative to magnets tend to counteract that motion.

eddy diffusion [FL MECH] Diffusion which occurs in turbulent flow, by the rapid process of mixing of the swirling eddies of fluid. Also known as turbulent diffusion.

eddy diffusivity [FL MECH] The exchange coefficient for the diffusion of a conservative property by eddies in a turbulent flow. Also known as coefficient of eddy diffusion; eddy diffusion coefficient.

eddy flux [FL MECH] The rate of transport (or flux) of fluid properties such as momentum, mass heat, or suspended matter by means of eddies in a turbulent motion; the rate of turbulent exchange. Also known as moisture flux; turbulent flux.

eddy heat conduction [THERMO] The transfer of heat by means of eddies in turbulent flow, treated analogously to molecular conduction. Also known as eddy heat flux; eddy conduction.

eddy kinetic energy [FL MECH] The kinetic energy of that component of fluid flow which represents a departure from the average kinetic energy of the fluid, the mode of averaging depending on the particular problem. Also known as turbulence energy.

eddy resistance [FL MECH] Resistance or drag of a ship resulting from eddies that are shed from the hull or appendages of the ship and carry away energy.

eddy spectrum [FL MECH] **1.** The distribution of frequencies of rotation of eddies in a turbulent flow, or of those eddies having some range of sizes. **2.** The distribution of kinetic energy among eddies with various frequencies or sizes.

eddy velocity [FL MECH] The difference between the mean velocity of fluid flow and the instantaneous velocity at a point. Also known as fluctuation velocity.

eddy viscosity [FL MECH] The turbulent transfer of momentum by eddies giving rise to an internal fluid friction, in a manner analogous to the action of molecular viscosity in laminar flow, but taking place on a much larger scale.

effective angle of attack [AERO ENG] That part of a given angle of attack that lies between the chord of an airfoil and a line representing the resultant velocity of the disturbed airflow.

effective discharge area [DES ENG] A nominal or calculated area of flow through a pressure relief valve for use in flow formulas to determine valve capacity.

effective exhaust velocity [AERO ENG] A fictitious exhaust velocity that yields the observed value of jet thrust in calculations.

effective gun bore line [MECH] The line which a projectile should follow when the muzzle velocity of the antiaircraft gun is vectorially added to the aircraft velocity.

effective launcher line [MECH] The line along which the aircraft rocket would go if it were not affected by gravity.

effective pitch [AERO ENG] The distance traveled by an airplane along its flight path for one complete turn of the propeller.

effective rake [MECH ENG] The angular relationship between the plane of the tooth face of the cutter and the line through the tooth point measured in the direction of chip flow.

effective thrust [AERO ENG] In a rocket motor or engine, the theoretical thrust less the effects of incomplete combustion and friction flow in the nozzle.

effector [CONT SYS] A motor, solenoid, or hydraulic piston that turns commands to a teleoperator into specific manipulative actions.

efficiency Abbreviated eff. [THERMO] The ratio of the work done by a heat engine to the heat energy absorbed by it. Also known as thermal efficiency.

Ehrenfest's equations [THERMO] Equations which state that for the phase curve $P(T)$ of a second-order phase transition the derivative of pressure P with respect to temperature T is equal to $(C_p^f - C_p^i)/TV(\gamma^f - \gamma^i) = (\gamma^f - \gamma^i)/(K^f - K^i)$, where i and f

refer to the two phases, γ is the coefficient of volume expansion, K is the compressibility, C_p is the specific heat at constant pressure, and V is the volume.

ejection capsule [AERO ENG] In an aircraft or spacecraft, a detachable compartment (serving as a cockpit or cabin) or a payload capsule which may be ejected as a unit and parachuted to the ground.

ejection seat [AERO ENG] Emergency device which expels the pilot safely from a high-speed airplane.

ejector condenser [MECH ENG] A type of direct-contact condenser in which vacuum is maintained by high-velocity injection water; condenses steam and discharges water, condensate, and noncondensables to the atmosphere.

Ekman sucking [FL MECH] A boundary-layer phenomenon in which fluid near the bottom of a spinning vessel is drawn toward the edge of the vessel along the bottom.

elastic [MECH] Capable of sustaining deformation without permanent loss of size or shape.

elastica [MECH] The elastic curve formed by a uniform rod that is originally straight, then is bent in a principal plane by applying forces, and couples only at its ends.

elastic aftereffect [MECH] The delay of certain substances in regaining their original shape after being deformed within their elastic limits. Also known as elastic lag.

elastic axis [MECH] The lengthwise line of a beam along which transverse loads must be applied in order to produce bending only, with no torsion of the beam at any section.

elastic body [MECH] A solid body for which the additional deformation produced by an increment of stress completely disappears when the increment is removed. Also known as elastic solid.

elastic buckling [MECH] An abrupt increase in the lateral deflection of a column at a critical load while the stresses acting on the column are wholly elastic.

elastic center [MECH] That point of a beam in the plane of the section lying midway between the flexural center and the center of twist in that section.

elastic collision [MECH] A collision in which the sum of the kinetic energies of translation of the participating systems is the same after the collision as before.

elastic curve [MECH] The curved shape of the longitudinal centroidal surface of a beam when the transverse loads acting on it produced wholly elastic stresses.

elastic deformation [MECH] Reversible alteration of the form or dimensions of a solid body under stress or strain.

elastic failure [MECH] Failure of a body to recover its original size and shape after a stress is removed.

elastic flow [MECH] Return of a material to its original shape following deformation.

elastic force [MECH] A force arising from the deformation of a solid body which depends only on the body's instantaneous deformation and not on its previous history, and which is conservative.

elastic hysteresis [MECH] Phenomenon exhibited by some solids in which the deformation of the solid depends not only on the stress applied to the solid but also on the previous history of this stress; analogous to magnetic hysteresis, with magnetic field strength and magnetic induction replaced by stress and strain respectively.

elasticity [MECH] 1. The property whereby a solid material changes its shape and size under action of opposing forces, but recovers its original configuration when the forces are removed. 2. The existence of forces which tend to restore to its original position any part of a medium (solid or fluid) which has been displaced.

elasticity number 1 [FL MECH] A dimensionless number which is a measure of the ratio of elastic forces to inertial forces on a viscoelastic fluid flowing in a pipe, and is equal to the product of the fluid's relaxation time and its dynamic viscosity, divided by the product of the fluid's density and the square of the radius of the pipe. Symbolized N_{El1}.

elasticity number 2 [FL MECH] A dimensionless number used in studying the effect of elasticity on a flow process, equal to the fluid's density times its specific heat at constant pressure, divided by the product of its coefficient of bulk expansion and its bulk modulus. Symbolized K_E.

elastic limit [MECH] The maximum stress a solid can sustain without undergoing permanent deformation.

elasticoviscosity [FL MECH] That property of a fluid whose rate of deformation under stress is the sum of a part corresponding to a viscous Newtonian fluid and a part obeying Hooke's law.

elastic potential energy [MECH] Capacity that a body has to do work by virtue of its deformation.

elastic ratio [MECH] The ratio of the elastic limit to the ultimate strength of a solid.

elastic recovery [MECH] That fraction of a given deformation of a solid which behaves elastically.

elastic scattering [MECH] Scattering due to an elastic collision.

elastic strain energy [MECH] The work done in deforming a solid within its elastic limit.

elastic theory [MECH] Theory of the relations between the forces acting on a body and the resulting changes in dimensions.

elastic vibration [MECH] Oscillatory motion of a solid body which is sustained by elastic forces and the inertia of the body.

elastodynamics [MECH] The study of the mechanical properties of elastic waves.

elastoplasticity [MECH] State of a substance subjected to a stress greater than its elastic limit but not so great as to cause it to rupture, in which it exhibits both elastic and plastic properties.

elbow [DES ENG] 1. A fitting that connects two pipes at an angle, often of 90°. 2. A sharp corner in a pipe.

electric boiler [MECH ENG] A steam generator using electric energy, in immersion, resistor, or electrode elements, as the source of heat.

electric brake [MECH ENG] An actuator in which the actuating force is supplied by current flowing through a solenoid, or through an electromagnet which is thereby attracted to disks on the rotating member, actuating the brake shoes; this force is counteracted by the force of a compression spring. Also known as electromagnetic brake.

electric coupling [MECH ENG] Magnetic-field coupling between the shafts of a driver and a driven machine.

electric drive [MECH ENG] A mechanism which transmits motion from one shaft to another and controls the velocity ratio of the shafts by electrical means.

electric engine [AERO ENG] A rocket engine in which the propellant is accelerated by some electric device. Also known as electric propulsion system; electric rocket.

electric hammer [MECH ENG] An electric-powered hammer; often used for riveting or caulking.

electric ignition [MECH ENG] Ignition of a charge of fuel vapor and air in an internal combustion engine by passing a high-voltage electric current between two electrodes in the combustion chamber.

electric locomotive [MECH ENG] A locomotive operated by electric power picked up from a system of continuous overhead wires, or, sometimes, from a third rail mounted alongside the track.

electric power generation [MECH ENG] The large-scale production of electric power for industrial, residential, and rural use, generally in stationary plants designed for that purpose.

electric power plant [MECH ENG] A power plant that converts a form of raw energy into electricity, for example, a hydro, steam, diesel, or nuclear generating station for stationary or transportation service.

electric power system [MECH ENG] A complex assemblage of equipment and circuits for generating, transmitting, transforming, and distributing electric energy.

electric propulsion [AERO ENG] A general term encompassing all the various types of propulsion in which the propellant consists of electrically charged particles which are accelerated by electric or magnetic fields, or both.

electric railroad [MECH ENG] A railroad which has a system of continuous overhead wires or a third rail mounted alongside the track to supply electric power to the locomotive and cars.

electric rocket *See* electric engine.

electric stacker [MECH ENG] A stacker whose carriage is raised and lowered by a winch powered by electric storage batteries.

electric typewriter [MECH ENG] A typewriter having an electric motor that provides power for all operations initiated by the touching of the keys.

electric vehicle [MECH ENG] Any ground vehicle whose original source of energy is electric power, such as an electric car or electric locomotive.

electrochemical thermodynamics [THERMO] The application of the laws of thermodynamics to electrochemical systems.

electrodrill [MECH ENG] A drilling machine driven by electric power.

electrofluid [FL MECH] Newtonian (or shear-thinning) fluid whose rheological or flow properties are changed into those of a viscoplastic type by the addition of electric-field modulation.

electrolytic grinding [MECH ENG] A combined grinding and machining operation in which the abrasive, cathodic grinding wheel is in contact with the anodic workpiece beneath the surface of an electrolyte.

electromachining [MECH ENG] The application of electric or ultrasonic energy to a workpiece to effect removal of material.

electromagnetic brake *See* electric brake.

electromagnetic clutch [MECH ENG] A clutch based on magnetic coupling between conductors, such as a magnetic fluid and powder clutch, an eddy-current clutch, or a hysteresis clutch.

electromagnetic propulsion [AERO ENG] Motive power for flight vehicles produced by electromagnetic acceleration of a plasma fluid.

electromechanical [MECH ENG] Pertaining to a mechanical device, system, or process which is electrostatically or electromagnetically actuated or controlled.

electromechanics [MECH ENG] The technology of mechanical devices, systems, or processes which are electrostatically or electromagnetically actuated or controlled.

electronic flame safeguard [MECH ENG] An electrode used in a burner system which detects the main burner flame and interrupts fuel flow if the flame is not detected.

electropulse engine [AERO ENG] An engine, for propelling a flight vehicle, that is based on the use of spark discharges through which intense electric and magnetic fields are established for periods ranging from microseconds to a few milliseconds; a resulting electromagnetic force drives the plasma along the leads and away from the spark gap.

electrostatic atomization [MECH ENG] Atomization in which a liquid jet or film is exposed to an electric field, and forces leading to atomization arise from either free charges on the surface or liquid polarization.

electrostriction [MECH] A form of elastic deformation of a dielectric induced by an electric field, associated with those components of strain which are independent of reversal of field direction, in contrast to the piezoelectric effect. Also known as electrostrictive strain.

electrostrictive strain *See* electrostriction.

electrothermal propulsion [AERO ENG] Propulsion of spacecraft by using an electric arc or other electric heater to bring hydrogen gas or other propellant to the high temperature required for maximum thrust; an arc-jet engine is an example.

electroviscous effect [FL MECH] Change in a liquid's viscosity induced by a strong electrostatic field.

elements of the trajectory [MECH] The various features of the trajectory such as the angle of departure, maximum ordinate, angle of fall, and so on.

elevating conveyor *See* bucket conveyor.

elevating machine *See* elevator.

elevator [AERO ENG] The hinged rear portion of the longitudinal stabilizing surface or tail plane of an aircraft, used to obtain longitudinal or pitch-control moments. [MECH ENG] Also known as elevating machine. **1.** Vertical, continuous-belt, or chain device with closely spaced buckets, scoops, arms, or trays to lift or elevate powders, granules, or solid objects to a higher level. **2.** Pneumatic device in which air or gas is used to elevate finely powdered materials through a closed conduit. **3.** An enclosed platform or car that moves up and down in a shaft for transporting people or materials. Also known as lift.

elevator angle [AERO ENG] The angular displacement of the elevator from its neutral position; it is positive when the trailing edge of the elevator is below the neutral position, and negative when it is above.

elevator dredge [MECH ENG] A dredge which has a chain of buckets, usually flattened across the front and mounted on a nearly vertical ladder; used principally for excavation of sand and gravel beds under bodies of water.

elevon [AERO ENG] The hinged rear portion of an aircraft wing, moved in the same direction on each side of the aircraft to obtain longitudinal control and differentially to obtain lateral control; elevon is a combination of the words elevator and aileron to denote that an elevon combines the functions of aircraft elevators and ailerons.

elliptical orbit [MECH] The path of a body moving along an ellipse, such as that described by either of two bodies revolving under their mutual gravitational attraction but otherwise undisturbed.

elliptic gear [MECH ENG] A change gear composed of two elliptically shaped gears, each rotating about one of its focal points.

elliptic spring [DES ENG] A spring made of laminated steel plates, arched to resemble an ellipse.

elongation [MECH] The fractional increase in a material's length due to stress in tension or to thermal expansion.

emagram [THERMO] A graph of the logarithm of the pressure of a substance versus its temperature, when it is held at constant volume; in meteorological investigations, the potential temperature is often the parameter.

embrittlement [MECH] Reduction or loss of ductility or toughness in a metal or plastic with little change in other mechanical properties.

emergency brake [MECH ENG] A brake that can be set by hand and, once set, continues to hold until released; used as a parking brake in an automobile.

emery wheel [DES ENG] A grinding wheel made of or having a surface of emery powder; used for grinding and polishing.

emissive power *See* emittance.

emissivity [THERMO] The ratio of the radiation emitted by a surface to the radiation emitted by a perfect blackbody radiator at the same temperature. Also known as thermal emissivity.

emittance [THERMO] The power radiated per unit area of a radiating surface. Also known as emissive power; radiating power.

empennage [AERO ENG] The assembly at the rear of an aircraft; it comprises the horizontal and vertical stabilizers. Also known as tail assembly.

end-feed centerless grinding [MECH ENG] Centerless grinding in which the piece is fed through grinding and regulating wheels to an end stop.

end lap [DES ENG] A joint in which two joining members are made to overlap by removal of half the thickness of each.

end loader [MECH ENG] A platform elevator at the rear of a truck.

end mill [MECH ENG] A machine which has a rotating shank with cutting teeth at the end and spiral blades on the peripheral surface; used for shaping and cutting metal.

end-milled keyway *See* profiled keyway.

end play [MECH ENG] Axial movement in a shaft-and-bearing assembly resulting from clearances between the components.

endurance limit *See* fatigue limit.

endurance ratio *See* fatigue ratio.

endurance strength *See* fatigue strength.

energy conversion efficiency [MECH ENG] The efficiency with which the energy of the working substance is converted into kinetic energy.

energy head [FL MECH] The elevation of the hydraulic grade line at any section of a waterway plus the velocity head of the mean velocity of the water in that section.

energy integral [MECH] A constant of integration resulting from integration of Newton's second law of motion in the case of a conservative force; equal to the sum of the kinetic energy of the particle and the potential energy of the force acting on it.

energy management [AERO ENG] In rocketry, the monitoring of the expenditure of fuel for flight control and navigation.

engine [MECH ENG] A machine in which power is applied to do work by the conversion of various forms of energy into mechanical force and motion.

engine balance [MECH ENG] Arrangement and construction of moving parts in reciprocating or rotating machines to reduce dynamic forces which may result in undesirable vibrations.

engine block   *See* cylinder block.

engine cooling   [MECH ENG] Controlling the temperature of internal combustion engine parts to prevent overheating and to maintain all operating dimensions, clearances, and alignment by a circulating coolant, oil, and a fan.

engine cycle   [THERMO] Any series of thermodynamic phases constituting a cycle for the conversion of heat into work; examples are the Otto cycle, Stirling cycle, and Diesel cycle.

engine cylinder   [MECH ENG] A cylindrical chamber in an engine in which the energy of the working fluid, in the form of pressure and heat, is converted to mechanical force by performing work on the piston. Also known as cylinder.

engine displacement   [MECH ENG] Volume displaced by each piston moving from bottom dead center to top dead center multiplied by the number of cylinders.

engine efficiency   [MECH ENG] Ratio between the energy supplied to an engine to the energy output of the engine.

engine inlet   [MECH ENG] A place of entrance for engine fuel.

engine knock   [MECH ENG] In spark ignition engines, the sound and other effects associated with ignition and rapid combustion of the last part of the charge to burn, before the flame front reaches it. Also known as combustion knock.

engine lathe   [MECH ENG] A manually operated lathe equipped with a headstock of the back-geared, cone-driven type or of the geared-head type.

engine performance   [MECH ENG] Relationship between power output, revolutions per minute, fuel or fluid consumption, and ambient conditions in which an engine operates.

enlargement loss   [FL MECH] Energy loss by friction in a flowing fluid when it moves into a cross-sectional area of sudden enlargement.

entering angle   [MECH ENG] The angle between the side-cutting edge of a tool and the machined surface of the work; angle is 90° for a tool with 0° side-cutting edge angle effective.

enthalpy   [THERMO] The sum of the internal energy of a system plus the product of the system's volume multiplied by the pressure exerted on the system by its surroundings. Also known as heat content; sensible heat; total heat.

enthalpy-entropy chart   [THERMO] A graph of the enthalpy of a substance versus its entropy at various values of temperature, pressure, or specific volume; useful in making calculations about a machine or process in which this substance is the working medium.

enthalpy of vaporization   *See* heat of vaporization.

entrained fluid   [FL MECH] Fluid in the form of mist, fog, or droplets that is carried out of a column or vessel by a rising gas or vapor stream.

entrance loss   [FL MECH] Energy loss by friction in a flowing fluid when it moves into a cross-sectional area of sudden contraction, as at the entrance of a pipe or a suddenly reduced area of a duct.

entropy   [THERMO] Function of the state of a thermodynamic system whose change in any differential reversible process is equal to the heat absorbed by the system from its surroundings divided by the absolute temperature of the system. Also known as thermal charge.

entry ballistics   [MECH] That branch of ballistics which pertains to the entry of a missile, spacecraft, or other object from outer space into and through an atmosphere.

entry corridor   [AERO ENG] Depth of the region between two trajectories which define the design limits of a vehicle about to enter a planetary atmosphere, or define the desired landing area (footprint).

environmental stress cracking [MECH] The susceptibility of a material to crack or craze in the presence of surface-active agents or other factors.

environmental survey satellite [AERO ENG] One of a series of meteorological satellites which completely photographs the earth each day. Abbreviated ESSA.

eon [MECH] A unit of time, equal to 10^9 years.

Eötvös rule [THERMO] The rule that the rate of change of molar surface energy with temperature is a constant for all liquids; deviations are encountered in practice.

epicyclic gear [MECH ENG] A system of gears in which one or more gears travel around the inside or the outside of another gear whose axis is fixed.

epicyclic train [MECH ENG] A combination of epicyclic gears, usually connected by an arm, in which some or all of the gears have a motion compounded of rotation about an axis and a translation or revolution of that axis.

equal-arm balance [MECH] A simple balance in which the distances from the point of support of the balance-arm beam to the two pans at the end of the beam are equal.

equaling file [DES ENG] A slightly bulging double-cut file used in fine toolmaking.

equalizer [MECH ENG] 1. A bar to which one attaches a vehicle's whiffletrees to make the pull of draft animals equal. Also known as equalizing bar. 2. A bar which joins a pair of axle springs on a railway locomotive or car for equalization of weight. Also known as equalizing bar. 3. A device which distributes braking force among independent brakes of an automotive vehicle. Also known as equalizer brake. 4. A machine which saws wooden stock to equal lengths.

equalizer brake *See* equalizer.

equalizing bar *See* equalizer.

equation of motion [FL MECH] One of a set of hydrodynamical equations representing the application of Newton's second law of motion to a fluid system; the total acceleration on an individual fluid particle is equated to the sum of the forces acting on the particle within the fluid. [MECH] 1. Equation which specifies the coordinates of particles as functions of time. 2. A differential equation, or one of several such equations, from which the coordinates of particles as functions of time can be obtained if the initial positions and velocities of the particles are known.

equation of piezotropy [THERMO] An equation obeyed by certain fluids which states that the time rate of change of the fluid's density equals the product of a function of the thermodynamic variables and the time rate of change of the pressure.

equatorial plane [MECH] A plane perpendicular to the axis of rotation of a rotating body and equidistant from the intersections of this axis with the body's surface, provided that the body is symmetric about the axis of rotation and is symmetric under reflection through this plane.

equilibrium [MECH] Condition in which a particle, or all the constituent particles of a body, are at rest or in unaccelerated motion in an inertial reference frame. Also known as static equilibrium.

equipollent [MECH] Of two systems of forces, having the same vector sum and the same total torque about an arbitrary point.

equipotential surface [MECH] A surface which is always normal to the lines of force of a field and on which the potential is everywhere the same.

equivalent airspeed [AERO ENG] The product of the true airspeed and the square root of the density ratio; used in structural design work to designate various design conditions.

equivalent bending moment [MECH] A bending moment which, acting alone, would produce in a circular shaft a normal stress of the same magnitude as the maximum normal stress produced by a given bending moment and a given twisting moment acting simultaneously.

equivalent blackbody temperature   [THERMO] For a surface, the temperature of a blackbody which emits the same amount of radiation per unit area as does the surface.

equivalent evaporation   [FL MECH] The amount of water, usually in pounds per hour, evaporated from a temperature of 212°F (100°C) to saturated steam at the same temperature.

equivalent nitrogen pressure   [MECH] The pressure that would be indicated by a device if the gas inside it were replaced by nitrogen of equivalent molecular density.

equivalent orifice   [MECH ENG] An expression of fan performance as the theoretical sharp-edge orifice area which would offer the same resistance to flow as the system resistance itself.

equivalent temperature   [THERMO] A term used in British engineering for that temperature of a uniform enclosure in which, in still air, a sizable blackbody at 75°F (23.9°C) would lose heat at the same rate as in the environment.

equivalent twisting moment   [MECH] A twisting moment which, if acting alone, would produce in a circular shaft a shear stress of the same magnitude as the shear stress produced by a given twisting moment and a given bending moment acting simultaneously.

equivalent viscous damping   [MECH] An assumed value of viscous damping used in analyzing a vibratory motion, such that the dissipation of energy per cycle at resonance is the same for the assumed or the actual damping force.

erection stress   [MECH] The internal forces exerted on a structural member during construction.

Ericsson cycle   [THERMO] An ideal thermodynamic cycle consisting of two isobaric processes interspersed with processes which are, in effect, isothermal, but each of which consists of an infinite number of alternating isentropic and isobaric processes.

error coefficient   [CONT SYS] The steady-state value of the output of a control system, or of some derivative of the output, divided by the steady-state actuating signal. Also known as error constant.

error constant   *See* error coefficient.

error signal   [CONT SYS] In an automatic control device, a signal whose magnitude and sign are used to correct the alignment between the controlling and the controlled elements.

ERTS   *See* earth resources technology satellite.

escalator   [MECH ENG] A continuously moving stairway and handrail.

escapement   [MECH ENG] A ratchet device that permits motion in one direction slowly.

escape rocket   [AERO ENG] A small rocket engine attached to the leading end of an escape tower, to provide additional thrust to the capsule in an emergency; it helps separate the capsule from the booster vehicle and carries it to an altitude where parachutes can be deployed.

escape tower   [AERO ENG] A trestle tower placed on top of a space capsule, connecting the capsule to the escape rocket on top of the tower; used for emergencies.

escape velocity   [AERO ENG] In space flight, the speed at which an object is able to overcome the gravitational pull of the earth.

escort fighter   [AERO ENG] A fighter designed or equipped for long-range missions, usually to accompany heavy bombers on raids.

escutcheon [DES ENG] An ornamental shield, flange, or border used around a dial, window, control knob, or other panel-mounted part. Also known as escutcheon plate.

escutcheon plate *See* escutcheon.

ESSA *See* environmental survey satellite.

established flow [FL MECH] The flow when the boundary layer of a fluid flowing in a duct completely fills the duct; that is, when the effect of the wall shearing stress extends completely across the duct.

Euler angles [MECH] Three angular parameters that specify the orientation of a body with respect to reference axes.

Euler equation [MECH] Expression for the energy removed from a gas stream by a rotating blade system (as a gas turbine), independent of the blade system (as a radial- or axial-flow system).

Euler equations of motion [MECH] A set of three differential equations expressing relations between the force moments, angular velocities, and angular accelerations of a rotating rigid body.

Euler force [MECH] The greatest load that a long, slender column can carry without buckling, according to the Euler formula for long columns.

Euler formula for long columns [MECH] A formula which gives the greatest axial load that a long, slender column can carry without buckling, in terms of its length, Young's modulus, and the moment of inertia about an axis along the center of the column.

Eulerian coordinates [FL MECH] Any system of coordinates in which properties of a fluid are assigned to points in space at each given time, without attempt to identify individual fluid parcels from one time to the next; a sequence of synoptic charts is a Eulerian representation of the data.

Eulerian correlation [FL MECH] The correlation between the properties of a flow at various points in space at a single instant of time. Also known as synoptic correlation.

Eulerian equation [FL MECH] A mathematical representation of the motions of a fluid in which the behavior and the properties of the fluid are described at fixed points in a coordinate system.

Euler method [MECH] A method of studying fluid motion and the mechanics of deformable bodies in which one considers volume elements at fixed locations in space, across which material flows; the Euler method is in contrast to the Lagrangian method.

Euler number 1 [FL MECH] A dimensionless number used in the study of fluid friction in conduits, equal to the pressure drop due to friction divided by the product of the fluid density and the square of the fluid velocity.

Euler number 2 [FL MECH] A dimensionless number equal to two times the Fanning friction factor.

Euler-Rodrigues parameter [MECH] One of four numbers which may be used to specify the orientation of a rigid body; they are components of a quaternion.

Euler's expansion [FL MECH] The transformation of a derivative (d/dt) describing the behavior of a moving particle with respect to time, into a local derivative $(\delta/\delta t)$ and three additional terms that describe the changing motion of a fluid as it passes through a fixed point.

evaporative condenser [MECH ENG] An apparatus in which vapor is condensed within tubes that are cooled by the evaporation of water flowing over the outside of the tubes.

evaporative cooling tower *See* wet cooling tower.

evaporator [MECH ENG] Any of many devices in which liquid is changed to the vapor state by the addition of heat, for example, distiller, still, dryer, water purifier, or refrigeration system element where evaporation proceeds at low pressure and consequent low temperature.

even pitch [DES ENG] The pitch of a screw in which the number of threads per inch is a multiple (or submultiple) of the threads per inch of the lead screw of the lathe on which the screw is cut.

excavator [MECH ENG] A machine for digging and removing earth.

excess coefficient [MECH ENG] The ratio $(A - R)/R$, where A is the amount of air admitted in the combustion of fuel and R is the amount required.

exchange coefficient [FL MECH] A coefficient of eddy flux in turbulent flow, defined in analogy to those coefficients of the kinetic theory of gases. Also known as austausch coefficient; eddy coefficient; interchange coefficient.

excitation [CONT SYS] The application of energy to one portion of a system or apparatus in a manner that enables another portion to carry out a specialized function; a generalization of the electricity and electronics definitions.

exhaust [MECH ENG] 1. The working substance discharged from an engine cylinder or turbine after performing work on the moving parts of the machine. 2. The phase of the engine cycle concerned with this discharge. 3. A duct for the escape of gases, fumes, and odors from an enclosure, sometimes equipped with an arrangement of fans.

exhaust deflecting ring [MECH ENG] A type of jetavator consisting of a ring so mounted at the end of a nozzle as to permit it to be rotated into the exhaust stream.

exhaust gas [MECH ENG] Spent gas leaving an internal combustion engine or gas turbine.

exhaust manifold [MECH ENG] A branched system of pipes to carry waste emissions away from the piston chambers of an internal combustion engine.

exhaust nozzle [AERO ENG] The terminal portion of a jet engine tail pipe.

exhaust pipe [MECH ENG] The duct through which engine exhaust is discharged.

exhaust stream [AERO ENG] The stream of matter or radiation emitted from the nozzle of a rocket or other reaction engine.

exhaust stroke [MECH ENG] The stroke of an engine, pump, or compressor that expels the fluid from the cylinder.

exhaust suction stroke [MECH ENG] A stroke of an engine that simultaneously removes used fuel and introduces fresh fuel to the cylinder.

exhaust valve [MECH ENG] The valve on a cylinder in an internal combustion engine which controls the discharge of spent gas.

exhaust velocity [FL MECH] The velocity of gaseous or other particles in the exhaust stream of the nozzle of a reaction engine, relative to the nozzle.

exline correction [FL MECH] Calculation of fluid-flow friction loss through annular sections with a correction for the flow eccentricity in the laminar-flow range.

expandable space structure [AERO ENG] A structure which can be packaged in a small volume for launch and then erected to its full size and shape outside the earth's atmosphere.

expanding brake [MECH ENG] A brake that operates by moving outward against the inside rim of a drum or wheel.

expansion [MECH ENG] Increase in volume of working material with accompanying drop in pressure of a gaseous or vapor fluid, as in an internal combustion engine or steam engine cylinder.

expansion bolt [DES ENG] A bolt having an end which, when embedded into masonry or concrete, expands under a pull on the bolt, thereby providing anchorage.

expansion chucking reamer [DES ENG] A machine reamer with an expansion screw at the end which increases the diameter.

expansion coefficient *See* coefficient of cubical expansion.

expansion cooling [MECH ENG] Cooling of a substance by having it undergo adiabatic expansion.

expansion engine [MECH ENG] Piston-cylinder device that cools compressed air via sudden expansion; used in production of pure gaseous oxygen via the Claude cycle.

expansion fit [DES ENG] A condition of optimum clearance between certain mating parts in which the cold inner member is placed inside the warmer outer member and the temperature is allowed to equalize.

expansion joint [MECH ENG] 1. A joint between parts of a structure or machine to avoid distortion when subjected to temperature change. 2. A pipe coupling which, under temperature change, allows movement of a piping system without hazard to associated equipment.

expansion ratio [FL MECH] For the calculation of the mass flow of a gas out of a nozzle or other expanding duct, the ratio of the nozzle exit section area to the nozzle throat area, or the ratio of final to initial volume.   [MECH ENG] In a reciprocating piston engine, the ratio of cylinder volume with piston at bottom dead center to cylinder volume with piston at top dead center.

expansion shield [DES ENG] An anchoring device that expands as it is driven into masonry or concrete, pressing against the sides of the hole.

expansion valve [MECH ENG] A valve in which fluid flows under falling pressure and increasing volume.

expansion wave [FL MECH] A pressure wave or shock wave that decreases the density of air as the air passes through it.

expansive bit [DES ENG] A bit in which the cutting blade can be set at various sizes.

expansivity *See* coefficient of cubical expansion.

Explorer program [AERO ENG] A series of earth satellites begun by the U.S. Army and continued by NASA; *Explorer 1*, the first U.S. satellite, went into orbit on January 31, 1958.

explosion door [MECH ENG] A door in a furnace which is designed to open at a predetermined pressure.

explosion method [THERMO] Method of measuring the specific heat of a gas at constant volume by enclosing the gas with an explosive mixture, whose heat of reaction is known, in a chamber closed with a corrugated steel membrane which acts as a manometer, and by deducing the maximum temperature reached on ignition of the mixture from the pressure change.

explosion rupture disk device [MECH ENG] A protective device used where the pressure rise in the vessel occurs at a rapid rate.

explosive bolt [AERO ENG] A bolt designed to contain a remote-initiated explosive charge which, upon detonation, will shear the bolt or cause it to fail otherwise; applicable to such uses as stage separation of rockets, jettison of expended fuel tanks, and ejection of parachutes.

explosive decompression [AERO ENG] A sudden loss of pressure in a pressurized cabin, cockpit, or the like, so rapid as to be explosive, as when punctured by gunfire.

extended area [DES ENG] An engineering surface that has been extended areawise without increasing diameter, as by using pleats (as in filter cartridges) or fins (as in heat exchangers).

extensibility [MECH] The amount to which a material can be stretched or distorted without breaking.

extension bolt [DES ENG] A vertical bolt that can be slid into place by a long extension rod; used at the top of doors.

extension ladder [DES ENG] A ladder of two or more nesting sections which can be extended to almost the combined length of the sections.

extension spring [DES ENG] A tightly coiled spring designed to resist a tensile force.

exterior ballistics [MECH] The science concerned with behavior of a projectile after leaving the muzzle of the firing weapon.

external aileron [AERO ENG] An aileron offset from the wing; that is, not forming a part of the wing.

external brake [MECH ENG] A brake that operates by contacting the outside of a brake drum.

external centerless grinding [MECH ENG] A process by which a metal workpiece is finished on its external surface by supporting the piece on a blade while it is advanced between a regulating wheel and grinding wheel.

external combustion engine [MECH ENG] An engine in which the generation of heat is effected in a furnace or reactor outside the engine cylinder.

external force [MECH] A force exerted on a system or on some of its components by an agency outside the system.

external grinding [MECH ENG] Grinding the outer surface of a rotating piece of work.

external header [MECH ENG] Manifold connecting sections of a cast iron boiler.

externally fired boiler [MECH ENG] A boiler that has refractory or cooling tubes surrounding its furnace.

external shoe brake [MECH ENG] A friction brake operated by the application of externally contracting elements.

external thread [DES ENG] A screw thread cut on an outside surface.

external wave [FL MECH] 1. A wave in fluid motion having its maximum amplitude at an external boundary such as a free surface. 2. Any surface wave on the free surface of a homogeneous incompressible fluid is an external wave.

external work [THERMO] The work done by a system in expanding against forces exerted from outside.

extraction parachute [AERO ENG] An auxiliary parachute designed to release and extract cargo from aircraft in flight and to deploy cargo parachutes.

extraction turbine [MECH ENG] A steam turbine equipped with openings through which partly expanded steam is bled at one or more stages.

extravehicular activity [AERO ENG] Activity conducted outside a spacecraft during space flight.

eyebar [DES ENG] A metal bar having a hole or eye through each enlarged end.

eyebolt [DES ENG] A bolt with a loop at one end.

eyelet [DES ENG] A small ring or barrel-shaped piece of metal inserted into a hole for reinforcement.

eye screw [DES ENG] A screw with an open loop head.

Eyring formula [FL MECH] A formula, based on the Eyring theory of rate processes, which relates shear stress acting on a liquid and the resulting rate of shear.

Eyring molecular system [FL MECH] Theory to account for liquid properties; assumes that each liquid molecule can move freely within a certain free volume. Also known as Eyring theory.

Eyring theory *See* Eyring molecular system.

f *See* Fanning friction factor.

F-6 *See* Skyray.

F-11 *See* Tiger.

F-84 *See* Thunderstreak.

F-86 *See* Sabrejet.

F-89 *See* Scorpion.

F-100 *See* Super Sabre.

F-105 *See* Thunderchief.

face-discharge bit [MECH ENG] A liquid-coolant bit designed for drilling in soft formations and for use on a double-tube core barrel, the inner tube of which fits snugly into a recess cut into the inside wall of the bit directly above the inside reaming stones; the coolant flows through the bit and is ejected at the cutting face. Also known as bottom-discharge bit; face-ejection bit.

face-ejection bit *See* face-discharge bit.

face gear [DES ENG] A gear having teeth cut on the face.

face milling [MECH ENG] Milling flat surfaces perpendicular to the rotational axis of the cutting tool.

facing [MECH ENG] Machining the end of a flat rotating surface by applying a tool perpendicular to the axis of rotation in a spiral planar path.

factor of safety [MECH] 1. The ratio between the breaking load on a member, appliance, or hoisting rope and the safe permissible load on it. Also known as safety factor. 2. *See* factor of stress intensity.

factor of stress concentration [MECH] Any irregularity producing localized stress in a structural member subject to load. Also known as fatigue-strength reduction factor.

factor of stress intensity [MECH] The ratio of the maximum stress to which a structural member can be subjected, to the maximum stress to which it is likely to be subjected. Also known as factor of safety.

Fahrenheit scale [THERMO] A temperature scale; the temperature in degrees Fahrenheit (°F) is the sum of 32 plus 9/5 the temperature in degrees Celsius; water at 1 atmosphere (101, 325 newtons per square meter) pressure freezes very near 32°F and boils very near 212°F.

failure [MECH] Condition caused by collapse, break, or bending, so that a structure or structural element can no longer fulfill its purpose.

faired cable [DES ENG] A trawling cable covered by streamlined surfaces to reduce hydrodynamic drag.

fairing [AERO ENG] A structure or surface on an aircraft or rocket that functions to reduce drag, such as the streamlined nose of a satellite-launching rocket.

fairlead [AERO ENG] A guide through which an airplane antenna or control cable passes. [MECH ENG] A group of pulleys or rollers used in conjunction with a winch or similar apparatus to permit the cable to be reeled from any direction.

Fales-Stuart windmill [MECH ENG] A windmill developed for farm use from the two-blade airfoil propeller. Also known as Stuart windmill.

Falk flexible coupling [MECH ENG] A spring coupling in which a continuous steel spring is threaded back and forth through axial slots in the periphery of two hubs on the shaft ends.

fall [MECH ENG] The rope or chain of a hoisting tackle.

fallaway section [AERO ENG] A section of a rocket vehicle that is cast off from the vehicle during flight, especially such a section that falls back to the earth.

fall block [MECH ENG] A pulley block that rises and falls with the load on a lifting tackle.

faller [MECH ENG] A machine part whose operation depends on a falling action.

falling body [MECH] A body whose motion is accelerated toward the center of the earth by the force of gravity, other forces acting on it being negligible by comparison.

falling film [FL MECH] A theoretical liquid film that moves downward in even flow on a vertical surface in laminar flow; the concept is used for heat- and mass-transfer calculations.

fall velocity *See* settling velocity.

fan [MECH ENG] **1.** A device, usually consisting of a rotating paddle wheel or an airscrew, with or without a casing, for producing currents in order to circulate, exhaust, or deliver large volumes of air or gas. **2.** A vane to keep the sails of a windmill facing the direction of the wind.

fan brake [MECH ENG] A fan used to provide a load for a driving mechanism.

fan efficiency [MECH ENG] The ratio obtained by dividing a fan's useful power output by the power input (the power supplied to the fan shaft); it is expressed as a percentage.

fang bolt [DES ENG] A bolt having a triangular nut with sharp projections at its corners; used to attach metal pieces to wood.

fanjet [AERO ENG] A turbojet engine whose performance has been improved by the addition of a fan which operates in an annular duct surrounding the engine.

Fanning friction factor [FL MECH] A dimensionless number used in studying fluid friction in pipes, equal to the pipe diameter times the drop in pressure in the fluid due to friction as it passes through the pipe, divided by the product of the pipe length and the kinetic energy of the fluid per unit volume. Symbolized f.

Fanning's equation [FL MECH] The equation expressing that frictional pressure drop of fluid flowing in a pipe is a function of the Reynolds number, rate of flow, acceleration due to gravity, and length and diameter of the pipe.

Fanno flow [FL MECH] An ideal flow used to study the flow of fluids in long pipes; the flow obeys the same simplifying assumptions as Rayleigh flow except that the assumption there is no friction is replaced by the requirement the flow be adiabatic.

fan rating [MECH ENG] The head, quantity, power, and efficiency expected from a fan operating at peak efficiency.

fan ring [DES ENG] Circular metallic collar encircling (but spaced away from) the tips of the fan blade in process equipment, such as air-cooled heat exchangers; ring design is critical to the efficiency of fan performance.

fan shaft [DES ENG] The spindle on which a fan impeller is mounted.

fan static pressure [MECH ENG] The total pressure rise diminished by the velocity pressure in the fan outlet.

fan test [MECH ENG] Observations of the quantity, total pressure, and power of air circulated by a fan running at a known constant speed.

fan total head [MECH ENG] The sum of the fan static head and the velocity head at the fan discharge corresponding to a given quantity of airflow.

fan total pressure [MECH ENG] The algebraic difference between the mean total pressure at the fan outlet and the mean total pressure at the fan inlet.

fan velocity pressure [MECH ENG] The velocity pressure corresponding to the average velocity at the fan outlet.

fast coupling [MECH ENG] A flexible geared coupling that uses two interior hubs on the shafts with circumferential gear teeth surrounded by a casing having internal gear teeth to mesh and connect the two hubs.

fastener [DES ENG] 1. A device for joining two separate parts of an article or structure. 2. A device for holding closed a door, gate, or similar structure.

fastening [DES ENG] A spike, bolt, nut, or other device to connect rails to ties.

fast-spiral drill *See* high-helix drill.

fatigue [MECH] Failure of a material by cracking resulting from repeated or cyclic stress.

fatigue life [MECH] The number of applied repeated stress cycles a material can endure before failure.

fatigue limit [MECH] The maximum stress that a material can endure for an infinite number of stress cycles without breaking. Also known as endurance limit.

fatigue ratio [MECH] The ratio of the fatigue limit or fatigue strength to the static tensile strength. Also known as endurance ratio.

fatigue strength [MECH] The maximum stress a material can endure for a given number of stress cycles without breaking. Also known as endurance strength.

fatigue-strength reduction factor *See* factor of stress concentration.

feasible method *See* interaction prediction method.

feather [MECH ENG] To change the pitch on a propeller.

featheredge [DES ENG] A wood tool with a level edge used to straighten angles in the finish coat of plaster.

feathering [MECH ENG] A pitch position in a controllable-pitch propeller; it is used in the event of engine failure to stop the windmilling action, and occurs when the blade angle is about 90° to the plane of rotation. Also known as full feathering.

feathering propeller [MECH ENG] A variable-pitch marine or airscrew propeller capable of increasing pitch beyond the normal high pitch value to the feathered position.

feed [MECH ENG] Forward motion imparted to the cutters or drills of cutting or drilling machinery.

feedback branch [CONT SYS] A branch in a signal-flow graph that belongs to a feedback loop.

feedback compensation [CONT SYS] Improvement of the response of a feedback control system by placing a compensator in the feedback path, in contrast to cascade compensation. Also known as parallel compensation.

feedback control loop   *See* feedback loop.

feedback control signal   [CONT SYS] The portion of an output signal which is retransmitted as an input signal.

feedback control system   [CONT SYS] A system in which the value of some output quantity is controlled by feeding back the value of the controlled quantity and using it to manipulate an input quantity so as to bring the value of the controlled quantity closer to a desired value. Also known as closed-loop control system.

feedback loop   [CONT SYS] A closed transmission path or loop that includes an active transducer and consists of a forward path, a feedback path, and one or more mixing points arranged to maintain a prescribed relationship between the loop input signal and the loop output signal. Also known as feedback control loop.

feedback regulator   [CONT SYS] A feedback control system that tends to maintain a prescribed relationship between certain system signals and other predetermined quantities.

feedback transfer function   [CONT SYS] In a feedback control loop, the transfer function of the feedback path.

feed-control valve   [MECH ENG] A small valve, usually a needle valve, on the outlet of the hydraulic-feed cylinder on the swivel head of a diamond drill, used to control minutely the speed of the hydraulic piston travel and hence the rate at which the bit is made to penetrate the rock.

feeder   [MECH ENG] 1. A conveyor adapted to control the rate of delivery of bulk materials, packages, or objects, or a control device which separates or assembles objects. 2. A device for delivering materials to a processing unit.

feeder-breaker   [MECH ENG] A unit that breaks and feeds ore or crushed rock to a materials-handling system at a required rate.

feeder conveyor   [MECH ENG] A short auxiliary conveyor designed to transport materials to another conveyor. Also known as stage loader.

feedforward control   [CONT SYS] Process control in which changes are detected at the process input and an anticipating correction signal is applied before process output is affected.

feed nut   [MECH ENG] The threaded sleeve fitting around the feed screw on a gear-feed drill swivel head, which is rotated by means of paired gears driven from the spindle or feed shaft.

feed pipe   [MECH ENG] The pipe which conducts water to a boiler drum.

feed pitch   [DES ENG] The distance between the centers of adjacent feed holes in punched paper tape.

feed pressure   [MECH ENG] Total weight or pressure, expressed in pounds or tons, applied to the drilling stem to make the drill bit cut and penetrate the geologic, rock, or ore formation.

feed pump   [MECH ENG] A pump used to supply water to a steam boiler.

feed rate   *See* cutting speed.

feed ratio   [MECH ENG] The number of revolutions a drill stem and bit must turn to advance the drill bit 1 inch when the stem is attached to and rotated by a screw- or gear-feed type of drill swivel head with a particular pair of the set of gears engaged. Also known as feed speed.

feed screw   [MECH ENG] The externally threaded drill-rod drive rod in a screw- or gear-feed swivel head on a diamond drill; also used on percussion drills, lathes, and other machinery.

feed shaft [MECH ENG] A short shaft or countershaft in a diamond-drill gear-feed swivel head which is rotated by the drill motor through gears or a fractional drive and by means of which the engaged pair of feed gears is driven.

feed speed *See* feed ratio.

feed travel [MECH ENG] The distance a drilling machine moves the steel shank in traveling from top to bottom of its feeding range.

feed trough [MECH ENG] A receptacle into which feedwater overflows from a boiler drum.

feedwater [MECH ENG] The water supplied to a boiler or still.

feedwater heater [MECH ENG] An apparatus that utilizes steam extracted from an engine or turbine to heat boiler feedwater.

feeler pin [MECH ENG] A pin that allows a duplicating machine to operate only when there is a supply of paper.

female fitting [DES ENG] In a paired pipe or an electrical or mechanical connection, the portion (fitting) that receives, contrasted to the male portion (fitting) that inserts.

femtometer [MECH] A unit of length, equal to 10^{-15} meter; used particularly in measuring nuclear distances. Abbreviated fm. Also known as fermi.

fence [AERO ENG] A stationary plate or vane projecting from the upper surface of an airfoil, substantially parallel to the airflow, used to prevent spanwise flow.

fermi *See* femtometer.

ferrule [DES ENG] 1. A metal ring or cap attached to the end of a tool handle, post, or other device to strengthen and protect it. 2. A bushing inserted in the end of a boiler flue to spread and tighten it.

fiber stress [MECH] 1. The tensile or compressive stress on the fibers of a fiber metal or other fibrous material, especially when fiber orientation is parallel with the neutral axis. 2. Local stress through a small area (a point or line) on a section where the stress is not uniform, as in a beam under bending load.

fibrous fracture [MECH] Failure of a material resulting from a ductile crack; broken surfaces are dull and silky. Also known as ductile fracture.

fiducial temperature [THERMO] Any of the temperatures assigned to a number of reproducible equilibrium states on the International Practical Temperature Scale; standard instruments are calibrated at these temperatures.

field excitation [MECH ENG] Control of the speed of a series motor in an electric or diesel-electric locomotive by changing the relation between the armature current and the field strength, either through a reduction in field current by shunting the field coils with resistance, or through the use of field taps.

fifteen-degrees calorie *See* calorie.

fighter aircraft [AERO ENG] A military aircraft designed primarily to destroy other aircraft in the air; may also be used to bomb military targets; it is maneuverable and has a high rate of climb.

fighter bomber [AERO ENG] A fighter aircraft that is designed to have bombs, or rockets, added to it so that it may be used as a bomber.

fighter interceptor [AERO ENG] A fighter aircraft designed to intercept and shoot down enemy aircraft.

filar micrometer [DES ENG] An instrument used to measure small distances in the field of an eyepiece by using two parallel wires, one of which is fixed while the other is moved at right angles to its length by means of an accurately cut screw. Also known as bifilar micrometer.

file [DES ENG] A steel bar or rod with cutting teeth on its surface; used as a smoothing or forming tool.

fillet [DES ENG] A concave transition surface between two otherwise intersecting surfaces.

fillet gage [DES ENG] A gage for measuring convex or concave surfaces.

fill factor [MECH ENG] The approximate load that the dipper of a shovel is carrying, expressed as a percentage of the rated capacity.

film coefficient [THERMO] For a fluid confined in a vessel, the rate of flow of heat out of the fluid, per unit area of vessel wall divided by the difference between the temperature in the interior of the fluid and the temperature at the surface of the wall. Also known as convection coefficient.

film cooling [THERMO] The cooling of a body or surface, such as the inner surface of a rocket combustion chamber, by maintaining a thin fluid layer over the affected area.

film transport [MECH ENG] 1. The mechanism for moving photographic film through the region where light strikes it in recording film tracks or sound tracks of motion pictures. 2. The mechanism which moves the film print past the area where light passes through it in reproduction of picture and sound.

filter *See* compensator.

filter pump [MECH ENG] An aspirator or vacuum pump which creates a negative pressure on the filtrate side of the filter to hasten the process of filtering.

fin [AERO ENG] A fixed or adjustable vane or airfoil affixed longitudinally to an aerodynamically or ballistically designed body for stabilizing purposes. [DES ENG] A projecting flat plate or structure, as a cooling fin.

final filter *See* afterfilter.

final mass [AERO ENG] The mass of a rocket after its propellants are consumed.

fine grinding [MECH ENG] Grinding performed in a mill rotating on a horizontal axis in which the material undergoes final size reduction, to −100 mesh.

fineness ratio [AERO ENG] The ratio of the length of a streamlined body, as that of a fuselage or airship hull, to its maximum diameter.

finger bit [DES ENG] A steel rock-cutting bit having fingerlike, fixed or replaceable steel-cutting points.

finish grinding [MECH ENG] The last action of a grinding operation to achieve a good finish and accurate dimensions.

finishing nail [DES ENG] A wire nail with a small head that can easily be concealed.

finish plate [DES ENG] A plate which covers and protects the cylinder setscrews; it is fastened to the underplate and forms part of the armored front for a mortise lock.

finish turning [MECH ENG] The operation of machining a surface to accurate size and producing a smooth finish.

finned surface [MECH ENG] A tubular heat-exchange surface with extended projections on one side.

firebox [MECH ENG] The furnace of a locomotive or similar type of fire-tube boiler.

fire pump [MECH ENG] A pump for fire protection purposes usually driven by an independent, reliable prime mover and approved by the National Board of Fire Underwriters.

fireroom [MECH ENG] That portion of a fossil fuel–burning plant which contains the furnace and associated equipment.

fire-tube boiler [MECH ENG] A steam boiler in which hot gaseous products of combustion pass through tubes surrounded by boiler water.

firing chamber *See* combustion chamber.

firing machine [MECH ENG] A mechanical stoker used to feed coal to a boiler furnace.

firing pressure [MECH ENG] The highest pressure in an engine cylinder during combustion.

firing rate [MECH ENG] The rate at which fuel feed to a burner occurs, in terms of volume, heat units, or weight per unit time.

firmer chisel [DES ENG] A small hand chisel with a flat blade; used in woodworking.

firm-joint caliper [DES ENG] An outside or inside caliper whose legs are jointed together at the top with a nut and which must be opened and closed by hand pressure.

firmoviscosity [MECH] Property of a substance in which the stress is equal to the sum of a term proportional to the substance's deformation, and a term proportional to its rate of deformation.

first law of motion *See* Newton's first law.

first law of thermodynamics [THERMO] The law that heat is a form of energy, and the total amount of energy of all kinds in an isolated system is constant; it is an application of the principle of conservation of energy.

first-level controller [CONT SYS] A controller that is associated with one of the subsystems into which a large-scale control system is partitioned by plant decomposition, and acts to satisfy local objectives and constraints. Also known as local controller.

first-order transition [THERMO] A change in state of aggregation of a system accompanied by a discontinuous change in enthalpy, entropy, and volume at a single temperature and pressure.

fishtail bit [DES ENG] A drilling bit shaped like the tail of a fish.

fit [DES ENG] The dimensional relationship between mating parts, such as press, shrink, or sliding fit.

fixed end moment *See* fixing moment.

fixed-feed grinding [MECH ENG] Feeding processed material to a grinding wheel, or vice versa, in predetermined increments or at a given rate.

fixed fin [AERO ENG] A nonadjustable vane or airfoil affixed longitudinally to an aerodynamically or ballistically designed body for stabilizing purposes.

fixing moment [MECH] The bending moment at the end support of a beam necessary to fix it and prevent rotation. Also known as fixed end moment.

fixture [MECH ENG] A device used to hold and position a piece of work without guiding the cutting tool.

flaking mill [MECH ENG] A machine for converting material to flakes.

flame bucket [AERO ENG] A deep, cavelike construction built beneath a rocket launchpad, open at the top to receive the hot gases of the rocket, and open on one or three sides below, with a thick metal fourth side bent toward the open side or sides so as to deflect the exhaust gases.

flame deflector [AERO ENG] 1. In a vertical launch, any of variously designed obstructions that intercept the hot gases of the rocket engine so as to deflect them away from the ground or from a structure. 2. In a captive test, an elbow in the exhaust conduit or flame bucket that deflects the flame into the open.

flame detector [MECH ENG] A sensing device which indicates whether or not a fuel is burning, or if ignition has been lost, by transmitting a signal to a control system.

flameholder [AERO ENG] A device that sustains combustion in a flowing mixture within the combustion chamber of some types of jet engines.

flameout [AERO ENG] The extinguishing of the flame in a reaction engine, especially in a jet engine.

flanged pipe [DES ENG] A pipe with flanges at the ends; can be bolted end to end to another pipe.

flank [DES ENG] 1. The end surface of a cutting tool, adjacent to the cutting edge. 2. The side of a screw thread.

flank angle [DES ENG] The angle made by the flank of a screw thread with a line perpendicular to the axis of the screw.

flap [AERO ENG] 1. Any control surface, such as a speed brake, dive brake, or dive-recovery brake, used primarily to increase the lift or drag on an airplane, or to aid in recovery from a dive. 2. Any rudder attached to a rocket and acting either in the air or within the jet stream.

flaperon [AERO ENG] A control surface used both as a flap and as an aileron.

flap hinge *See* backflap hinge.

flap valve [MECH ENG] A valve fitted with a hinged flap or disk that swings in one direction only.

flare [AERO ENG] To descend in a smooth curve, making a transition from a relatively steep descent to a direction substantially parallel to the surface, when landing an aircraft. [DES ENG] An expansion at the end of a cylindrical body, as at the base of a rocket.

flareout [AERO ENG] That portion of the approach path of an aircraft in which the vertical component is modified to lessen the impact of landing.

flash boiler [MECH ENG] A boiler with hot tubes of small capacity; designed to immediately convert small amounts of water to superheated steam.

flat belt [DES ENG] A power transmission belt, in the form of leather belting, used where high-speed motion rather than power is the main concern.

flat-belt conveyor [MECH ENG] A conveyor belt in which the carrying run is supported by flat-belt idlers or pulleys.

flat-belt pulley [DES ENG] A smooth, flat-faced pulley made of cast iron, fabricated steel, wood, and paper and used with a flat-belt drive.

flat-blade turbine [MECH ENG] An impeller with flat blades attached to the margin.

flat-bottom crown *See* flat-face bit.

flat chisel [DES ENG] A steel chisel used to obtain a flat and finished surface.

flat crank [DES ENG] A crankshaft having one flat bearing journal.

flat drill [DES ENG] A type of rotary drill constructed from a flat piece of material.

flat edge trimmer [MECH ENG] A machine designed to trim the notched edges of metal shells.

flat-face bit [DES ENG] A diamond core bit whose face in cross section is square. Also known as flat-bottom crown; flat-nose bit; square-nose bit.

flat form tool [DES ENG] A tool having a square or rectangular cross section with the form along the end.

flathead rivet [DES ENG] A small rivet with a flat manufactured head used for general-purpose riveting.

flat-nose bit *See* flat-face bit.

flat rope [DES ENG] A steel or fiber rope having a flat cross section and composed of a number of loosely twisted ropes placed side by side, the lay of the adjacent strands being in opposite directions to secure uniformity in wear and to prevent twisting during winding.

flat spin [MECH] Motion of a projectile with a slow spin and a very large angle of yaw, happening most frequently in fin-stabilized projectiles with some spin-producing moment, when the period of revolution of the projectile coincides with the period of its oscillation; sometimes observed in bombs and in unstable spinning projectiles.

flat spring *See* leaf spring.

flat trajectory [MECH] A trajectory which is relatively flat, that is, described by a projectile of relatively high velocity.

flat-turret lathe [MECH ENG] A lathe with a low, flat turret on a power-fed cross-sliding headstock.

fl dr *See* fluid dram.

fleet [MECH ENG] Sidewise movement of a rope or cable when winding on a drum.

fleet angle [MECH ENG] In hoisting gear, the included angle between the rope, in its position of greatest travel across the drum, and a line drawn perpendicular to the drum shaft, passing through the center of the head sheave or lead sheave groove.

Flettner windmill [MECH ENG] An inefficient windmill with four arms, each consisting of a rotating cylinder actuated by a Savonius rotor.

flexibility [MECH] The quality or state of being able to be flexed or bent repeatedly.

flexible coupling [MECH ENG] A coupling used to connect two shafts and to accommodate their misalignment.

flexible shaft [MECH ENG] 1. A shaft that transmits rotary motion at any angle up to about 90°. 2. A shaft made of flexible material or of segments. 3. A shaft whose bearings are designed to accommodate a small amount of misalignment.

flexural modulus [MECH] A measure of the resistance of a beam of specified material and cross section to bending, equal to the product of Young's modulus for the material and the square of the radius of gyration of the beam about its neutral axis.

flexural rigidity [MECH] The ratio of the sideward force applied to one end of a beam to the resulting displacement of this end, when the other end is clamped.

flexural strength [MECH] Strength of a material in blending, that is, resistance to fracture.

flexure [MECH] 1. The deformation of any beam subjected to a load. 2. Any deformation of an elastic body in which the points originally lying on any straight line are displaced to form a plane curve.

flexure theory [MECH] Theory of the deformation of a prismatic beam having a length at least 10 times its depth and consisting of a material obeying Hooke's law, in response to stresses within the elastic limit.

flicker control [AERO ENG] Control of an aircraft, rocket, or such in which the control surfaces are deflected to their maximum degree with only a slight motion of the controller.

flight [AERO ENG] The movement of an object through the atmosphere or through space, sustained by aerodynamic reaction or other forces.

flight characteristic [AERO ENG] A characteristic exhibited by an aircraft, rocket, or the like in flight, such as a tendency to stall or to yaw, or an ability to remain stable at certain speeds.

flight conveyor [MECH ENG] A conveyor in which paddles, attached to single or double strands of chain, drag or push pulverized or granulated solid materials along a trough. Also known as drag conveyor.

flight deck [AERO ENG] In certain airplanes, an elevated compartment occupied by the crew for operating the airplane in flight.

flight dynamics [AERO ENG] The study of the motion of an aircraft or missile; concerned with transient or short-term effects relating to the stability and control of the vehicle, rather than to calculating such performance as altitude or velocity.

flight envelope [AERO ENG] The boundary depicting, for a specific aircraft, the limits of speed, altitude, and acceleration which that aircraft cannot safely exceed.

flight feeder [MECH ENG] Short-length flight conveyor used to feed solids materials to a process vessel or other receptacle at a preset rate.

flight instrument [AERO ENG] An aircraft instrument used in the control of the direction of flight, attitude, altitude, or speed of an aircraft, for example, the artificial horizon, airspeed indicator, altimeter, compass, rate-of-climb indicator, accelerometer, turn-and-bank indicator, and so on.

flight level [AERO ENG] A surface of constant atmospheric pressure which is related to the standard pressure datum.

flight path [AERO ENG] The path made or followed in the air or in space by an aircraft, rocket, or such.

flight-path angle [AERO ENG] The angle between the horizontal (or some other reference angle) and a tangent to the flight path at a point. Also known as flight-path slope.

flight-path slope *See* flight-path angle.

flight profile [AERO ENG] A graphic portrayal or plot of the flight path of an aeronautical vehicle in the vertical plane.

flight science [AERO ENG] The sum total of all knowledge that enables humans to accomplish flight; it is compounded of both science and engineering, and is concerned with airplanes, missiles, and crewed and uncrewed space vehicles.

flight simulator [AERO ENG] A training device or apparatus that simulates certain conditions of actual flight or of flight operations.

flight stability [AERO ENG] The property of an aircraft or missile to maintain its attitude and to resist displacement, and, if displaced, to tend to restore itself to the original attitude.

flight test [AERO ENG] **1.** A test by means of actual or attempted flight to see how an aircraft, spacecraft, space-air vehicle, or missile flies. **2.** A test of a component part of a flying vehicle, or of an object carried in such a vehicle, to determine its suitability or reliability in terms of its intended function by making it endure actual flight.

flint mill [MECH ENG] A mill employing pebbles to pulverize materials (for example, in cement manufacture).

flip-flop amplifier *See* wall-attachment amplifier.

float [DES ENG] A file which has a single set of parallel teeth.

float-cut file [DES ENG] A coarse file used on soft materials.

floating axle [MECH ENG] A live axle used to turn the wheels of an automotive vehicle; the weight of the vehicle is borne by housings at the ends of a fixed axle.

floating block *See* traveling block.

floating lever [MECH ENG] A horizontal brake lever with a movable fulcrum; used under railroad cars.

floating scraper [MECH ENG] A balanced scraper blade that rests lightly on a drum filter; removes solids collected on the rotating drum surface by riding on the drum's surface contour.

flood [MECH ENG] To supply an excess of fuel to a carburetor so that the level rises above the nozzle.

flooring saw [DES ENG] A pointed saw with teeth on both edges; cuts its own entrance into a material.

flow [FL MECH] The forward continuous movement of a fluid, such as gases, vapors, or liquids, through closed or open channels or conduits.

flowability [FL MECH] Capability of a liquid or loose particulate solid to move by flow.

flow coefficient [FL MECH] An experimentally determined proportionality constant, relating the actual velocity of fluid flow in a pipe, duct, or open channel to the theoretical velocity expected under certain assumptions. [MECH ENG] A dimensionless number used in studying the power required by fans, equal to the volumetric flow rate through the fan divided by the product of the rate of rotation of the fan and the cube of the impeller diameter.

flow distribution *See* flow field.

flow equation [FL MECH] Equation for the calculation of fluid (gas, vapor, liquid) flow through conduits or channels; consists of an interrelation of fluid properties (such as density or viscosity), environmental conditions (such as temperature or pressure), and conduit or channel geometry and conditions (such as diameter, cross-sectional shape, or surface roughness).

flow field [FL MECH] The velocity and the density of a fluid as functions of position and time. Also known as flow distribution.

flowing-temperature factor [THERMO] Calculation correction factor for gases flowing at temperatures other than that for which a flow equation is valid, that is, other than 60°F (15.5°C).

flow mixer [MECH ENG] Liquid-liquid mixing device in which the mixing action occurs as the liquids pass through it; includes jet nozzles and agitator vanes. Also known as line mixer.

flow net [FL MECH] A diagram used in studying the flow of a fluid through a permeable substance (such as water through a soil structure) having two nests of curves, one representing the flow lines, which follow the path of the fluid, and the other the equipotential lines, which connect points of equal head.

flow pattern [FL MECH] Pattern of two-phase flow in a conduit or channel pipe, taking into consideration the ratio of gas to liquid and conditions of flow resistance and liquid holdup.

flow rate [FL MECH] Also known as rate of flow. **1.** Time required for a given quantity of flowable material to flow a measured distance. **2.** Weight or volume of flowable material flowing per unit time.

flow-rating pressure [MECH ENG] The value of inlet static pressure at which the relieving capacity of a pressure-relief device is established.

flow resistance [FL MECH] **1.** Any factor within a conduit or channel that impedes the flow of fluid, such as surface roughness or sudden bends, contractions, or expansions. **2.** *See* viscosity.

flow separation *See* boundary-layer separation.

flow stress [MECH] The stress along one axis at a given value of strain that is required to produce plastic deformation.

fl oz *See* fluid ounce.

fluctuation velocity   *See* eddy velocity.

fluid clutch   *See* fluid drive.

fluid-controlled valve   [MECH ENG] A valve for which the valve operator is activated by a fluid energy, in contrast to electrical, pneumatic, or manual energy.

fluid coupling   [MECH ENG] A device for transmitting rotation between shafts by means of the acceleration and deceleration of a fluid such as oil. Also known as hydraulic coupling.

fluid density   [FL MECH] The mass of a fluid per unit volume.

fluid die   [MECH ENG] A die for shaping parts by liquid pressure; a plunger forces the liquid against the part to be shaped, making the part conform to the shape of a die.

fluid dram   [MECH] Abbreviated fl dr. **1.** A unit of volume used in the United States for measurement of liquid substances, equal to 1/8 fluid ounce, or 3.6966911953125 $\times$ 10^{-6} cubic meter. **2.** A unit of volume used in the United Kingdom for measurement of liquid substances and occasionally of solid substances, equal to 1/8 fluid ounce or approximately 3.55163 $\times$ 10^{-6} cubic meter.

fluid drive   [MECH ENG] A power coupling operated on a hydraulic turbine principle in which the engine flywheel has a set of turbine blades which are connected directly to it and which are driven in oil, thereby turning another set of blades attached to the transmission gears of the automobile. Also known as fluid clutch; hydraulic clutch.

fluid dynamics   [FL MECH] The science of fluids in motion.

fluid-film bearing   [MECH ENG] An antifriction bearing in which rubbing surfaces are kept apart by a film of lubricant such as oil.

fluid friction   [FL MECH] Conversion of mechanical energy in fluid flow into heat energy.

fluidity   [FL MECH] The reciprocal of viscosity; expresses the ability of a substance to flow.

fluidized-bed combustion   [MECH ENG] A method of burning particulate fuel, such as coal, in which the amount of air required for combustion far exceeds that found in conventional burners; the fuel particles are continually fed into a bed of mineral ash in the proportions of 1 part fuel to 200 parts ash, while a flow of air passes up through the bed, causing it to act like a turbulent fluid.

fluid mechanics   [MECH] The science concerned with fluids, either at rest or in motion, and dealing with pressures, velocities, and accelerations in the fluid, including fluid deformation and compression or expansion.

fluid ounce   [MECH] Abbreviated fl oz. **1.** A unit of volume that is used in the United States for measurement of liquid substances, equal to 1/16 liquid pint, or 231/128 (approximately 1.804) cubic inches, or 2.95735295625 $\times$ 10^{-5} cubic meter. **2.** A unit of volume used in the United Kingdom for measurement of liquid substances, and occasionally of solid substances, equal to 1/20 pint or approximately 2.84130 $\times$ 10^{-5} cubic meter.

fluid resistance   [FL MECH] The force exerted by a gas or liquid opposing the motion of a body through it. Also known as resistance.

fluid statics   [FL MECH] The determination of pressure intensities and forces exerted by fluids at rest.

fluid stress   [MECH] Stress associated with plastic deformation in a solid material.

fluid ton   [MECH] A unit of volume equal to 32 cubic feet or approximately 9.0614 $\times$ 10^{-2} cubic meter; used for many hydrometallurgical, hydraulic, and other industrial purposes.

fluid transmission [MECH ENG] Automotive transmission with fluid drive.

flute [DES ENG] A groove having a curved section, especially when parallel to the main axis, as on columns, drills, and other cylindrical or conical shaped pieces.

fluted chucking reamer [DES ENG] A machine reamer with a straight or tapered shank and with straight or spiral flutes; the ends of the teeth are ground on a slight chamfer for end cutting.

flute length [DES ENG] On a twist drill, the length measured from the outside corners of the cutting lips to the farthest point at the back end of the flutes.

fluting [MECH ENG] A machining operation whereby flutes are formed parallel to the main axis of cylindrical or conical parts.

flutter *See* aeronautical flutter.

fly [MECH ENG] A fan with two or more blades used in timepieces or light machinery to govern speed by air resistance.

fly-by-wire system [AERO ENG] A flight control system that uses electric wiring instead of mechanical or hydraulic linkages to control the actuators for the ailerons, flaps, and other control surfaces of an aircraft.

fly cutter [MECH ENG] A cutting tool that revolves with the arbor of a lathe.

fly cutting [MECH ENG] Cutting with a milling cutter provided with only one tooth.

flying angle [AERO ENG] The acute angle between the longitudinal axis of an aircraft and the horizontal axis in normal level flight, or the angle of attack of a wing in normal level flight.

flying boat [AERO ENG] A seaplane with a fuselage that acts as a hull and is the means of the plane's support on water.

flying-crane helicopter [AERO ENG] A heavy-lift helicopter used in rapid loading and unloading of, for example, cargo ships.

flywheel [MECH ENG] A rotating element attached to the shaft of a machine for the maintenance of uniform angular velocity and revolutions per minute. Also known as balance wheel.

fm *See* femtometer.

Fo$_f$ *See* Fourier number.

foam [FL MECH] A froth of bubbles on the surface of a liquid, often stabilized by organic contaminants, as found at sea or along shore.

folding fin [AERO ENG] A fin hinged at its base to lie flat, especially a fin on a rocket that lies flat until the rocket is in flight.

foot [MECH] The unit of length in the British systems of units, equal to exactly 0.3048 meter. Abbreviated ft.

foot-pound [MECH] **1.** Unit of energy or work in the English gravitational system, equal to the work done by 1 pound of force when the point at which the force is applied is displaced 1 foot in the direction of the force; equal to approximately 1.355818 joule. Abbreviated ft-lb; ft-lbf. **2.** Unit of torque in the English gravitational system, equal to the torque produced by 1 pound of force acting at a perpendicular distance of 1 foot from an axis of rotation. Also known as pound-foot. Abbreviated lbf-ft.

foot-poundal [MECH] **1.** A unit of energy or work in the English absolute system, equal to the work done by a force of magnitude 1 poundal when the point at which the force is applied is displaced 1 foot in the direction of the force; equal to approximately 0.04214011 joules. Abbreviated ft-pdl. **2.** A unit of torque in the English absolute system, equal to the torque produced by a force of magnitude 1 poundal acting at a perpendicular distance of 1 foot from the axis of rotation. Also known as poundal-foot. Abbreviated pdl-ft.

foot section [MECH ENG] In both belt and chain conveyors that portion of the conveyor at the extreme opposite end from the delivery point.

footstock [MECH ENG] A device containing a center which supports the workpiece on a milling machine; usually used in conjunction with a dividing head.

foot valve [MECH ENG] A valve in the bottom of the suction pipe of a pump which prevents backward flow of water.

force [MECH] That influence on a body which causes it to accelerate; quantitatively it is a vector, equal to the body's time rate of change of momentum.

force constant [MECH] The ratio of the force to the deformation of a system whose deformation is proportional to the applied force.

forced-air heating [MECH ENG] A warm-air heating system in which positive air circulation is provided by means of a fan or a blower.

forced circulation [MECH ENG] The use of a pump or other fluid-movement device in conjunction with liquid-processing equipment to move the liquid through pipes and process vessels; contrasted to gravity or thermal circulation.

forced-circulation boiler [MECH ENG] A once-through steam generator in which water is pumped through successive parts.

forced convection [THERMO] Heat convection in which fluid motion is maintained by some external agency.

forced draft [MECH ENG] Air under positive pressure produced by fans at the point where air or gases enter a unit, such as a combustion furnace.

forced oscillation [MECH] An oscillation produced in a simple oscillator or equivalent mechanical system by an external periodic driving force. Also known as forced vibration.

forced ventilation [MECH ENG] A system of ventilation in which air is forced through ventilation ducts under pressure.

forced vibration *See* forced oscillation.

forced wave [FL MECH] Any wave which is required to fit irregularities at the boundary of a system or satisfy some impressed force within the system; the forced wave will not in general be a characteristic mode of oscillation of the system.

force feedback [CONT SYS] A method of error detection in which the force exerted on the effector is sensed and fed back to the control, usually by mechanical, hydraulic, or electric transducers.

force polygon [MECH] A closed polygon whose sides are vectors representing the forces acting on a body in equilibrium.

forceps [DES ENG] A pincerlike instrument for grasping objects.

force pump [MECH ENG] A pump fitted with a solid plunger and a suction valve which draws and forces a liquid to a considerable height above the valve or puts the liquid under a considerable pressure.

fore pump *See* backing pump.

forklift [MECH ENG] A machine, usually powered by hydraulic means, consisting of two or more prongs which can be raised and lowered and are inserted under heavy materials or objects for hoisting and moving them.

forklift truck *See* fork truck.

fork truck [MECH ENG] A vehicle equipped with a forklift. Also known as forklift truck.

form cutter   *See* formed cutter.

form drag   [FL MECH]   **1.** The drag from all causes resulting from the particular shape of a body relative to its direction of motion, as of fuselage, wing, or nacelle.   **2.** At supersonic speed, the drag caused by losses due to shock waves, exclusive of losses due to skin friction.

formed cutter   [MECH]   A cutting tool shaped to make surfaces with irregular geometry. Also known as form cutter.

form factor   [MECH]   The theoretical stress concentration factor for a given shape, for a perfectly elastic material.

form grinding   [MECH ENG]   Grinding by use of a wheel whose cutting face is contoured to the reverse shape of the desired form.

forming press   [MECH ENG]   A punch press for forming metal parts.

forming rolls   [MECH ENG]   Rolls contoured to give a desired shape to parts passing through them.

forming tool   [DES ENG]   A nonrotating tool that produces its inverse form on the workpiece.

fors   *See* G; gram-force.

forward path   [CONT SYS]   The transmission path from the loop actuating signal to the loop output signal in a feedback control loop.

forward transfer function   [CONT SYS]   In a feedback control loop, the transfer function of the forward path.

Foster's reactance theorem   [CONT SYS]   The theorem that the most general driving point impedance or admittance of a network, in which every mesh contains independent inductance and capacitance, is a meromorphic function whose poles and zeros are all simple and occur in conjugate pairs on the imaginary axis, and in which these poles and zeros alternate.

Foucault pendulum   [MECH]   A swinging weight supported by a long wire, so that the wire's upper support restrains the wire only in the vertical direction, and the weight is set swinging with no lateral or circular motion; the plane of the pendulum gradually changes, demonstrating the rotation of the earth on its axis.

fountain effect   [FL MECH]   The effect occurring when two containers of superfluid helium are connected by a capillary tube and one of them is heated, so that helium flows through the tube in the direction of higher temperature.

four-bar linkage   [MECH ENG]   A plane linkage consisting of four links pinned tail to head in a closed loop with lower, or closed, joints.

Fourdrinier machine   [MECH ENG]   A papermaking machine; a paper web is formed on an endless wire screen; the screen passes through presses and over dryers to the calenders and reels.

fourier   *See* thermal ohm.

Fourier heat equation   *See* Fourier law of heat conduction; heat equation.

Fourier law of heat conduction   [THERMO]   The law that the rate of heat flow through a substance is proportional to the area normal to the direction of flow and to the negative of the rate of change of temperature with distance along the direction of flow. Also known as Fourier heat equation.

Fourier number   [FL MECH]   A dimensionless number used in unsteady-state flow problems, equal to the product of the dynamic viscosity and a characteristic time divided by the product of the fluid density and the square of a characteristic length. Symbolized Fo_f.   [THERMO]   A dimensionless number used in the study of unsteady-state heat transfer, equal to the product of the thermal conductivity and a characteristic time, divided by the product of the density, the specific heat at constant pressure, and the distance from the midpoint of the body through which heat is passing to the surface. Symbolized N_{Foh}.

four-stroke cycle　[MECH ENG]　An internal combustion engine cycle completed in four piston strokes; includes a suction stroke, compression stroke, expansion stroke, and exhaust stroke.

four-way valve　[MECH ENG]　A valve at the junction of four waterways which allows passage between any two adjacent waterways by means of a movable element operated by a quarter turn.

four-wheel drive　[MECH ENG]　An arrangement in which the drive shaft acts on all four wheels of the automobile.

fox lathe　[MECH ENG]　A lathe with chasing bar and leaders for cutting threads; used for turning brass.

fracture strength　*See* fracture stress.

fracture stress　[MECH]　The minimum tensile stress that will cause fracture. Also known as fracture strength.

fracture wear　[MECH]　The wear on individual abrasive grains on the surface of a grinding wheel caused by fracture.

framing square　[DES ENG]　A graduated carpenter's square used for cutting off and making notches.

Francis formula　[FL MECH]　An equation for the calculation of water flow rate over a rectangular weir in terms of length and head.

Francis turbine　[MECH ENG]　A reaction hydraulic turbine of relatively medium speed with radial flow of water in the runner.

frangible　[MECH]　Breakable, fragile, or brittle.

free balloon　[AERO ENG]　A balloon that ascends without a tether, propulsion or guidance; it is made to descend by the release of gas.

free convection　*See* natural convection.

free convection number　*See* Grashof number.

free energy　[THERMO]　1. The internal energy of a system minus the product of its temperature and its entropy. Also known as Helmholtz free energy; Helmholtz function; Helmholtz potential; thermodynamic potential at constant volume; work function. 2. *See* Gibbs free energy.

free enthalpy　*See* Gibbs free energy.

free fall　[MECH]　The ideal falling motion of a body acted upon only by the pull of the earth's gravitational field.

free falling　[MECH ENG]　In ball milling, the peripheral speed at which part of the crop load breaks clear on the ascending side and falls clear to the toe of the charge.

free fit　[DES ENG]　A fit between mating pieces where accuracy is not essential or where large variations in temperature may occur.

free flight　[MECH]　Unconstrained or unassisted flight.

free-flight angle　[MECH]　The angle between the horizontal and a line in the direction of motion of a flying body, especially a rocket, at the beginning of free flight.

free-flight trajectory　[MECH]　The path of a body in free fall.

free-piston engine [MECH ENG] A prime mover utilizing free-piston motion controlled by gas pressure in the cylinders.

free-stream Mach number [AERO ENG] The Mach number of the total airframe (entire aircraft) as contrasted with the local Mach number of a section of the airframe.

free surface [FL MECH] A boundary between two homogeneous fluids.

free turbine [MECH ENG] In a turbine engine, a turbine wheel that drives the output shaft and is not connected to the shaft driving the compressor.

free vector [MECH] A vector whose direction in space is prescribed but whose point of application and line of application are not prescribed.

free vortex [FL MECH] Two-dimensional fluid flow in which the fluid moves in concentric circles at speeds inversely proportional to the radii of the circles.

freezer [MECH ENG] An insulated unit, compartment, or room in which perishable foods are quick-frozen and stored.

freeze-up [MECH ENG] Abnormal operation of a refrigerating unit because ice has formed at the expansion device.

french [MECH] A unit of length used to measure small diameters, especially those of fiber optic bundles, equal to $\frac{1}{3}$ millimeter.

french coupling [DES ENG] A coupling having both right- and left-handed threads.

frequency domain [CONT SYS] Pertaining to a method of analysis, particularly useful for fixed linear systems in which one does not deal with functions of time explicitly, but with their Laplace or Fourier transforms, which are functions of frequency.

frequency locus [CONT SYS] The path followed by the frequency transfer function or its inverse, either in the complex plane or on a graph of amplitude against phase angle; used in determining zeros of the describing function.

frequency-response trajectory [CONT SYS] The path followed by the frequency-response phasor in the complex plane as the frequency is varied.

frequency transformation [CONT SYS] A transformation used in synthesizing a band-pass network from a low-pass prototype, in which the frequency variable of the transfer function is replaced by a function of the frequency. Also known as low-pass band-pass transformation.

friction [MECH] A force which opposes the relative motion of two bodies whenever such motion exists or whenever there exist other forces which tend to produce such motion.

frictional grip [MECH] The adhesion between the wheels of a locomotive and the rails of the railroad track.

frictional secondary flow *See* secondary flow.

friction bearing [MECH ENG] A solid bearing that directly contacts and supports an axle end.

friction brake [MECH ENG] A brake in which the resistance is provided by friction.

friction clutch [MECH ENG] A clutch in which torque is transmitted by pressure of the clutch faces on each other.

friction coefficient *See* coefficient of friction.

friction damping [MECH] The conversion of the mechanical vibrational energy of solids into heat energy by causing one dry member to slide on another.

friction drive [MECH ENG] A drive that operates by the friction forces set up when one rotating wheel is pressed against a second wheel.

friction factor [FL MECH] Any of several dimensionless numbers used in studying fluid friction in pipes, equal to the Fanning friction factor times some dimensionless constant.

friction flow [FL MECH] Fluid flow in which a significant amount of mechanical energy is dissipated into heat by action of viscosity.

friction gear [MECH ENG] Gearing in which motion is transmitted through friction between two surfaces in rolling contact.

friction head [FL MECH] The head lost by the flow in a stream or conduit due to frictional disturbances set up by the moving fluid and its containing conduit and by intermolecular friction.

friction horsepower [MECH ENG] Power dissipated in a machine through friction.

frictionless flow *See* inviscid flow.

friction loss [MECH] Mechanical energy lost because of mechanical friction between moving parts of a machine.

friction saw [MECH ENG] A toothless circular saw used to cut materials by fusion due to frictional heat.

friction sawing [MECH ENG] A burning process to cut stock to length by using a blade saw operating at high speed; used especially for the structural parts of mild steel and stainless steel.

friction torque [MECH] The torque which is produced by frictional forces and opposes rotational motion, such as that associated with journal or sleeve bearings in machines.

frigorie [THERMO] A unit of rate of extraction of heat used in refrigeration, equal to 1000 fifteen-degree calories per hour, or 1.16264 ± 0.00014 watt.

Frise aileron [AERO ENG] A type of aileron having its leading edge projecting well ahead of the hinge axis.

frog [DES ENG] A hollow on one or both of the larger faces of a brick or block; reduces weight of the brick or block; may be filled with mortar.

frontal passage [AERO ENG] The transit of an aircraft through a frontal zone.

front-end loader [MECH ENG] An excavator consisting of an articulated bucket mounted on a series of movable arms at the front of a crawler or rubber-tired tractor.

Froude number 1 [FL MECH] A dimensionless number used in studying the motion of a body floating on a fluid with production of surface waves and eddies; equal to the ratio of the square of the relative speed to the product of the acceleration of gravity and a characteristic length of the body. Symbolized N_{Fr1}.

Froude number 2 [FL MECH] A dimensionless number, equal to the ratio of the speed of flow of a fluid in an open channel to the speed of very small gravity waves, the latter being equal to the square root of the product of the acceleration of gravity and a characteristic length. Symbolized N_{Fr2}.

ft *See* foot.

ft-lb *See* foot-pound.

ft-lbf *See* foot-pound.

ft-pdl *See* foot-poundal.

fuel bed [MECH ENG] A layer of burning fuel, as on a furnace grate or a cupola.

fuel injection [MECH ENG] The delivery of fuel to an internal combustion engine cylinder by pressure from a mechanical pump.

fuel injector [MECH ENG] A pump mechanism that sprays fuel into the cylinder of an internal combustion engine at the appropriate part of the cycle.

fuel pump [MECH ENG] A pump for drawing fuel from a storage tank and delivering it to an engine or furnace.

fuel shutoff [AERO ENG] 1. The action of shutting off the flow of liquid fuel into a combustion chamber or of stopping the combustion of a solid fuel. 2. The event or time marking this action.

fuel structure ratio *See* fuel-weight ratio.

fuel system [MECH ENG] A system which stores fuel for present use and delivers it as needed.

fuel tank [MECH ENG] The operating, fuel-storage component of a fuel system.

fuel-weight ratio [AERO ENG] The ratio of the weight of a rocket's fuel to the weight of the unfueled rocket. Also known as fuel structure ratio.

fugacity [THERMO] A function used as an analog of the partial pressure in applying thermodynamics to real systems; at a constant temperature it is proportional to the exponential of the ratio of the chemical potential of a constituent of a system divided by the product of the gas constant and the temperature, and it approaches the partial pressure as the total pressure of the gas approaches zero.

fulcrum [MECH] The rigid point of support about which a lever pivots.

full feathering *See* feathering.

full-gear [MECH ENG] The condition of a steam engine when the valve is operated to the maximum extent by the link motion.

full-track vehicle [MECH ENG] A vehicle entirely supported, driven, and steered by an endless belt, or track, on each side; for example, a tank.

funal *See* sthène.

functional decomposition [CONT SYS] The partitioning of a large-scale control system into a nested set of generic control functions, namely the regulatory or direct control function, the optimizing control function, the adaptive control function, and the self-organizing function.

funicular polygon [MECH] 1. The figure formed by a light string hung between two points from which weights are suspended at various points. 2. A force diagram for such a string, in which the forces (weights and tensions) acting on points of the string from which weights are suspended are represented by a series of adjacent triangles.

funnel [DES ENG] A tube with one conical end that sometimes holds a filter; the function is to direct flow of a liquid or, if a filter is present, to direct a flow that was filtered.

furlong [MECH] A unit of length, equal to 1/8 mile, 660 feet, or 201.168 meters.

fuselage [AERO ENG] In an airplane, the central structure to which wings and tail are attached; it accommodates flight crew, passengers, and cargo.

fusibility [THERMO] The quality or degree of being capable of being liquefied by heat.

fusing disk [MECH ENG] A rapidly spinning disk that cuts metal by melting it.

G

g *See* gram; gravity vector.

G [MECH] A unit of acceleration equal to the standard acceleration of gravity, 9.80665 meters per second per second, or approximately 32.1740 feet per second per second. Also known as fors; grav.

gage block [DES ENG] A chrome steel block having two flat, parallel surfaces with the parallel distance between them being the size marked on the block to a guaranteed accuracy of a few millionths of an inch; used as the standard of precise lineal measurement for most manufacturing processes. Also known as precision block; size block.

gage point [DES ENG] A point used to position a part in a jig, fixture, or qualifying gage.

gage pressure [MECH ENG] The amount by which the total absolute pressure exceeds the ambient atmospheric pressure.

gage Also spelled gauge. [DES ENG] 1. A device for determining the relative shape or size of an object. 2. The thickness of a metal sheet, a rod, or a wire.

gain asymptotes [CONT SYS] Asymptotes to a logarithmic graph of gain as a function of frequency.

gain-crossover frequency [CONT SYS] The frequency at which the magnitude of the loop ratio is unity.

gain margin [CONT SYS] The reciprocal of the magnitude of the loop ratio at the phase crossover frequency, frequently expressed in decibels.

gain scheduling [CONT SYS] A method of eliminating influences of variations in the process dynamics of a control system by changing the parameters of the regulator as functions of auxiliary variables which correlate well with those dynamics.

gal [MECH] 1. The unit of acceleration in the centimeter-gram-second system, equal to 1 centimeter per second squared; commonly used in geodetic measurement. Formerly known as galileo. Symbolized Gal. 2. *See* gallon.

Gal *See* gal.

Galilean transformation [MECH] A mathematical transformation used to relate the space and time variables of two uniformly moving (inertial) reference systems in nonrelativistic kinematics.

galileo *See* gal.

Galileo number [FL MECH] A dimensionless number used in studying the circulation of viscous liquids, equal to the cube of a characteristic dimension, times the acceleration of gravity, times the square of the liquid's density, divided by the square of its viscosity. Symbol N_{Gal}.

Galileo's law of inertia *See* Newton's first law.

gallon [MECH] Abbreviated gal. 1. A unit of volume used in the United States for measurement of liquid substances, equal to 231 cubic inches, or to 3.785 411 784

$\times$ 10^{-3} cubic meter, or to 3.785 411 784 liters; equal to 128 fluid ounces. **2.** A unit of volume used in the United Kingdom for measurement of liquid and solid substances, usually the former; equal to the volume occupied by 10 pounds of weight of water of density 0.998859 gram per milliliter in air of density 0.001217 gram per milliliter against weights of density 8.136 grams per milliliter (the milliliter here has its old definition of 1.000028 $\times$ 10^{-6} cubic meter); approximately equal to 277.420 cubic inches, or to 4.54609 $\times$ 10^{-3} cubic meter, or to 4.54609 liters; equal to 160 fluid ounces.

gamma [MECH] A unit of mass equal to 10^{-6} gram or 10^{-9} kilogram.

gang drill [MECH ENG] A set of drills operated together in the same machine; used in rock drilling.

gang saw [MECH ENG] A steel frame in which thin, parallel saws are arranged to operate simultaneously in cutting logs.

gantry crane [MECH ENG] A bridgelike hoisting machine having fixed supports or arranged for running along tracks on ground level.

gap-framepress [MECH ENG] A punch press whose frame is open at bed level so that wide work or strip work can be inserted.

gap lathe [MECH ENG] An engine lathe with a sliding bed providing enough space for turning large-diameter work.

garnet hinge [DES ENG] A hinge with a vertical bar and horizontal strap.

garter spring [DES ENG] A closed ring formed of helically wound wire.

gas bearing [MECH ENG] A journal or thrust bearing lubricated with gas. Also known as gas-lubricated bearing.

gas-compression cycle [MECH ENG] A refrigeration cycle in which hot, compressed gas is cooled in a heat exchanger, then passes into a gas expander which provides an exhaust stream of cold gas to another heat exchanger that handles the sensible-heat refrigeration effect and exhausts the gas to the compressor.

gas compressor [MECH ENG] A machine that increases the pressure of a gas or vapor by increasing the gas density and delivering the fluid against the connected system resistance.

gas constant [THERMO] The constant of proportionality appearing in the equation of state of an ideal gas, equal to the pressure of the gas times its molar volume divided by its temperature. Also known as gas-law constant.

gas cycle [THERMO] A sequence in which a gaseous fluid undergoes a series of thermodynamic phases, ultimately returning to its original state.

gas cylinder [MECH ENG] The chamber in which a piston moves in a positive displacement engine or compressor.

gas-deviation factor *See* compressibility factor.

gas engine [MECH ENG] An internal combustion engine that uses gaseous fuel.

gas generator [MECH ENG] An apparatus that supplies a high-pressure gas flow to drive compressors, airscrews, and other machines.

gas heater [MECH ENG] A unit heater designed to supply heat by forced convection, using gas as a heat source.

gas injection [MECH ENG] Injection of gaseous fuel into the cylinder of an internal combustion engine at the appropriate part of the cycle.

gas kinematics [FL MECH] The motion of a gas considered by itself, without regard for the causes of motion.

gas law [THERMO] Any law relating the pressure, volume, and temperature of a gas.

gas-law constant *See* gas constant.

gas-lubricated bearing *See* gas bearing.

gas mechanics [FL MECH] The action of forces on gases.

gasoline engine [MECH ENG] An internal combustion engine that uses a mixture of air and gasoline vapor as a fuel.

gasoline pump [MECH ENG] A device that pumps and measures the gasoline supplied to a motor vehicle, as at a filling station.

gas pliers [DES ENG] Pliers for gripping round objects such as pipes, tubes, and circular rods.

gas slippage [FL MECH] Phenomenon of gas bypassing liquids that occurs when the diameter of capillary openings approaches the mean free path of the gas; occurs not only in capillary tubing, but in porous oil-reservoir formations.

gas turbine [MECH ENG] A heat engine that converts energy of fuel into work by using compressed, hot gas as the working medium and that usually delivers its mechanical output through a rotating shaft. Also known as combustion turbine.

gas-turbine nozzle [MECH ENG] The component of a gas turbine in which the hot, high-pressure gas expands and accelerates to high velocity.

gas viscosity [FL MECH] The internal fluid function of a gas.

Gates crusher [MECH ENG] A gyratory crusher which has a cone or mantle that is moved eccentrically by the lower bearing sleeve.

gate valve [DES ENG] A valve with a disk-shaped closing element that fits tightly over an opening through which water passes.

Gauss' principle of least constraint [MECH] The principle that the motion of a system of interconnected material points subjected to any influence is such as to minimize the constraint on the system; here the constraint, during an infinitesimal period of time, is the sum over the points of the product of the mass of the point times the square of its deviation from the position it would have occupied at the end of the time period if it had not been connected to other points.

g-cm *See* gram-centimeter.

gear [DES ENG] A toothed machine element used to transmit motion between rotating shafts when the center distance of the shafts is not too large. [MECH ENG] **1.** A mechanism performing a specific function in a machine. **2.** An adjustment device of the transmission in a motor vehicle which determines mechanical advantage, relative speed, and direction of travel.

gearbox *See* transmission.

gear case [MECH ENG] An enclosure, usually filled with lubricating fluid, in which gears operate.

gear cutter [MECH ENG] A machine or tool for cutting teeth in a gear.

gear cutting [MECH ENG] The cutting or forming of a uniform series of toothlike projections on the surface of a workpiece.

gear down [MECH ENG] To arrange gears so the driven part rotates at a slower speed than the driving part.

gear drive [MECH ENG] Transmission of motion or torque from one shaft to another by means of direct contact between toothed wheels.

geared turbine [MECH ENG] A turbine connected to a set of reduction gears.

gear forming [MECH ENG] A method of gear cutting in which the desired tooth shape is produced by a tool whose cutting profile matches the tooth form.

gear generating [MECH ENG] A method of gear cutting in which the tooth is produced by the conjugate or total cutting action of the tool plus the rotation of the workpiece.

gear grinding [MECH ENG] A gear-cutting method in which gears are shaped by formed grinding wheels and by generation; primarily a finishing operation.

gear hobber [MECH ENG] A machine that mills gear teeth; the rotational speed of the hob has a precise relationship to that of the work.

gearing [MECH ENG] A set of gear wheels.

gearing chain [MECH ENG] A continuous chain used to transmit motion from one toothed wheel, or sprocket, to another.

gearless traction [MECH ENG] Direct drive, without reduction gears.

gear level [MECH ENG] To arrange gears so that the driven part and driving part turn at the same speed.

gear loading [MECH ENG] The power transmitted or the contact force per unit length of a gear.

gearmotor [MECH ENG] A motor combined with a set of speed-reducing gears.

gear pump [MECH ENG] A rotary pump in which two meshing gear wheels contrarotate so that the fluid is entrained on one side and discharged on the other.

gear ratio [MECH ENG] The ratio of the angular speed of the driving member of a gear train or similar mechanism to that of the driven member; specifically, the number of revolutions made by the engine per revolution of the rear wheels of an automobile.

gear shaper [MECH ENG] A machine that makes gear teeth by means of a reciprocating cutter that rotates slowly with the work.

gear-shaving machine [MECH ENG] A finishing machine that removes excess metal from machined gears by the axial sliding motion of a straight-rack cutter or a circular gear cutter.

gearshift [MECH ENG] A device for engaging and disengaging gears.

gear teeth [DES ENG] Projections on the circumference or face of a wheel which engage with complementary projections on another wheel to transmit force and motion.

gear train [MECH ENG] A combination of two or more gears used to transmit motion between two rotating shafts or between a shaft and a slide.

gear up [MECH ENG] To arrange gears so that the driven part rotates faster than the driving part.

gear wheel [MECH ENG] A wheel that meshes gear teeth with another part.

geepound *See* slug.

gel strength [FL MECH] Of a colloid, the ability or a measure of its ability to form gels.

GEM *See* ground-effect machine.

generalized coordinates [MECH] A set of variables used to specify the position and orientation of a system, in principle defined in terms of cartesian coordinates of the system's particles and of the time in some convenient manner; the number of such coordinates equals the number of degrees of freedom of the system Also known as Lagrangian coordinates.

generalized force [MECH] The generalized force corresponding to a generalized coordinate is the ratio of the virtual work done in an infinitesimal virtual displacement, which alters that coordinate and no other, to the change in the coordinate.

generalized momentum   *See* conjugate momentum.

generalized velocity   [MECH] The derivative with respect to time of one of the generalized coordinates of a particle. Also known as Lagrangian generalized velocity.

generating flow   [FL MECH] For a liquid allowed to flow smoothly into a duct, the flow while the boundary layer, which starts at the entrance and grows until it fills the duct, is growing.

generating plant   *See* generating station.

generating station   [MECH ENG] A stationary plant containing apparatus for large-scale conversion of some form of energy (such as hydraulic, steam, chemical, or nuclear energy) into electrical energy. Also known as generating plant; power station.

geodetic satellite   [AERO ENG] An artificial earth satellite used to obtain data for geodetic triangulation calculations.

geographical mile   [MECH] The length of 1 minute of arc of the Equator, or 6087.08 feet (1855.34 meters), which approximates the length of the nautical mile.

geometrical pitch   [AERO ENG] The distance a component of an airplane propeller would move forward in one complete turn of the propeller if the path it was moving along was a helix that had an angle equal to an angle between a plane perpendicular to the axis of the propeller and the chord of the component.

geometrical similarity   [FL MECH] Property of two fluid flows for which a simple alteration of scales of length and velocity transforms one into the other.

geostationary satellite   [AERO ENG] A satellite that orbits the earth from west to east at such a speed as to remain fixed over a given place on the earth's equator at approximately 35,900 kilometers altitude; makes one revolution in 24 hours, synchronous with the earth's rotation. Also known as synchronous satellite.

Gerstner wave   [FL MECH] A rotational gravity wave of finite amplitude.

gewel hinge   [DES ENG] A hinge consisting of a hook inserted in a loop.

gf   *See* gram-force.

Giaque's temperature scale   [THERMO] The internationally accepted scale of absolute temperature, in which the triple point of water is defined to have a temperature of 273.16 K.

Gibbs free energy   [THERMO] The thermodynamic function $G = H - TS$, where H is enthalpy, T absolute temperature, and S entropy; Also known as free energy; free enthalpy; Gibbs function.

Gibbs function   *See* Gibbs free energy.

Gibbs-Helmholtz equation   [THERMO] 1. Either of two thermodynamic relations that are useful in calculating the internal energy U or enthalpy H of a system; they may be written $U = F - T(\partial F/\partial T)_V$ and $H = G - T(\partial G/\partial T)_P$, where F is the free energy, G is the Gibbs free energy, T is the absolute temperature, V is the volume, and P is the pressure. 2. Any of the similar equations for changes in thermodynamic potentials during an isothermal process.

gill   [MECH] 1. A unit of volume used in the United States for the measurement of liquid substances, equal to ¼ U.S. liquid pint, or to $1.1829411825 \times 10^{-4}$ cubic meter. 2. A unit of volume used in the United Kingdom for the measurement of liquid substances, and occasionally of solid substances, equal to ¼ U.K. pint, or to approximately 1.42065×10^{-4} cubic meter.

gimbaled nozzle   [MECH ENG] A nozzle supported on a gimbal.

gimlet   [DES ENG] A small tool consisting of a threaded tip, grooved shank, and a cross handle; used for boring holes in wood.

gimlet bit [DES ENG] A bit with a threaded point and spiral flute; used for drilling small holes in wood.

gin [MECH ENG] A hoisting machine in the form of a tripod with a windlass, pulleys, and ropes.

gin pole [MECH ENG] A hand-operated derrick which has a nearly vertical pole supported by guy ropes; the load is raised on a rope that passes through a pulley at the top and over a winch at the foot. Also known as guyed-mast derrick; pole derrick; standing derrick.

gin tackle [MECH ENG] A tackle made for use with a gin.

Gleason bevel gear system [DES ENG] The standard for bevel gear designs in the United States; employs a basic pressure angle of 20° with long and short addenda for ratios other than 1:1 to avoid undercut pinions and to increase strength.

glide [AERO ENG] Descent of an aircraft at a normal angle of attack, with little or no thrust.

glide angle *See* gliding angle.

glide path [AERO ENG] 1. The flight path of an aeronautical vehicle in a glide, seen from the side. Also known as glide trajectory. 2. The path used by an aircraft or spacecraft in a landing approach procedure.

glider [AERO ENG] A fixed-wing aircraft designed to glide, and sometimes to soar; usually does not have a power plant.

glide slope [AERO ENG] *See* gliding angle.

glide trajectory *See* glide path.

gliding angle [AERO ENG] The angle between the horizontal and the glide path of an aircraft. Also known as glide angle; glide slope.

glow plug [MECH ENG] A small electric heater, located inside a cylinder of a diesel engine, that preheats the air and aids the engine in starting.

glue-joint ripsaw [MECH ENG] A heavy-gage ripsaw used on straight-line or self-feed rip machines; the cut is smooth enough to permit gluing of joints from the saw.

glug [MECH] A unit of mass, equal to the mass which is accelerated by 1 centimeter per second per second by a force of 1 gram-force, or to 980.665 grams.

gm *See* gram.

goal coordination method [CONT SYS] A method for coordinating the subproblem solutions in plant decomposition, in which Lagrange multipliers enter into the subsystem cost functions as shadow prices, and these are adjusted by the second-level controller in an iterative procedure which culminates (if the method is applicable) in the satisfaction of the subsystem coupling relationships. Also known as interaction balance method; nonfeasible method.

Goertler parameter [FL MECH] A dimensionless number used in studying boundary-layer flow on curved surfaces, equal to the Reynolds number, where the characteristic length is the boundary-layer momentum thickness, times the square root of this thickness, divided by the square root of the surface's radius of curvature.

go gage [DES ENG] A test device that just fits a part if it has the proper dimensions (often used in pairs with a "no go" gage to establish maximum and minimum dimensions).

Goldberg-Mohn friction [FL MECH] A force proportional to the velocity of a current and the density of the medium; used as a first approximation in estimating frictional effects in the atmosphere and the ocean.

gold point [THERMO] The temperature of the freezing point of gold at a pressure of 1 standard atmosphere (101,325 newtons per square meter); used to define the International Practical Temperature Scale of 1968, on which it is assigned a value of 1337.58 K or 1064.43°C.

gooseneck [DES ENG] **1.** A pipe, bar, or other device having a curved or bent shape resembling that of the neck of a goose. **2.** *See* water swivel.

gouge [DES ENG] A curved chisel for wood, bone, stone, and so on.

governor [MECH ENG] A device, especially one actuated by the centrifugal force of whirling weights opposed by gravity or by springs, used to provide automatic control of speed or power of a prime mover.

gr *See* grain.

grabbing crane [MECH ENG] An excavator made up of a crane carrying a large grab or bucket in the form of a pair of half scoops, hinged to dig into the earth as they are lifted.

grab bucket [MECH ENG] A bucket with hinged jaws or teeth that is hung from cables on a crane or excavator and is used to dig and pick up materials.

grab dredger [MECH ENG] Dredging equipment comprising a grab or grab bucket that is suspended from the jib head of a crane. Also known as grapple dredger.

grabhook [DES ENG] A hook used for grabbing, as in lifting blocks of stone, in which case the hooks are used in pairs connected with a chain, and are so constructed that the tension of the chain causes them to adhere firmly to the rock.

Gradall [MECH ENG] A hydraulic backhoe equipped with an extensible boom that excavates, backfills, and grades.

gradeability [MECH ENG] The performance of earthmovers on various inclines, measured in percent grade.

grader [MECH ENG] A high-bodied, wheeled vehicle with a leveling blade mounted between the front and rear wheels; used for fine-grading relatively loose and level earth.

Graetz number [THERMO] A dimensionless number used in the study of streamline flow, equal to the mass flow rate of a fluid times its specific heat at constant pressure divided by the product of its thermal conductivity and a characteristic length. Also spelled Grätz number. Symbolized N_{Gz}.

Graetz problem [FL MECH] The problem of determining the steady-state temperature field in a fluid flowing in a circular tube when the wall of the tube is held at a uniform temperature and the fluid enters the tube at a different uniform temperature.

Graham's law of diffusion [FL MECH] The law that the rate of diffusion of a gas is inversely proportional to the square root of its density.

grain [MECH] A unit of mass in the United States and United Kingdom, common to the avoirdupois, apothecaries', and troy systems, equal to 1/7000 of a pound, or to 6.479891×10^{-5} kilogram. Abbreviated gr.

grain spacing [DES ENG] Relative location of abrasive grains on the surface of a grinding wheel.

gram [MECH] The unit of mass in the centimeter-gram-second system of units, equal to 0.001 kilogram. Abbreviated g; gm.

gram-calorie *See* calorie.

gram-centimeter [MECH] A unit of energy in the centimeter-gram-second gravitational system, equal to the work done by a force of magnitude 1 gram force when the point at which the force is applied is displaced 1 centimeter in the direction of the force. Abbreviated g-cm.

gram-force [MECH] A unit of force in the centimeter-gram-second gravitational system, equal to the gravitational force on a 1-gram mass at a specified location. Abbreviated gf. Also known as fors; gram-weight; pond.

gram-weight *See* gram-force.

graphical statics [MECH] A method of determining forces acting on a rigid body in equilibrium, in which forces are represented on a diagram by straight lines whose lengths are proportional to the magnitudes of the forces.

graphic panel [CONT SYS] A master control panel which indicates the status of equipment and operations in a system, and their relationships.

grapnel [DES ENG] An implement with claws used to recover a lost core, drill fittings, and junk from a borehole or for other grappling operations. Also known as grapple.

grapple dredger *See* grab dredger.

grapple hook [DES ENG] An iron hook used on the end of a rope to snag lines, to hold one ship alongside another, or as a fishing tool. Also known as grappling iron.

grappling iron *See* grapple hook.

Grashof formula [FL MECH] A formula, $m = 0.0165A_2p_1^{0.97}$, used to express the discharge m of saturated steam, where A_2 is the area of the orifice in square inches, and p_1 is reservoir pressure in pounds per square inch.

Grashof number [FL MECH] A dimensionless number used in the study of free convection of a fluid caused by a hot body, equal to the product of the fluid's coefficient of thermal expansion, the temperature difference between the hot body and the fluid, the cube of a typical dimension of the body and the square of the fluid's density, divided by the square of the fluid's dynamic viscosity. Also known as free convection number.

grasshopper linkage [MECH ENG] A straight-line mechanism used in some early steam engines.

grav *See* G.

gravel pump [MECH ENG] A centrifugal pump with renewable impellers and lining, used to pump a mixture of gravel and water.

gravitational constant [MECH] The constant of proportionality in Newton's law of gravitation, equal to the gravitational force between any two particles times the square of the distance between them, divided by the product of their masses. Also known as constant of gravitation.

gravitational displacement [MECH] The gravitational field strength times the gravitational constant. Also known as gravitational flux density.

gravitational energy *See* gravitational potential energy.

gravitational field [MECH] The field in a region in space in which a test particle would experience a gravitational force; quantitatively, the gravitational force per unit mass on the particle at a particular point.

gravitational flux density *See* gravitational displacement.

gravitational force [MECH] The force on a particle due to its gravitational attraction to other particles.

gravitational instability [MECH] Instability of a dynamic system in which gravity is the restoring force.

gravitational potential [MECH] The amount of work which must be done against gravitational forces to move a particle of unit mass to a specified position from a reference position, usually a point at infinity.

gravitational potential energy [MECH] The energy that a system of particles has by virtue of their positions, equal to the work that must be done against gravitational forces to assemble the particles from some reference configuration, such as mutually infinite separation. Also known as gravitational energy.

gravitational pressure *See* hydrostatic pressure.

gravitational systems of units [MECH] Systems in which length, force, and time are regarded as fundamental, and the unit of force is the gravitational force on a standard body at a specified location on the earth's surface.

gravity [MECH] The gravitational attraction at the surface of a planet or other celestial body.

gravity-gradient attitude control [AERO ENG] A device that regulates automatically attitude or orientation of an aircraft or spacecraft by responding to changes in gravity acting on the craft.

gravity vector [MECH] The force of gravity per unit mass at a given point. Symbolized g.

gravity wave [FL MECH] 1. A wave at a gas-liquid interface which depends primarily upon gravitational forces, surface tension and viscosity being of secondary importance. 2. A wave in a fluid medium in which restoring forces are provided primarily by buoyancy (that is, gravity) rather than by compression.

gravity wheel conveyor [MECH ENG] A downward-sloping conveyor trough with closely spaced axle-mounted wheel units on which flat-bottomed containers or objects are conveyed from point to point by gravity pull.

graybody [THERMO] An energy radiator which has a blackbody energy distribution, reduced by a constant factor, throughout the radiation spectrum or within a certain wavelength interval. Also known as nonselective radiator.

grid [DES ENG] A network of equally spaced lines forming squares, used for determining permissible locations of holes on a printed circuit board or a chassis.

Griffith's criterion [MECH] A criterion for the fracture of a brittle material under biaxial stress, based on the theory that the strength of such a material is limited by small cracks.

Griffiths' method [THERMO] A method of measuring the mechanical equivalent of heat in which the temperature rise of a known mass of water is compared with the electrical energy needed to produce this rise.

grinder [MECH ENG] Any device or machine that grinds, such as a pulverizer or a grinding wheel.

grinding [MECH ENG] 1. Reducing a material to relatively small particles. 2. Removing material from a workpiece with a grinding wheel.

grinding burn [MECH ENG] Overheating a localized area of the work in grinding operations.

grinding mill [MECH ENG] A machine consisting of a rotating cylindrical drum, that reduces the size of particles of ore or other materials fed into it; three main types are ball, rod, and tube mills.

grinding ratio [MECH ENG] Ratio of the volume of ground material removed from the workpiece to the volume removed from the grinding wheel.

grinding stress [MECH] Residual tensile or compressive stress, or a combination of both, on the surface of a material due to grinding.

grinding wheel [DES ENG] A wheel or disk having an abrasive material such as alumina or silicon carbide bonded to the surface.

grit size [DES ENG] Size of the abrasive particles on a grinding wheel.

grizzly crusher [MECH ENG] A machine with a series of parallel rods or bars for crushing rock and sorting particles by size.

grommet nut [DES ENG] A blind nut with a round head; used with a screw to attach a hinge to a door.

groove [DES ENG] A long, narrow channel in a surface.

grooved drum [DES ENG] Drum with a grooved surface to support and guide a rope.

grooving saw [MECH ENG] A circular saw for cutting grooves.

gross ton *See* ton.

ground [AERO ENG] To forbid (an aircraft or individual) to fly, usually for a relatively short time.

ground effect [AERO ENG] Increase in the lift of an aircraft operating close to the ground caused by reaction between high-velocity downwash from its wing or rotor and the ground.

ground-effect machine *See* air-cushion vehicle.

ground handling equipment *See* ground support equipment.

ground support equipment [AERO ENG] That equipment on the ground, including all implements, tools, and devices (mobile or fixed), required to inspect, test, adjust, calibrate, appraise, gage, measure, repair, overhaul, assemble, disassemble, transport, safeguard, record, store, or otherwise function in support of a rocket, space vehicle, or the like, either in the research and development phase or in an operational phase, or in support of the guidance system used with the missile, vehicle, or the like. Abbreviated GSE. Also known as ground handling equipment.

growth factor [AERO ENG] The additional weight of fuel and structural material required by the addition of 1 pound of payload to the original payload.

grub screw [DES ENG] A headless screw with a slot at one end to receive a screwdriver.

GSE *See* ground support equipment.

guard circle [DES ENG] The closed loop at the end of a grooved record.

guard ring [THERMO] A device used in heat flow experiments to ensure an even distribution of heat, consisting of a ring that surrounds the specimen and is made of a similar material.

guidance system [AERO ENG] The control devices used in guidance of an aircraft or spacecraft.

guide bearing [MECH ENG] A plain bearing used to guide a machine element in its lengthwise motion, usually without rotation of the element.

guide idler [MECH ENG] An idler roll with its supporting structure mounted on a conveyor frame to guide the belt in a defined horizontal path, usually by contact with the edge of the belt.

guides [MECH ENG] **1.** Pulleys to lead a driving belt or rope in a new direction or to keep it from leaving its desired direction. **2.** Tracks that support and determine the path of a skip bucket and skip bucket bail. **3.** Tracks guiding the chain or buckets of a bucket elevator. **4.** The runway paralleling the path of the conveyor which limits the conveyor or parts of a conveyor to movement in a defined path.

Gukhman number [THERMO] A dimensionless number used in studying convective heat transfer in evaporation, equal to $(t_0 - t_m)/T_0$, where t_0 is the temperature of a hot gas stream, t_m is the temperature of a moist surface over which it is flowing, and T_0 is the absolute temperature of the gas stream. Symbolized Gu; N_{Gu}.

gun reaction   [MECH] The force exerted on the gun mount by the rearward movement of the gun resulting from the forward motion of the projectile and hot gases. Also known as recoil.

gust-gradient distance   [AERO ENG] The horizontal distance along an aircraft flight path from the "edge" of a gust to the point at which the gust reaches its maximum speed.

gust load   [MECH] The wind load on an antenna due to gusts.

gust tunnel   [AERO ENG] A type of wind tunnel that has an enclosed space and is used to test the effect of gusts on an airplane model in free flight to determine how atmospheric gusts affect the flight of an airplane.

guy derrick   [MECH ENG] A derrick having a vertical pole supported by guy ropes to which a boom is attached by rope or cable suspension at the top and by a pivot at the foot.

guyed-mast derrick   *See* gin pole.

gyratory breaker   *See* gyratory crusher.

gyratory crusher   [MECH ENG] A primary breaking machine in the form of two cones, an outer fixed cone and a solid inner erect cone mounted on an eccentric bearing. Also known as gyratory breaker.

gyratory screen   [MECH ENG] Boxlike machine with a series of horizontal screens nested in a vertical stack with downward-decreasing mesh-opening sizes; near-circular motion causes undersized material to sift down through each screen in succession.

gyropendulum   [MECH ENG] A gravity pendulum attached to a rapidly spinning gyro wheel.

gyroplane   [AERO ENG] A rotorcraft whose rotors are not power-driven.

gyroscopic couple   [MECH ENG] The turning moment which opposes any change of the inclination of the axis of rotation of a gyroscope.

gyroscopic precession   [MECH] The turning of the axis of spin of a gyroscope as a result of an external torque acting on the gyroscope; the axis always turns toward the direction of the torque.

gyroscopics   [MECH] The branch of mechanics concerned with gyroscopes and their use in stabilization and control of ships, aircraft, projectiles, and other objects.

gyro wheel   [MECH ENG] The rapidly spinning wheel in a gyroscope, which resists being disturbed.

h See hour.

ha See hectare.

Hagen-Poiseuille law [FL MECH] In the case of laminar flow of fluid through a circular pipe, the loss of head due to fluid friction is 32 times the product of the fluid's viscosity, the pipe length, and the fluid velocity, divided by the product of the acceleration of gravity, the fluid density, and the square of the pipe diameter.

hairpin tube [DES ENG] A boiler tube bent into a hairpin, or U, shape.

half-dog setscrew [DES ENG] A setscrew with a short, blunt point.

half nut [DES ENG] A nut split lengthwise so that it can be clamped around a screw.

half-round file [DES ENG] A file that is flat on one side and convex on the other.

half-track [MECH ENG] **1.** A chain-track drive system for a vehicle; consists of an endless metal belt on each side of the vehicle driven by one of two inside sprockets and running on bogie wheels; the revolving belt lays down on the ground a flexible track of cleated steel or hard-rubber plates; the front end of the vehicle is supported by a pair of wheels. **2.** A motor vehicle equipped with half-tracks.

Hamiltonian function [MECH] A function of the generalized coordinates and momenta of a system, equal in value to the sum over the coordinates of the product of the generalized momentum corresponding to the coordinate, and the coordinate's time derivative, minus the Lagrangian of the system; it is numerically equal to the total energy if the Lagrangian does not depend on time explicitly; the equations of motion of the system are determined by the functional dependence of the Hamiltonian on the generalized coordinates and momenta.

Hamilton-Jacobi theory [MECH] A theory that provides a means for discussing the motion of a dynamic system in terms of a single partial differential equation of the first order, the Hamilton-Jacobi equation.

Hamilton's equations of motion [MECH] A set of first-order, highly symmetrical equations describing the motion of a classical dynamical system, namely $\dot{q}_j = \partial H/\partial \dot{p}_j$, $p_j = -\partial H/\partial q_j$; here q_j ($j = 1, 2,...$) are generalized coordinates of the system, p_j is the momentum conjugate to q_j, and H is the Hamiltonian. Also known as canonical equations of motion.

Hamilton's principle [MECH] A variational principle which states that the path of a conservative system in configuration space between two configurations is such that the integral of the Lagrangian function over time is a minimum or maximum relative to nearby paths between the same end points and taking the same time.

hammer [DES ENG] **1.** A hand tool used for pounding and consisting of a solid metal head set crosswise on the end of a handle. **2.** An arm with a striking head for sounding a bell or gong. [MECH ENG] A power tool with a metal block or a drill for the head.

hammer drill [MECH ENG] Any of three types of fast-cutting, compressed-air rock drills (drifter, sinker, and stoper) in which a hammer strikes rapid blows on a loosely held piston, and the bit remains against the rock in the bottom of the hole, rebounding slightly at each blow, but does not reciprocate.

hammerhead [DES ENG] The striking part of a hammer.

hammerhead crane [MECH ENG] A crane with a horizontal jib that is counterbalanced.

hammer mill [MECH ENG] **1.** A type of impact mill or crusher in which materials are reduced in size by hammers revolving rapidly in a vertical plane within a steel casing. Also known as beater mill. **2.** A grinding machine which pulverizes feed and other products by several rows of thin hammers revolving at high speed.

hammer milling [MECH ENG] Crushing or fracturing materials in a hammer mill.

hand auger [DES ENG] A hand tool resembling a large carpenters' bit or comprising a short cylindrical container with cutting lips attached to a rod; used to bore shallow holes in the soil to obtain samples of it and other relatively unconsolidated near-surface materials.

hand brake [MECH ENG] A manually operated brake.

handcar [MECH ENG] A small, four-wheeled, hand-pumped car used on railroad tracks to transport men and equipment for construction or repair work; other cars for the same purpose are motor-operated.

hand drill [DES ENG] A small, portable drilling machine which is operated by hand.

handle [MECH ENG] The arm connecting the bucket with the boom in a dipper shovel or hoe.

hand punch [DES ENG] A hand-held device for punching holes in paper or cards.

handsaw [DES ENG] A saw operated by hand, with a backward and forward arm movement.

handset [DES ENG] A combination of a telephone-type receiver and transmitter, designed for holding in one hand.

handset bit [DES ENG] A bit in which the diamonds are manually set into holes that are drilled into a malleable-steel bit blank and shaped to fit the diamonds.

hand winch [MECH ENG] A winch that is operated by hand.

hanger bolt [DES ENG] A bolt with a machine-screw thread on one end and a lag-screw thread on the other.

hanging-drop atomizer [MECH ENG] An atomizing device used in gravitational atomization; functions by quasi-static emission of a drop from a wetted surface. Also known as pendant atomizer.

hanging load [MECH ENG] **1.** The weight that can be suspended on a hoist line or hook device in a drill tripod or derrick without causing the members of the derrick or tripod to buckle. **2.** The weight suspended or supported by a bearing.

Hardinge feeder-weigher [MECH ENG] A pivoted, short belt conveyor which controls the rate of material flow from a hopper by weight per cubic foot.

Hardinge mill [MECH ENG] A tricone type of ball mill; the cones become steeper from the feed end toward the discharge end.

hard-laid [DES ENG] Pertaining to rope with strands twisted at a 45° angle.

hard landing [AERO ENG] A landing made without deceleration, as by impact on the moon.

hardwood bearing [MECH ENG] A fluid-film bearing made of lignum vitae which has a natural gum, or of hard maple which is impregnated with oil, grease, or wax.

harmonic motion [MECH] A periodic motion that is a sinusoidal function of time, that is, motion along a line given by the equation $x = a \cos(kt + \theta)$, where t is the time parameter, and a, k, and θ are constants. Also known as harmonic vibration; simple harmonic motion (SHM).

harmonic oscillator [MECH] Any physical system that is bound to a position of stable equilibrium by a restoring force or torque proportional to the linear or angular displacement from this position.

harmonic speed changer [MECH ENG] A mechanical-drive system used to transmit rotary, linear, or angular motion at high ratios and with positive motion.

harmonic synthesizer [MECH] A machine which combines elementary harmonic constituents into a single periodic function; a tide-predicting machine is an example.

harmonic vibration *See* harmonic motion.

harness [AERO ENG] **1.** Straps arranged to hold an occupant of a spacecraft or aircraft in the seat. **2.** Straps worn by a parachutist or used to suspend a load from a parachute.

harpoon [DES ENG] A barbed spear used to catch whales.

Hartford loop [MECH ENG] A condensate return arrangement for low-pressure, steam-heating systems featuring a steady water line in the boiler.

hatchet [DES ENG] A small ax with a short handle and a hammerhead in addition to the cutting edge.

Hayward grab bucket [MECH ENG] A clamshell type of grab bucket used for handling coal, sand, gravel, and other flowable materials.

Hayward orange peel [MECH ENG] A grab bucket that operates like the clamshell type but has four blades pivoted to close.

H bit [DES ENG] A core bit manufactured and used in Canada having inside and outside diameters of 2.875 and 3.875 inches (73.025 and 98.425 millimeters), respectively; the matching reaming shell has an outside diameter of 3.906 inches (99.2124 millimeters).

head *See* pressure head.

header [MECH ENG] A machine used for gathering or upsetting materials; used for screw, rivet, and bolt heads.

head loss [FL MECH] The drop in the sum of pressure head, velocity head, and potential head between two points along the path of a flowing fluid, due to causes such as fluid friction.

head motion [MECH ENG] The vibrator on a reciprocating table concentrator which imparts motion to the deck.

head pulley [MECH ENG] The pulley at the discharge end of a conveyor belt; may be either an idler or a drive pulley.

head-pulley-drive conveyor [MECH ENG] A conveyor having the belt driven by the head pulley without a snub pulley.

head shaft [MECH ENG] The shaft driven by a chain and mounted at the delivery end of a chain conveyor; it serves as the mount for a sprocket which drives the drag chain.

headstock [MECH ENG] **1.** The device on a lathe for carrying the revolving spindle. **2.** The movable head of certain measuring machines. **3.** The device on a cylindrical grinding machine for rotating the work.

heat [THERMO] Energy in transit due to a temperature difference between the source from which the energy is coming and a sink toward which the energy is going; other types of energy in transit are called work.

heat balance [THERMO] The equilibrium which is known to exist when all sources of heat gain and loss for a given region or body are accounted for.

heat barrier *See* thermal barrier.

heat budget [THERMO] The statement of the total inflow and outflow of heat for a planet, spacecraft, biological organism, or other entity.

heat capacity [THERMO] The quantity of heat required to raise a system one degree in temperature in a specified way, usually at constant pressure or constant volume. Also known as thermal capacity.

heat conduction [THERMO] The flow of thermal energy through a substance from a higher- to a lower-temperature region.

heat conductivity *See* thermal conductivity.

heat content *See* enthalpy.

heat convection [THERMO] The transfer of thermal energy by actual physical movement from one location to another of a substance in which thermal energy is stored. Also known as thermal convection.

heat cycle *See* thermodynamic cycle.

heat death [THERMO] The condition of any isolated system when its entropy reaches a maximum, in which matter is totally disordered and at a uniform temperature, and no energy is available for doing work.

heat dump *See* heatsink.

heat engine [MECH ENG] A machine that converts heat into work (mechanical energy). [THERMO] A thermodynamic system which undergoes a cyclic process during which a positive amount of work is done by the system; some heat flows into the system and a smaller amount flows out in each cycle.

heat equation [THERMO] A parabolic second-order differential equation for the temperature of a substance in a region where no heat source exists: $\partial t/\partial \tau = (k/\rho c)(\partial^2 t/\partial x^2 + \partial^2 t/\partial y^2 + \partial t^2/\partial z^2)$, where x, y, and z are space coordinates, τ is the time, $t(x,y,z,\tau)$ is the temperature, k is the thermal conductivity of the body, ρ is its density, and c is its specific heat; this equation is fundamental to the study of heat flow in bodies. Also known as Fourier heat equation; heat flow equation.

heat flow [THERMO] Heat thought of as energy flowing from one substance to another; quantitatively, the amount of heat transferred in a unit time. Also known as heat transmission.

heat flow equation *See* heat equation.

heat flux [THERMO] The amount of heat transferred across a surface of unit area in a unit time. Also known as thermal flux.

heat of ablation [THERMO] A measure of the effective heat capacity of an ablating material, numerically the heating rate input divided by the mass loss rate which results from ablation.

heat of adsorption [THERMO] The increase in enthalpy when 1 mole of a substance is adsorbed upon another at constant pressure.

heat of aggregation [THERMO] The increase in enthalpy when an aggregate of matter, such as a crystal, is formed at constant pressure.

heat of compression [THERMO] Heat generated when air is compressed.

heat of condensation [THERMO] The increase in enthalpy accompanying the conversion of 1 mole of vapor into liquid at constant pressure and temperature.

heat of cooling [THERMO] Increase in enthalpy during cooling of a system at constant pressure, resulting from an internal change such as an allotropic transformation.

heat of crystallization [THERMO] The increase in enthalpy when 1 mole of a substance is transformed into its crystalline state at constant pressure.

heat of evaporation *See* heat of vaporization.

heat of fusion [THERMO] The increase in enthalpy accompanying the conversion of 1 mole, or a unit mass, of a solid to a liquid at its melting point at constant pressure and temperature. Also known as latent heat of fusion.

heat of mixing [THERMO] The difference between the enthalpy of a mixture and the sum of the enthalpies of its components at the same pressure and temperature.

heat of solidification [THERMO] The increase in enthalpy when 1 mole of a solid is formed from a liquid or, less commonly, a gas at constant pressure and temperature.

heat of sublimation [THERMO] The increase in enthalpy accompanying the conversion of 1 mole, or unit mass, of a solid to a vapor at constant pressure and temperature. Also known as latent heat of sublimation.

heat of transformation [THERMO] The increase in enthalpy of a substance when it undergoes some phase change at constant pressure and temperature.

heat of vaporization [THERMO] The quantity of energy required to evaporate 1 mole, or a unit mass, of a liquid, at constant pressure and temperature. Also known as enthalpy of vaporization; heat of evaporation; latent heat of vaporization.

heat of wetting [THERMO] 1. The heat of adsorption of water on a substance. 2. The additional heat required, above the heat of vaporization of free water, to evaporate water from a substance in which it has been absorbed.

heat pump [MECH ENG] A device which transfers heat from a cooler reservoir to a hotter one, expending mechanical energy in the process, especially when the main purpose is to heat the hot reservoir rather than refrigerate the cold one.

heat quantity [THERMO] A measured amount of heat; units are the small calorie, normal calorie, mean calorie, and large calorie.

heat radiation [THERMO] The energy radiated by solids, liquids, and gases in the form of electromagnetic waves as a result of their temperature. Also known as thermal radiation.

heat rate [MECH ENG] An expression of the conversion efficiency of a thermal power plant or engine, as heat input per unit of work output; for example, Btu/kWh.

heat release [THERMO] The quantity of heat released by a furnace or other heating mechanism per second, divided by its volume.

heat resistance *See* thermal resistance.

heatsink [AERO ENG] 1. A type of protective device capable of absorbing heat and used as a heat shield. 2. In nuclear propulsion, any thermodynamic device, such as a radiator or condenser, that is designed to absorb the excess heat energy of the working fluid. Also known as heat dump. [THERMO] Any (gas, solid, or liquid) region where heat is absorbed.

heat source [THERMO] Any device or natural body that supplies heat.

heat transfer [THERMO] The movement of heat from one body to another (gas, liquid, solid, or combinations thereof) by means of radiation, convection, or conduction.

heat-transfer coefficient [THERMO] The amount of heat which passes through a unit area of a medium or system in a unit time when the temperature difference between the boundaries of the system is 1 degree.

heat transmission *See* heat flow.

heat transport [THERMO] Process by which heat is carried past a fixed point or across a fixed plane, as in a warm current.

heavier-than-air craft [AERO ENG] Any aircraft weighing more than the air it displaces.

heavy bomber [AERO ENG] Any large bomber considered to be relatively heavy, such as a bomber having a gross weight, including bomb load, of 250,000 pounds (113,000 kilograms) or more, as the B-36 and the B-52.

heavy-duty car [MECH ENG] A railway motorcar weighing more than 1400 pounds (635 kilograms), propelled by an engine of 12–30 horsepower (8900–22,400 watts), and designed for hauling heavy equipment and for hump-yard service.

heavy-duty tool block *See* open-side tool block.

heavy force fit [DES ENG] A fit for heavy steel parts or shrink fits in medium sections.

heavy section car [MECH ENG] A railway motorcar weighing 1200–1400 pounds (544–635 kilograms) and propelled by an 8–12 horsepower (6000–8900 watts) engine.

hectare [MECH] A unit of area in the metric system equal to 100 ares or 10,000 square meters. Abbreviated ha.

hectogram [MECH] A unit of mass equal to 100 grams. Abbreviated hg.

hectoliter [MECH] A metric unit of volume equal to 100 liters or to 0.1 cubic meter. Abbreviated hl.

hectometer [MECH] A unit of length equal to 100 meters. Abbreviated hm.

heel *See* heel block.

heel block [MECH ENG] A block or plate that is usually fixed on the die shoe to minimize deflection of a punch or cam. Also known as heel.

helical angle [MECH] In the study of torsion, the angular displacement of a longitudinal element, originally straight on the surface of an untwisted bar, which becomes helical after twisting.

helical conveyor [MECH ENG] A conveyor for the transport of bulk materials which consists of a horizontal shaft with helical paddles or ribbons rotating inside a stationary tube.

helical-flow turbine [MECH ENG] A steam turbine in which the steam is directed tangentially and radially inward by nozzles against buckets milled in the wheel rim; the steam flows in a helical path, reentering the buckets one or more times. Also known as tangential helical-flow turbine.

helical gear [MECH ENG] Gear wheels running on parallel axes, with teeth twisted oblique to the gear axis.

helical milling [MECH ENG] Milling in which the work is simultaneously rotated and translated.

helical rake angle [DES ENG] The angle between the axis of a reamer and a plane tangent to its helical cutting edge; also applied to milling cutters.

helical spline broach [MECH ENG] A broach used to produce internal helical splines having a straight-sided or involute form.

helical spring [DES ENG] A bar or wire of uniform cross section wound into a helix.

helicopter [AERO ENG] An aircraft fitted to sustain itself by motor-driven horizontal rotating blades (rotors) that accelerate the air downward, providing a reactive lift force, or accelerate the air at an angle to the vertical, providing lift and thrust.

helium refrigerator [MECH ENG] A refrigerator which uses liquid helium to cool substances to temperatures of 4 K or less.

helix angle [DES ENG] That angle formed by the helix of the thread at the pitch-diameter line and a line at right angles to the axis.

Helmholtz flow [FL MECH] Flow with free streamlines or vortex sheets.

Helmholtz free energy *See* free energy.

Helmholtz function *See* free energy.

Helmholtz instability [FL MECH] The hydrodynamic instability arising from a shear, or discontinuity, in current speed at the interface between two fluids in two-dimensional motion; the perturbation gains kinetic energy at the expense of that of the basic currents. Also known as shearing instability.

Helmholtz potential *See* free energy.

Helmholtz's theorem [FL MECH] The theorem that in the isentropic flow of a nonviscous fluid which is not subject to body forces, individual vortices always consist of the same fluid particles.

Helmholtz wave [FL MECH] An unstable wave in a system of two homogeneous fluids with a velocity discontinuity at the interface.

hemming [MECH ENG] Forming of an edge by bending the metal back on itself.

hemp-core cable *See* standard wire rope.

hereditary mechanics [MECH] A field of mechanics in which quantities, such as stress, depend not only on other quantities, such as strain, at the same instant but also on integrals involving the values of such quantities at previous times.

hermaphrodite caliper [DES ENG] A layout tool having one leg pointed and the other like that of an inside caliper; used to locate the center of irregularly shaped stock or to lay out a line parallel to an edge.

herpolhode [MECH] The curve traced out on the invariable plane by the point of contact between the plane and the inertia ellipsoid of a rotating rigid body not subject to external torque.

herpolhode cone *See* space cone.

herringbone gear [MECH ENG] The equivalent of two helical gears of opposite hand placed side by side.

Hertz's law [MECH] A law which gives the radius of contact between a sphere of elastic material and a surface in terms of the sphere's radius, the normal force exerted on the sphere, and Young's modulus for the material of the sphere.

heterogeneous fluid [FL MECH] A fluid within which the density varies from point to point; for most purposes the atmosphere must be treated as heterogeneous, particularly with regard to the decrease of density with height.

hexagonal-head bolt [DES ENG] A standard wrench head bolt with a hexagonal head.

hexagonal nut [DES ENG] A plain nut in hexagon form. Also known as hex nut.

hex nut *See* hexagonal nut.

hg *See* hectogram.

higher pair [MECH ENG] A link in a mechanism in which the mating parts have surface (instead of line or point) contact.

high-front shovel [MECH ENG] A power shovel with a dipper stick mounted high on the boom for stripping and overburden removal.

high heat [THERMO] Heat absorbed by the cooling medium in a calorimeter when products of combustion are cooled to the initial atmospheric (ambient) temperature.

high-helix drill [DES ENG] A two-flute twist drill with a helix angle of 35–40°; used for drilling deep holes in metals, such as aluminum, copper, hard brass, and soft steel. Also known as fast-spiral drill.

high-intensity atomizer [MECH ENG] A type of atomizer used in electrostatic atomization, based on stress sufficient to overcome tensile strength of the liquid.

high-lift truck [MECH ENG] A forklift truck with a fixed or telescoping mast to permit high elevation of a load.

high-speed machine [MECH ENG] A diamond drill capable of rotating a drill string at a minimum of 2500 revolutions per minute, as contrasted with the normal maximum speed of 1600–1800 revolutions per minute attained by the average diamond drill.

high-temperature water boiler [MECH ENG] A boiler which provides hot water, under pressure, for space heating of large areas.

Hildebrand function [THERMO] The heat of vaporization of a compound as a function of the molal concentration of the vapor; it is nearly the same for many compounds.

hill-climbing [MECH ENG] Adjustment, either continuous or periodic, of a self-regulating system to achieve optimum performance.

Hindley screw [DES ENG] An endless screw or worm of hourglass shape that fits a part of the circumference of a worm wheel so as to increase the bearing area and thus diminish wear. Also known as hourglass screw; hourglass worm.

hinge [DES ENG] A pair of metal leaves forming a jointed device on which a swinging part turns.

hinge moment [AERO ENG] The tendency of an aerodynamic force to produce motion about the hinge line of a control surface.

hl *See* hectoliter.

hm *See* hectometer.

hob [DES ENG] A master model made from hardened steel which is used to press the shape of a plastics mold into a block of soft steel. [MECH ENG] A rotary cutting tool with its teeth arranged along a helical thread; used for generating gear teeth.

hobber *See* hobbing machine.

hobbing [DES ENG] In plastics manufacturing, the act of creating multiple mold cavities by pressing a hob into soft metal cavity blanks. [MECH ENG] Cutting evenly spaced forms, such as gear teeth, on the periphery of cylindrical workpieces.

hobbing machine [MECH ENG] A machine for cutting gear teeth in gear blanks or for cutting worm, spur, or helical gears. Also known as hobber.

hobnail [DES ENG] A short, large-headed, sharp-pointed nail; used to attach soles to heavy shoes.

hodograph method [FL MECH] A method for studying two-dimensional steady fluid flow in which the independent variables are taken as the components of the velocity with respect to cartesian or polar coordinates, rather than the coordinates themselves.

hoe [DES ENG] An implement consisting of a long handle with a thin, flat, straight-edged blade attached transversely to the end; used for cultivating and weeding.

hoe shovel [MECH ENG] A revolving shovel with a pull-type bucket rigidly attached to a stick hinged on the end of a live boom.

hohlraum *See* blackbody.

Hohmann orbit [AERO ENG] A minimum-energy-transfer orbit.

Hohmann trajectory [AERO ENG] The minimum-energy trajectory between two planetary orbits, utilizing only two propulsive impulses.

hoist [MECH ENG] **1.** To move or lift something by a rope-and-pulley device. **2.** A power unit for a hoisting machine, designed to lift from a position directly above the load and therefore mounted to facilitate mobile service.

hoist back-out switch [MECH ENG] A protective switch that permits hoist operation only in the reverse direction in case of overwind.

hoist cable [MECH ENG] A fiber rope, wire rope, or chain by means of which force is exerted on the sheaves and pulleys of a hoisting machine.

hoist hook [DES ENG] A swivel hook attached to the end of a hoist cable for securing a load.

hoisting [MECH ENG] **1.** Raising a load, especially by means of tackle. **2.** Either of two power-shovel operations: the raising or lowering of the boom, or the lifting or dropping of the dipper stick in relation to the boom.

hoisting drum *See* drum.

hoisting machine [MECH ENG] A mechanism for raising and lowering material with intermittent motion while holding the material freely suspended.

hoisting power [MECH ENG] The capacity of the hoisting mechanism on a hoisting machine.

hoist overspeed device [MECH ENG] A device used to prevent a hoist from operating at speeds greater than predetermined values by activating an emergency brake when the predetermined speed is exceeded.

hoist overwind device [MECH ENG] A device which can activate an emergency brake when a hoisted load travels beyond a predetermined point into a danger zone.

hoist slack-brake switch [MECH ENG] A device that automatically cuts off power to the hoist motor and sets the brake if the links in the brake rigging require tightening or if the brakes require relining.

holdback [MECH ENG] A brake on an inclined-belt conveyor system which is automatically activated in the event of power failure, thus preventing the loaded belt from running downward.

hole saw *See* crown saw.

hollander [MECH ENG] An elongate tube with a central midfeather and a cylindrical beater roll; formerly used for stock preparation in paper manufacture.

hollow drill [DES ENG] A drill rod or stem having an axial hole for the passage of water or compressed air to remove cuttings from a drill hole. Also known as hollow rod; hollow stem.

hollow mill [MECH ENG] A milling cutter with three or more cutting edges that revolve around the cylindrical workpiece.

hollow rod *See* hollow drill.

hollow-rod churn drill [MECH ENG] A churn drill with hollow rods instead of steel wire rope.

hollow shafting [MECH ENG] Shafting made from hollowed-out rods or hollow tubing to minimize weight, allow internal support, or permit other shafting to operate through the interior.

hollow stem *See* hollow drill.

holonomic constraints [MECH] An integrable set of differential equations which describe the restrictions on the motion of a system; a function relating several variables, in the form $f(x_1, \ldots, x_n) = 0$, in optimization or physical problems.

holonomic system [MECH] A system in which the constraints are such that the original coordinates can be expressed in terms of independent coordinates and possibly also the time.

homogenizer [MECH ENG] A machine that blends or emulsifies a substance by forcing it through fine openings against a hard surface.

hone [MECH ENG] A machine for honing that consists of a holding device containing several oblong stones arranged in a circular pattern.

honeycomb radiator [MECH ENG] A heat-exchange device utilizing many small cells, shaped like a bees' comb, for cooling circulating water in an automobile.

honing [MECH ENG] The process of removing a relatively small amount of material from a cylindrical surface by means of abrasive stones to obtain a desired finish or extremely close dimensional tolerance.

hood [DES ENG] An opaque shield placed above or around the screen of a cathode-ray tube to eliminate extraneous light.

hook [DES ENG] A piece of hard material, especially metal, formed into a curve for catching, holding, or pulling something.

hook-and-eye hinge [DES ENG] A hinge consisting of a hook (usually attached to a gate post) over which an eye (usually attached to the gate) is placed.

hook bolt [DES ENG] A bolt with a hook or L band at one end and threads at the other to fit a nut.

Hookean deformation [MECH] Deformation of a substance which is proportional to the force applied to it.

Hookean solid [MECH] An ideal solid which obeys Hooke's law exactly for all values of stress, however large.

Hooke number *See* Cauchy number.

Hooke's joint [MECH ENG] A simple universal joint; consists of two yokes attached to their respective shafts and connected by means of a spider. Also known as Cardan joint.

Hooke's law [MECH] The law that the stress of a solid is directly proportional to the strain applied to it.

hook wrench [DES ENG] A wrench with a hook for turning a nut or bolt.

horizontal auger [MECH ENG] A rotary drill, usually powered by a gasoline engine, for making horizontal blasting holes in quarries and opencast pits.

horizontal boiler [MECH ENG] A water-tube boiler having a main bank of straight tubes inclined toward the rear at an angle of 5 to 15° from the horizontal.

horizontal boring machine [MECH ENG] A boring machine adapted for work not conveniently revolved, for milling, slotting, drilling, tapping, boring, and reaming long holes and for making interchangeable parts that must be produced without jigs and fixtures.

horizontal broaching machine [MECH ENG] A pull-type broaching machine having the broach mounted on the horizontal plane.

horizontal crusher [MECH ENG] Rotary size reducer in which the crushing cone is supported on a horizontal shaft; needs less headroom than vertical models.

horizontal drilling machine [MECH ENG] A drilling machine in which the drill bits extend in a horizontal direction.

horizontal engine [MECH ENG] An engine with horizontal stroke.

horizontal firing [MECH ENG] The firing of fuel in a boiler furnace in which the burners discharge fuel and air into the furnace horizontally.

horizontal lathe [MECH ENG] A horizontally mounted lathe with which longitudinal and radial movements are applied to a workpiece that rotates.

horizontal milling machine [MECH ENG] A knee-type milling machine with a horizontal spindle and a swiveling table for cutting helices.

horizontal pendulum [MECH] A pendulum that moves in a horizontal plane, such as a compass needle turning on its pivot.

horizontal return tubular boiler [MECH ENG] A fire-tube boiler having tubes within a cylindrical shell that are attached to the end closures; products of combustion are transported under the lower half of the shell and back through the tubes.

horizontal screen [MECH ENG] Shaking screen with horizontal plates.

horizontal-tube evaporator [MECH ENG] A horizontally mounted tube-and-shell type of liquid evaporator, used most often for preparation of boiler feedwater.

horn socket [DES ENG] A cone-shaped fishing tool especially designed to recover lost collared drill rods, drill pipe, or tools in bored wells.

horsepower [MECH] The unit of power in the British engineering system, equal to 550 foot-pounds per second, approximately 745.7 watts. Abbreviated hp.

hose [DES ENG] Flexible tube used for conveying fluids.

hose clamp [DES ENG] Band or brace to attach the raw end of a hose to a water outlet.

hose coupling [DES ENG] Device to interconnect two or more pieces of hose.

hose fitting [DES ENG] Any attachment or accessory item for a hose.

hot-air engine [MECH ENG] A heat engine in which air or other gases, such as hydrogen, helium, or nitrogen, are used as the working fluid, operating on cycles such as the Stirling or Ericsson.

hot-air furnace [MECH ENG] An encased heating unit providing warm air to ducts for circulation by gravity convection or by fans.

hot-bulb [MECH ENG] Pertaining to an ignition method used in semidiesel engines in which the fuel mixture is ignited in a separate chamber kept above the ignition temperature by the heat of compression.

Hotchkiss drive [MECH ENG] An automobile rear suspension designed to take torque reactions through longitudinal leaf springs.

hot saw [MECH ENG] A power saw used to cut hot metal.

hotshot wind tunnel [AERO ENG] A wind tunnel in which electrical energy is discharged into a pressurized arc chamber, increasing the temperature and pressure in the arc chamber so that a diaphragm separating the arc chamber from an evacuated chamber is ruptured, and the heated gas from the arc chamber is then accelerated in a conical nozzle to provide flows with mach numbers of 10 to 27 for durations of 10 to 100 milliseconds.

hot strength *See* tensile strength.

hot-water heating [MECH ENG] A heating system for a building in which the heat-conveying medium is hot water and the heat-emitting means are radiators, convectors, or panel coils. Also known as hydronic heating.

hot well [MECH ENG] A chamber for collecting condensate, as in a steam condenser serving an engine or turbine.

hour [MECH] A unit of time equal to 3600 seconds. Abbreviated h; hr.

hourglass screw *See* Hindley screw.

hourglass worm *See* Hindley screw.

hour-out line *See* time front.

hovercraft *See* air-cushion vehicle.

hp *See* horsepower.

hr   *See* hour.

H rod   [DES ENG] A drill rod having an outside diameter of 3½ inches (8.89 centimeters).

hub   [DES ENG] 1. The cylindrical central part of a wheel, propeller, or fan. 2. A piece in a lock that is turned by the knob spindle, causing the bolt to move. 3. A short coupling that joins plumbing pipes.

hubcap   [DES ENG] A metal cap fastened or clamped to the end of an axle, as on motor vehicles.

humidifier   [MECH ENG] An apparatus for supplying moisture to the air and for maintaining desired humidity conditions.

Humpage gears   [MECH ENG] A train of bevel gears used for speed reduction.

Humphrey gas pump   [MECH ENG] A combined internal combustion engine and pump in which the metal piston has been replaced by a column of water.

Humphries equation   [THERMO] An equation which gives the ratio of specific heats at constant pressure and constant volume in moist air as a function of water vapor pressure.

hunt   [AERO ENG] 1. Of an aircraft or rocket, to weave about its flight path, as if seeking a new direction or another angle of attack; specifically, to yaw back and forth. 2. Of a control surface, to rotate up and down or back and forth without being detected by the pilot.

hunting   [CONT SYS] Undesirable oscillation of an automatic control system, wherein the controlled variable swings on both sides of the desired value.   [MECH ENG] Irregular engine speed resulting from instability of the governing device.

hunting tooth   [DES ENG] An extra tooth on the larger of two gear wheels so that the total number of teeth will not be an integral multiple of the number on the smaller wheel.

Huttig equation   [THERMO] An equation which states that the ratio of the volume of gas adsorbed on the surface of a nonporous solid at a given pressure and temperature to the volume of gas required to cover the surface completely with a unimolecular layer equals $(1 + r) c^r/(1 + c^r)$, where r is the ratio of the equilibrium gas pressure to the saturated vapor pressure of the adsorbate at the temperature of adsorption, and c is the product of a constant and the exponential of $(q - q_l)/RT$, where q is the heat of adsorption into a first layer molecule, q_l is the heat of liquefaction of the adsorbate, T is the temperature, and R is the gas constant.

hybrid propulsion   [AERO ENG] Propulsion utilizing energy released by a liquid propellant with a solid propellant in the same rocket engine.

hybrid rocket   [AERO ENG] A rocket with an engine utilizing a liquid propellant with a solid propellant in the same rocket engine.

hydraucone   [DES ENG] A conical, spreading type of draft tube used on hydraulic turbine installations.

hydraulic accumulator   [MECH ENG] A hydraulic flywheel that stores potential energy by accumulating a quantity of pressurized hydraulic fluid in a suitable enclosed vessel.

hydraulic actuator   [MECH ENG] A cylinder or fluid motor that converts hydraulic power into useful mechanical work; mechanical motion produced may be linear, rotary, or oscillatory.

hydraulic air compressor   [MECH ENG] A device in which water falling down a pipe entrains air which is released at the bottom under compression to do useful work.

hydraulic amplifier [CONT SYS] A device which increases the power of a signal in a hydraulic servomechanism or other system through the use of fixed and variable orifices. Also known as hydraulic intensifier.

hydraulic analog table [FL MECH] An experimental facility based on the hydraulic analogy; the water flows over a smooth horizontal surface and is bounded by vertical walls geometrically similar to the boundaries of the corresponding compressible gas flow; flow patterns are easily observed, and boundary changes may be made rapidly and inexpensively during exploratory studies.

hydraulic analogy [FL MECH] The analogy between the flow of a shallow liquid and the flow of a compressible gas; various phenomena such as shock waves occur in both systems; the analogy requires neglect of vertical accelerations in the liquid, and restrictions on the ratio of specific heats for the gas.

hydraulic backhoe [MECH ENG] A backhoe operated by a hydraulic mechanism.

hydraulic brake [MECH ENG] A brake in which the retarding force is applied through the action of a hydraulic press.

hydraulic circuit [MECH ENG] A circuit whose operation is analogous to that of an electric circuit except that electric currents are replaced by currents of water or other fluids, as in a hydraulic control.

hydraulic classifier [MECH ENG] A classifier in which particles are sorted by specific gravity in a stream of hydraulic water that rises at a controlled rate; heavier particles gravitate down and are discharged at the bottom, while lighter ones are carried up and out. Also known as hydrosizer.

hydraulic clutch *See* fluid drive.

hydraulic conductivity *See* permeability coefficient.

hydraulic conveyor [MECH ENG] A system for handling material, such as ash from a coal-fired furnace; refuse is flushed from a hopper or slag tank to a grinder which discharges to a pump for conveying to a disposal area or a dewatering bin.

hydraulic coupling *See* fluid coupling.

hydraulic cylinder [MECH ENG] The cylindrical chamber of a positive displacement pump.

hydraulic dredge [MECH ENG] A dredge consisting of a large suction pipe which is mounted on a hull and supported and moved about by a boom, a mechanical agitator or cutter head which churns up earth in front of the pipe, and centrifugal pumps mounted on a dredge which suck up water and loose solids.

hydraulic drill [MECH ENG] A rotary drill powered by hydrodynamic means and used to make shot-firing holes in coal or rock, or to make a well hole.

hydraulic drive [MECH ENG] A mechanism transmitting motion from one shaft to another, the velocity ratio of the shafts being controlled by hydrostatic or hydrodynamic means.

hydraulic elevator [MECH ENG] An elevator operated by water pressure. Also known as hydraulic lift.

hydraulic excavator digger [MECH ENG] An excavation machine which employs hydraulic pistons to actuate mechanical digging elements.

hydraulic friction [FL MECH] Resistance to flow which is exerted on the surface of contact between a stream and its conduit and which induces a loss of energy.

hydraulic grade line [FL MECH] 1. In a closed channel, a line joining the elevations that water would reach under atmospheric pressure. 2. The free water surface in an open channel.

hydraulic gradient [FL MECH] With regard to an aquifer, the rate of change of pressure head per unit of distance of flow at a given point and in a given direction.

hydraulic intensifier *See* hydraulic amplifier.

hydraulic jack [MECH ENG] A jack in which force is applied through the mechanism of a hydraulic press.

hydraulic jump [FL MECH] A steady-state, finite-amplitude disturbance in a channel, in which water passes turbulently from a region of (uniform) low depth and high velocity to a region of (uniform) high depth and low velocity; when applied to hydraulic jumps, the usual hydraulic formulas governing the relations of velocity and depth do not conserve energy.

hydraulic lift *See* hydraulic elevator.

hydraulic loss [FL MECH] The loss in fluid power due to flow friction within the system.

hydraulic machine [MECH ENG] A machine powered by a motor activated by the confined flow of a stream of liquid, such as oil or water under pressure.

hydraulic motor [MECH ENG] A motor activated by water or other liquid under pressure.

hydraulic nozzle [MECH ENG] An atomizing device in which fluid pressure is converted into fluid velocity.

hydraulic power system [MECH ENG] A power transmission system comprising machinery and auxiliary components which function to generate, transmit, control, and utilize hydraulic energy.

hydraulic press [MECH ENG] A combination of a large and a small cylinder connected by a pipe and filled with a fluid so that the fluid pressure created by a small force acting on the small-cylinder piston will result in a large force on the large piston. Also known as hydrostatic press.

hydraulic pump *See* hydraulic ram.

hydraulic radius [FL MECH] The ratio of the cross-sectional area of a conduit in which a fluid is flowing to the inner perimeter of the conduit.

hydraulic ram [MECH ENG] A device for forcing running water to a higher level by using the kinetic energy of flow; the flow of water in the supply pipeline is periodically stopped so that a small portion of water is lifted by the velocity head of a larger portion. Also known as hydraulic pump.

hydraulic rope-geared elevator [MECH ENG] An elevator hoisted by a system of ropes and sheaves attached to a piston in a hydraulic cylinder.

hydraulics [FL MECH] The branch of science and technology concerned with the mechanics of fluids, especially liquids.

hydraulic scale [MECH ENG] An industrial scale in which the load applied to the load-cell piston is converted to hydraulic pressure.

hydraulic separation [MECH ENG] Mechanical classification using a hydraulic classifier.

hydraulic shovel [MECH ENG] A revolving shovel in which hydraulic rams or motors are substituted for drums and cables.

hydraulic sprayer [MECH ENG] A machine that sprays large quantities of insecticide or fungicide on crops.

hydraulic stacker [MECH ENG] A tiering machine whose carriage is raised or lowered by a hydraulic cylinder.

hydraulic swivel head [MECH ENG] In a drill machine, a swivel head equipped with hydraulically actuated cylinders and pistons to exert pressure on and move the drill rod string longitudinally.

hydraulic turbine [MECH ENG] A machine which converts the energy of an elevated water supply into mechanical energy of a rotating shaft.

hydrocyclone [MECH ENG] A cyclone separator in which granular solids are removed from a stream of water and classified by centrifugal force.

hydrodynamic equations [FL MECH] Three equations which express the net acceleration of a unit water particle as the sum of the partial accelerations due to pressure gradient force, frictional force, earth's deflecting force, gravitational force, and other factors.

hydrodynamic pressure [FL MECH] The difference between the pressure of a fluid and the hydrostatic pressure; this concept is useful chiefly in problems of the steady flow of an incompressible fluid in which the hydrostatic pressure is constant for a given elevation (as when the fluid is bounded above by a rigid plate), so that the external force field (gravity) may be eliminated from the problem.

hydrodynamics [FL MECH] The study of the motion of a fluid and of the interactions of the fluid with its boundaries, especially in the incompressible inviscid case.

hydroelasticity [FL MECH] 1. Theory of elasticity of a fluid. 2. The interaction between the flow of water or other liquid and the elastic behavior of a body immersed in it.

hydroelectric generator [MECH ENG] An electric rotating machine that transforms mechanical power from a hydraulic turbine or water wheel into electric power.

hydroelectric plant [MECH ENG] A facility at which electric energy is produced by hydroelectric generators. Also known as hydroelectric power station.

hydroelectric power station *See* hydroelectric plant.

hydrokinematics [FL MECH] The study of the motion of a liquid apart from the cause of motion.

hydrokinetics [FL MECH] The study of the forces produced by a liquid as a consequence of its motion.

hydromechanics [FL MECH] The study of liquids, traditionally water, as a medium for the transmission of forces.

hydrometry [FL MECH] The science and technology of measuring specific gravities, particularly of liquids.

hydronic heating *See* hot-water heating.

hydropneumatic recoil system [MECH ENG] A recoil mechanism that absorbs the energy of recoil by the forcing of oil through orifices and returns the gun to battery by compressed gas.

hydroseparator [MECH ENG] A separator in which solids in suspension are agitated by hydraulic pressure or stirring devices.

hydrosizer *See* hydraulic classifier.

hydrostatic bearing [MECH ENG] A sleeve bearing in which high-pressure oil is pumped into the area between the shaft and the bearing so that the shaft is raised and supported by an oil film.

hydrostatic modulus *See* bulk modulus of elasticity.

hydrostatic press *See* hydraulic press.

hydrostatic pressure [FL MECH] 1. The pressure at a point in a fluid at rest due to the weight of the fluid above it. Also known as gravitational pressure. 2. The negative of the stress normal to a surface in a fluid.

hydrostatic roller conveyor [MECH ENG] A portion of a roller conveyor that has rolls weighted with liquid to control the speed of the moving objects.

hydrostatics [FL MECH] The study of liquids at rest and the forces exerted on them or by them.

hydrostatic strength [MECH] The ability of a body to withstand hydrostatic stress.

hydrostatic stress [MECH] The condition in which there are equal compressive stresses or equal tensile stresses in all directions, and no shear stresses on any plane.

hydrostatic weighing [FL MECH] A method of determining the density of a sample in which the sample is weighed in air, and then weighed in a liquid of known density; the volume of the sample is equal to the loss of weight in the liquid divided by the density of the liquid.

hyl *See* metric-technical unit of mass.

hyperbolic flareout [AERO ENG] A flareout obtained by changing the glide slope from a straight line to a hyperbolic curve at an appropriate distance from touchdown at an airport.

hyperbolic point [FL MECH] A singular point in a streamline field which constitutes the intersection of a convergence line and a divergence line; it is analogous to a col in the field of a single-valued scalar quantity. Also known as neutral point.

hyperbolic trajectory [AERO ENG] A trajectory entered by a spacecraft when its velocity exceeds the escape velocity of a planet, satellite, or star.

hyperoid axle [MECH ENG] A type of rear-axle drive gear set which generally carries the pinion 1.5–2 inches (38–51 millimeters) or more below the centerline of the gear.

hypersonic [FL MECH] Pertaining to hypersonic speeds, or air currents moving at hypersonic speeds.

hypersonic flight [AERO ENG] Flight at speeds well above the local velocity of sound; by convention, hypersonic regime starts at about five times the speed of sound and extends upward indefinitely.

hypersonic flow [FL MECH] Flow of a fluid over a body at hypersonic speeds, and in which shock waves start at a finite distance from the surface of the body.

hypersonic glider [AERO ENG] An unpowered vehicle, specifically a reentry vehicle, designed to fly at hypersonic speeds.

hypersonic inlet [FL MECH] An entrance or orifice for admission of fluids at hypersonic speeds.

hypersonic nozzle [FL MECH] A supersonic nozzle designed to accelerate a fluid to hypersonic speeds.

hypersonic speed [FL MECH] A speed of an object greater than about five times the speed of sound in the fluid through which the object is moving.

hypervelocity [MECH] **1.** Muzzle velocity of an artillery projectile of 3500 feet per second (1067 meters per second) or more. **2.** Muzzle velocity of a small-arms projectile of 5000 feet per second (1524 meters per second) or more. **3.** Muzzle velocity of a tank-cannon projectile in excess of 3350 feet per second (1021 meters per second).

hypoid gear [MECH ENG] Gear wheels connecting nonparallel, nonintersecting shafts, usually at right angles.

hypoid generator [MECH ENG] A gear-cutting machine for making hypoid gears.

hysteresis clutch [MECH ENG] A clutch in which torque is produced by attraction between induced poles in a magnetized iron ring and the control field.

hysteresis damping   [MECH] Damping of a vibration due to energy lost through mechanical hysteresis.

hysteretic damping   [MECH] Damping of a vibrating system in which the retarding force is proportional to the velocity and inversely proportional to the frequency of the vibration.

I

IAS *See* indicated airspeed.

ice pick [DES ENG] A hand tool for chipping ice.

ice tongs [DES ENG] Tongs for handling cubes or blocks of ice.

ID *See* inside diameter.

ideal aerodynamics [FL MECH] A branch of aerodynamics that deals with simplifying assumptions that help explain some airflow problems and provide approximate answers. Also known as ideal fluid dynamics.

ideal exhaust velocity [FL MECH] The theoretical maximum velocity, relative to the nozzle, of the gas flow as it passes from a given nozzle inlet temperature and pressure to a given ambient pressure, when the combustion gas has a given mean molecular weight.

ideal flow [FL MECH] 1. Fluid flow which is incompressible, two-dimensional, irrotational, steady, and nonviscous. 2. *See* inviscid flow.

ideal fluid [FL MECH] 1. A fluid which has ideal flow. 2. *See* inviscid fluid.

ideal fluid dynamics *See* ideal aerodynamics.

ideal gas [THERMO] Also known as perfect gas. 1. A gas whose molecules are infinitely small and exert no force on each other. 2. A gas that obeys Boyle's law (the product of the pressure and volume is constant at constant temperature) and Joule's law (the internal energy is a function of the temperature alone).

ideal gas law [THERMO] The equation of state of an ideal gas which is a good approximation to real gases at sufficiently high temperatures and low pressures; that is, $PV = RT$, where P is the pressure, V is the volume per mole of gas, T is the temperature, and R is the gas constant.

ideal radiator *See* blackbody.

ideal rocket [AERO ENG] A rocket motor or rocket engine that would have a velocity equal to the velocity of its jet gases.

identification [CONT SYS] The procedures for deducing a system's transfer function from its response to a step-function input or to an impulse.

idle [MECH ENG] To run without a load.

idler gear [MECH ENG] A gear situated between a driving gear and a driven gear to transfer motion, without any change of direction or of gear ratio.

idler pulley [MECH ENG] A pulley used to guide and tighten the belt or chain of a conveyor system.

idler wheel [MECH ENG] 1. A wheel used to transmit motion or to guide and support something. 2. A roller with a rubber surface used to transfer power by frictional means in a sound-recording or sound-reproducing system.

idling jet [MECH ENG] A carburetor part that introduces gasoline during minimum load or speed of the engine.

idling system [MECH ENG] A system to obtain adequate metering forces at low air-speeds and small throttle openings in an automobile carburetor in the idling position.

ignition delay *See* ignition lag.

ignition lag [MECH ENG] In the internal combustion engine, the time interval between the passage of the spark and the inflammation of the air-fuel mixture. Also known as ignition delay.

ignition system [MECH ENG] The system in an internal combustion engine that initiates the chemical reaction between fuel and air in the cylinder charge by producing a spark.

I-head cylinder [MECH ENG] The internal combustion engine construction having both inlet and exhaust valves located in the cylinder head.

ihp *See* indicated horsepower.

impact [MECH] A forceful collision between two bodies which is sufficient to cause an appreciable change in the momentum of the system on which it acts. Also known as impulsive force.

impact breaker [MECH ENG] A device that utilizes the energy from falling stones in addition to power from massive impellers for complete breaking up of stone. Also known as double impeller breaker.

impact crusher [MECH ENG] A machine for crushing large chunks of solid materials by sharp blows imposed by rotating hammers, or steel plates or bars; some crushers accept lumps as large as 28 inches (about 70 centimeters) in diameter, reducing them to ¼ inch (6 millimeters) and smaller.

impact energy [MECH] The energy necessary to fracture a material. Also known as impact strength.

impact force *See* set forward force.

impact grinding [MECH ENG] A technique used to break up particles by direct fall of crushing bodies on them.

impact loss [FL MECH] Loss of head in a flowing stream due to the impact of water particles upon themselves or some bounding surface.

impact mill [MECH ENG] A unit that reduces the size of rocks and minerals by the action of rotating blades projecting the material against steel plates.

impactor [MECH ENG] A machine or part whose operating principle is striking blows.

impact predictor [AERO ENG] A device which takes information from a trajectory measuring system and continuously computes the point (in real time) at which the rocket will strike the earth.

impact pressure *See* dynamic pressure.

impact roll [MECH ENG] An idler roll protected by a covering of a resilient material from the shock of the loading of material onto a conveyor belt, so as to reduce the damage to the belt.

impact screen [MECH ENG] A screen designed to swing or rock forward when loaded and to stop abruptly by coming in contact with a stop.

impact strength [MECH] 1. Ability of a material to resist shock loading. 2. *See* impact energy.

impact stress [MECH] Force per unit area imposed on a material by a suddenly applied force.

impact velocity [MECH] The velocity of a projectile or missile at the instant of impact. Also known as striking velocity.

impact wrench [MECH ENG] A compressed-air or electrically operated wrench that gives a rapid succession of sudden torques.

impeller [MECH ENG] The rotating member of a turbine, blower, fan, axial or centrifugal pump, or mixing apparatus. Also known as rotor.

impeller pump [MECH ENG] Any pump using a mechanical agency to provide continuous power to move liquids.

imperfect gas *See* real gas.

imperial pint *See* pint.

impregnated bit [DES ENG] A sintered, powder-metal matrix bit with fragmented bort or whole diamonds of selected screen sizes uniformly distributed throughout the entire crown section.

impulse [MECH] The integral of a force over an interval of time.

impulse modulation [CONT SYS] Modulation of a signal in which it is replaced by a series of impulses, equally spaced in time, whose strengths (integrals over time) are proportional to the amplitude of the signal at the time of the impulse.

impulse response [CONT SYS] The response of a system to an impulse which differs from zero for an infinitesimal time, but whose integral over time is unity; this impulse may be represented mathematically by a Dirac delta function.

impulse train [CONT SYS] An input consisting of an infinite series of unit impulses, equally separated in time.

impulse turbine [MECH ENG] A prime mover in which fluid under pressure enters a stationary nozzle where its pressure (potential) energy is converted to velocity (kinetic) energy and absorbed by the rotor.

impulsive force *See* impact.

in. *See* inch.

inch [MECH] A unit of length in common use in the United States and Great Britain, equal to $\frac{1}{12}$ foot or 2.54 centimeters. Abbreviated in.

inch of mercury [MECH] The pressure exerted by a 1-inch-high (2.54-centimeter-high) column of mercury that has a density of 13.5951 grams per cubic centimeter when the acceleration of gravity has the standard value of 9.80665 m/sec^2 or approximately 32.17398 ft/sec^2; equal to 3386.388640341 newtons per square meter; used as a unit in the measurement of atmospheric pressure.

inclined cableway [MECH ENG] A monocable arrangement in which the track cable has a slope sufficiently steep to allow the carrier to run down under its own weight.

inclined orbit [AERO ENG] A satellite orbit which is inclined with respect to the earth's equator.

inclined plane [MECH] A plane surface at an angle to some force or reference line.

incomplete lubrication [MECH ENG] Lubrication that takes place when the load on the rubbing surfaces is carried partly by a fluid viscous film and partly by areas of boundary lubrication; friction is intermediate between that of fluid and boundary lubrication.

incompressibility [MECH] Quality of a substance which maintains its original volume under increased pressure.

incompressibility condition [FL MECH] The condition prevailing when dp/dt, the time rate of change of the density of a fluid, is zero; this is a valid assumption for most problems in dynamic oceanography.

incompressible flow [FL MECH] Fluid motion without any change in density.

incompressible fluid [FL MECH] A fluid which is not reduced in volume by an increase in pressure.

indented bolt [DES ENG] A type of anchor bolt that has indentations to hold better in cemented grout.

independent chuck [DES ENG] A chuck for holding work by means of four jaws, each of which is moved independently of the others.

independent suspension [MECH ENG] In automobiles, a system of springs and guide links by which wheels are mounted independently on the chassis.

independent wire-rope core [DES ENG] A core of steel in a wire rope made in accordance with the best practice and design, either bright (uncoated) galvanized or drawn galvanized wire.

index center [MECH ENG] One of two machine-tool centers used to hold work and to rotate it by a fixed amount.

index chart [MECH ENG] **1.** A chart used in conjunction with an indexing or dividing head, which correlates the index plate, hole circle, and index crank motion with the desired angular subdivisions. **2.** A chart indicating the arrangement of levers in a machine to obtain desired output speed or fuel rate.

index crank [MECH ENG] The crank handle of an index head used to turn the spindle.

index head [MECH ENG] A headstock that can be affixed to the table of a milling machine, planer, or shaper; work may be mounted on it by a chuck or centers, for indexing.

indexing [MECH ENG] The process of providing discrete spaces, parts, or angles in a workpiece by using an index head.

indexing fixture [MECH ENG] A fixture that changes position with regular steplike movements.

index plate [DES ENG] A plate with circular graduations or holes arranged in circles, each circle with different spacing; used for indexing on machines.

indicated airspeed [AERO ENG] The airspeed as shown by a differential-pressure airspeed indicator, uncorrected for instrument and installation errors; a simple computation for altitude and temperature converts indicated airspeed to true airspeed. Abbreviated IAS.

indicated altitude [AERO ENG] The uncorrected reading of a barometric altimeter.

indicated horsepower [MECH ENG] The horsepower delivered by an engine as calculated from the average pressure of the working fluid in the cylinders and the displacement. Abbreviated ihp.

induced angle of attack [AERO ENG] The downward vertical angle between the horizontal and the velocity (relative to the wing of an aircraft) of the airstream passing over the wing.

induced draft [MECH ENG] A mechanical draft produced by suction stream jets or fans at the point where air or gases leave a unit.

induced-draft cooling tower [MECH ENG] A structure for cooling water by circulating air where the load is on the suction side of the fan.

induced drag [FL MECH] That part of the drag caused by the downflow or downwash of the airstream passing over the wing of an aircraft, equal to the lift times the tangent of the induced angle of attack.

induction pump [MECH ENG] Any pump operated by electromagnetic induction.

induction valve *See* inlet valve.

inelastic [MECH] Not capable of sustaining a deformation without permanent change in size or shape.

inelastic buckling [MECH] Sudden increase of deflection or twist in a column when compressive stress reaches the elastic limit but before elastic buckling develops.

inelastic collision [MECH] A collision in which the total kinetic energy of the colliding particles is not the same after the collision as before it.

inelastic stress [MECH] A force acting on a solid which produces a deformation such that the original shape and size of the solid are not restored after removal of the force.

inequality of Clausius *See* Clausius inequality.

inertia [MECH] That property of matter which manifests itself as a resistance to any change in the momentum of a body.

inertia ellipsoid [MECH] An ellipsoid used in describing the motion of a rigid body; it is fixed in the body, and the distance from its center to its surface in any direction is inversely proportional to the square root of the moment of inertia about the corresponding axis. Also known as momental ellipsoid; Poinsot ellipsoid.

inertia governor [MECH ENG] A speed-control device utilizing suspended masses that respond to speed changes by reason of their inertia.

inertial coordinate system *See* inertial reference frame.

inertial flow [FL MECH] Flow in which no external forces are exerted on a fluid.

inertial force [MECH] The fictitious force acting on a body as a result of using a non-inertial frame of reference; examples are the centrifugal and Coriolis forces that appear in rotating coordinate systems. Also known as effective force.

inertial instability [FL MECH] 1. Generally, instability in which the only form of energy transferred between the steady state and the disturbance in the fluid is kinetic energy. 2. The hydrodynamic instability arising in a rotating fluid mass when the velocity distribution is such that the kinetic energy of a disturbance grows at the expense of kinetic energy of the rotation. Also known as dynamic instability.

inertial mass [MECH] The mass of an object as determined by Newton's second law, in contrast to the mass as determined by the proportionality to the gravitational force.

inertial reference frame [MECH] A coordinate system in which a body moves with constant velocity as long as no force is acting on it. Also known as inertial coordinate system.

inertia matrix [MECH] A matrix M used to express the kinetic energy T of a mechanical system during small displacements from an equilibrium position, by means of the equation $T = \frac{1}{2}\dot{q}^T M \dot{q}$, where $\dot{q}$ is the vector whose components are the derivatives of the generalized coordinates of the system with respect to time, and $\dot{q}^T$ is the transpose of $\dot{q}$.

inertia starter [MECH ENG] A device utilizing inertial principles to start the rotator of an internal combustion engine.

inertia tensor [MECH] A tensor associated with a rigid body whose product with the body's rotation vector yields the body's angular momentum.

inertia wave [FL MECH] 1. Any wave motion in which no form of energy other than kinetic energy is present; in this general sense, Helmholtz waves, barotropic disturbances, Rossby waves, and so forth, are inertia waves. 2. More restrictedly, a wave motion in which the source of kinetic energy of the disturbance is the rotation of the fluid about some given axis; in the atmosphere a westerly wind system is such a source, the inertia waves here being, in general, stable.

inextensional deformation [MECH] A bending of a surface that leaves unchanged the length of any line drawn on the surface and the curvature of the surface at each point.

in-feed centerless grinding [MECH ENG] A metal-cutting process by which a cylindrical workpiece is ground to a prescribed surface smoothness and diameter by the insertion of the workpiece between a grinding wheel and a canted regulating wheel; the rotation of the regulating wheel controls the rotation and feed rate of the workpiece.

influence line [MECH] A graph of the shear, stress, bending moment, or other effect of a movable load on a structural member versus the position of the load.

infragravity wave [FL MECH] A gravity wave whose period ranges from 30 seconds to 5 minutes.

inhaul cable [MECH ENG] In a cable excavator, the line that pulls the bucket to dig and bring in soil. Also known as digging line.

inherent damping [MECH ENG] A method of vibration damping which makes use of the mechanical hysteresis of such materials as rubber, felt, and cork.

initial free space [MECH] In interior ballistics, the portion of the effective chamber capacity not displaced by propellant.

initial mass [AERO ENG] The mass of a rocket missile at the beginning of its flight.

initial shot start pressure [MECH] In interior ballistics, the pressure required to start the motion of the projectile from its initial loaded position; in fixed ammunition, it includes pressure required to separate projectile and cartridge case and to start engraving the rotating band.

initial-value problem [FL MECH] A dynamical problem whose solution determines the state of a system at all times subsequent to a given time at which the state of the system is specified by given initial conditions; the initial-value problem is contrasted with the steady-state problem, in which the state of the system remains unchanged in time. Also known as transient problem.

initial yaw [MECH] The yaw of a projectile the instant it leaves the muzzle of a gun.

injection [AERO ENG] The process of placing a spacecraft into a specific trajectory, such as an earth orbit or an encounter trajectory to Mars. [MECH ENG] The introduction of fuel, fuel and air, fuel and oxidizer, water, or other substance into an engine induction system or combustion chamber.

injection carburetor [MECH ENG] A carburetor in which fuel is delivered under pressure into a heated part of the engine intake system. Also known as pressure carburetor.

injection pump [MECH ENG] A pump that forces a measured amount of fuel through a fuel line and atomizing nozzle in the combustion chamber of an internal combustion engine.

injector [MECH ENG] 1. An apparatus containing a nozzle in an actuating fluid which is accelerated and thus entrains a second fluid, so delivering the mixture against a pressure in excess of the actuating fluid. 2. A plug with a valved nozzle through which fuel is metered to the combustion chambers in diesel- or full-injection engines. 3. A jet through which feedwater is injected into a boiler, or fuel is injected into a combustion chamber.

inlet box [MECH ENG] A closure at the fan inlet or inlets in a boiler for attachment of the fan to the duct system.

inlet valve [MECH ENG] The valve through which a fluid is drawn into the cylinder of a positive-displacement engine, pump, or compressor. Also known as induction valve.

in-line engine [MECH ENG] A multiple-cylinder engine with cylinders aligned in a row.

in-line linkage [MECH ENG] A power-steering linkage which has the control valve and actuator combined in a single assembly.

insert bit [DES ENG] A bit into which inset cutting points of various preshaped pieces of hard metal (usually a sintered tungsten carbide–cobalt powder alloy) are brazed or hand-peened into slots or holes cut or drilled into a blank bit. Also known as slug bit.

inserted-tooth cutter [DES ENG] A milling cutter in which the teeth can be replaced.

inside caliper [DES ENG] A caliper that has two legs with feet that turn outward; used to measure inside dimensions, as the diameter of a hole.

inside diameter [DES ENG] The length of a line which passes through the center of a hollow cylindrical or spherical object, and whose end points lie on the inner surface of the object. Abbreviated ID.

inside face [DES ENG] That part of the bit crown nearest to or parallel with the inside wall of an annular or coring bit.

inside gage [DES ENG] The inside diameter of a bit as measured between the cutting points, such as between inset diamonds on the inside-wall surface of a core bit.

inside micrometer [DES ENG] A micrometer caliper with the points turned outward for measuring the internal dimensions of an object.

instability [CONT SYS] A condition of a control system in which excessive positive feedback causes persistent, unwanted oscillations in the output of the system.

instantaneous axis [MECH] The axis about which a rigid body is carrying out a pure rotation at a given instant in time.

instantaneous center [MECH] A point about which a rigid body is rotating at a given instant in time. Also known as instant center.

instantaneous recovery [MECH] The immediate reduction in the strain of a solid when a stress is removed or reduced, in contrast to creep recovery.

instantaneous strain [MECH] The immediate deformation of a solid upon initial application of a stress, in contrast to creep strain.

intake manifold [MECH ENG] A system of pipes which feeds fuel to the various cylinders of a multicylinder internal combustion engine.

intake stroke [MECH ENG] The fluid admission phase or travel of a reciprocating piston and cylinder mechanism as, for example, in an engine, pump, or compressor.

intake valve [MECH ENG] The valve which opens to allow air or an air-fuel mixture to enter an engine cylinder.

integral action [CONT SYS] A control action in which the rate of change of the correcting force is proportional to the deviation.

integral compensation [CONT SYS] Use of a compensator whose output changes at a rate proportional to its input.

integral control [CONT SYS] Use of a control system in which the control signal changes at a rate proportional to the error signal.

integral-furnace boiler [MECH ENG] A type of steam boiler which incorporates furnace water-cooling in the circulatory system.

integral-mode controller [CONT SYS] A controller which produces a control signal proportional to the integral of the error signal.

integral network [CONT SYS] A compensating network which produces high gain at low input frequencies and low gain at high frequencies, and is therefore useful in achieving low steady-state errors. Also known as lagging network; lag network.

integral square error [CONT SYS] A measure of system performance formed by integrating the square of the system error over a fixed interval of time; this performance measure and its generalizations are frequently used in linear optimal control and estimation theory.

integral-type flange [DES ENG] A flange which is forged or cast with, or butt-welded to, a nozzle neck, pressure vessel, or piping wall.

interaction [FL MECH] With respect to wave components, the nonlinear action by which properties of fluid flow (such as momentum, energy, vorticity), are transferred from one portion of the wave spectrum to another, or viewed in another manner, between eddies of different size-scales.

interaction balance method *See* goal coordination method.

interaction prediction method [CONT SYS] A method for coordinating the subproblem solutions in plant decomposition, in which the interaction variables are specified by the second-level controller according to overall optimality conditions, and the subproblems are solved to satisfy local optimality conditions constrained by the specified values of the interaction variables. Also known as feasible method.

interceptor [AERO ENG] A crewed aircraft utilized for the identification or engagement of airborne objects.

interchange coefficient *See* exchange coefficient.

intercondenser [MECH ENG] A condenser between stages of a multistage steam jet pump.

intercooler [MECH ENG] A heat exchanger for cooling fluid between stages of a multistage compressor with consequent saving in power.

interface resistance [THERMO] 1. Impairment of heat flow caused by the imperfect contact between two materials at an interface. 2. Quantitatively, the temperature difference across the interface divided by the heat flux through it.

interference fit [DES ENG] A fit wherein one of the mating parts of an assembly is forced into a space provided by the other part in such a way that the condition of maximum metal overlap is achieved.

interior ballistics [MECH] The science concerned with the combustion of powder, development of pressure, and movement of a projectile in the bore of a gun.

interlocking cutter [DES ENG] A milling cutter assembly consisting of two mating sections with uniform or alternate overlapping teeth.

intermediate gear [MECH ENG] An idler gear interposed between a driver and driven gear.

intermittent firing [MECH ENG] Cyclic firing whereby fuel and air are burned in a furnace for frequent short time periods.

internal brake [MECH ENG] A friction brake in which an internal shoe follows the inner surface of the rotating brake drum, wedging itself between the drum and the point at which it is anchored; used in motor vehicles.

internal broaching [MECH ENG] The removal of material on internal surfaces, by means of a tool with teeth of progressively increasing size moving in a straight line or other prescribed path over the surface, other than for the origination of a hole.

internal combustion engine [MECH ENG] A prime mover in which the fuel is burned within the engine and the products of combustion serve as the thermodynamic fluid, as with gasoline and diesel engines.

internal energy [THERMO] A characteristic property of the state of a thermodynamic system, introduced in the first law of thermodynamics; it includes intrinsic energies of individual molecules, kinetic energies of internal motions, and contributions from interactions between molecules, but excludes the potential or kinetic energy of the system as a whole; it is sometimes erroneously referred to as heat energy.

internal floating-head exchanger [MECH ENG] Tube-and-shell heat exchanger in which the tube sheet (support for tubes) at one end of the tube bundle is free to move.

internal force [MECH] A force exerted by one part of a system on another.

internal friction [FL MECH] *See* viscosity. [MECH] Conversion of mechanical strain energy to heat within a material subjected to fluctuating stress.

internal furnace [MECH ENG] A boiler furnace having a firebox within a water-cooled heating surface.

internal gear [DES ENG] An annular gear having teeth on the inner surface of its rim.

internal grinder [MECH ENG] A machine designed for grinding the surfaces of holes.

internally fired boiler [MECH ENG] A fire-tube boiler containing an internal furnace which is water-cooled.

internal mix atomizer [MECH ENG] A type of pneumatic atomizer in which gas and liquid are mixed prior to the gas expansion through the nozzle.

internal stress [MECH] A stress system within a solid that is not dependent on external forces. Also known as residual stress.

internal thread [DES ENG] A screw thread cut on the inner surface of a hollow cylinder.

internal vibrator [MECH ENG] A vibrating device which is drawn vertically through placed concrete to achieve proper consolidation.

internal wave [FL MECH] A wave motion of a stably stratified fluid in which the maximum vertical motion takes place below the surface of the fluid.

internal work [THERMO] The work done in separating the particles composing a system against their forces of mutual attraction.

international practical temperature scale [THERMO] Temperature scale based on six points; the water triple point, the boiling points of oxygen, water, sulfur, and the solidification points of silver and gold; designated as °C, degrees Celsius, or t_{int}.

international table British thermal unit *See* British thermal unit.

international table calorie *See* calorie

international thread [DES ENG] A standardized metric system in which the pitch and diameter of the thread are related, with the thread having a rounded root and flat crest.

interplanetary flight [AERO ENG] Flight through the region of space between the planets, under the primary gravitational influence of the sun.

interplanetary probe [AERO ENG] An instrumented spacecraft that flies through the region of space between the planets.

interplanetary spacecraft [AERO ENG] A spacecraft designed for interplanetary flight.

interplanetary transfer orbit [AERO ENG] An elliptical trajectory tangent to the orbits of both the departure planet and the target planet.

interrupted screw [DES ENG] A screw with longitudinal grooves cut into the thread, and which locks quickly when inserted into a similar mating part.

interstellar probe   [AERO ENG] An instrumentated spacecraft propelled beyond the solar system to obtain specific information about interstellar environment.

interstellar travel   [AERO ENG] Space flight between stars.

intertube burner   [MECH ENG] A burner which utilizes a nozzle that discharges between adjacent tubes.

invariable line   [MECH] A line which is parallel to the angular momentum vector of a body executing Poinsot motion, and which passes through the fixed point in the body about which there is no torque.

invariable plane   [MECH] A plane which is perpendicular to the angular momentum vector of a rotating rigid body not subject to external torque, and which is always tangent to its inertia ellipsoid.

inverse cam   [MECH ENG] A cam that acts as a follower instead of a driver.

inverse feedback   *See* negative feedback.

inverse problem   [CONT SYS] The problem of determining, for a given feedback control law, the performance criteria for which it is optimal.

inversion   [MECH ENG] The conversion of basic four-bar linkages to special motion linkage, slider-crank mechanism, and slow-motion mechanism by successively holding fast, as ground link, members of a specific linkage (as linkages, such as parallelogram drag link).   [THERMO] A reversal of the usual direction of a variation or process, such as the change in sign of the expansion coefficient of water at 4°C, or a change in sign in the Joule-Thomson coefficient at a certain temperature.

inversion temperature   [THERMO] The temperature at which the Joule-Thomson effect of a gas changes sign.

inverted engine   [MECH ENG] An engine in which the cylinders are below the crankshaft.

inviscid flow   [FL MECH] Flow of an inviscid fluid. Also known as frictionless flow; ideal flow; nonviscous flow.

inviscid fluid   [FL MECH] A fluid which has no viscosity; it therefore can support no shearing stress, and flows without energy dissipation. Also known as ideal fluid; nonviscous fluid; perfect fluid.

involute gear tooth   [DES ENG] A gear tooth whose profile is established by an involute curve outward from the base circle.

involute spline   [DES ENG] A spline having the same general form as involute gear teeth, except that the teeth are one-half the depth and the pressure angle is 30°.

involute spline broach   [MECH ENG] A broach that cuts multiple keys in the form of internal or external involute gear teeth.

ion engine   [AERO ENG] An engine which provides thrust by expelling accelerated or high velocity ions; ion engines using energy provided by nuclear reactors are proposed for space vehicles.

ion propulsion   [AERO ENG] Vehicular motion caused by reaction from the high-speed discharge of a beam of electrically equally charged minute particles ejected behind the vehicle.

irreversible energy loss   [THERMO] Energy transformation process in which the resultant condition lacks the driving potential needed to reverse the process; the measure of this loss is expressed by the entropy increase of the system.

irreversible process   [THERMO] A process which cannot be reversed by an infinitesimal change in external conditions.

irreversible thermodynamics   *See* nonequilibrium thermodynamics.

irrotational flow [FL MECH] Fluid flow in which the curl of the velocity function is zero everywhere, so that the circulation of the velocity about any closed curve vanishes. Also known as acyclic motion; irrotational motion.

irrotational motion *See* irrotational flow.

isenthalpic expansion [THERMO] Expansion which takes place without any change in enthalpy.

isentrope [THERMO] A line of equal or constant entropy.

isentropic [THERMO] Having constant entropy; at constant entropy.

isentropic compression [THERMO] Compression which occurs without any change in entropy.

isentropic expansion [THERMO] Expansion which occurs without any change in entropy.

isentropic flow [THERMO] Fluid flow in which the entropy of any part of the fluid does not change as that part is carried along with the fluid.

isentropic process [THERMO] A change that takes place without any increase or decrease in entropy, such as a process which is both reversible and adiabatic.

isobaric [THERMO] Of equal or constant pressure, with respect to either space or time.

isobaric process [THERMO] A thermodynamic process of a gas in which the heat transfer to or from the gaseous system causes a volume change at constant pressure.

isochronism [MECH] The property of having a uniform rate of operation or periodicity, for example, of a pendulum or watch balance.

isochronous governor [MECH ENG] A governor that keeps the speed of a prime mover constant at all loads. Also known as astatic governor.

isodynamic [MECH] Pertaining to equality of two or more forces or to constancy of a force.

isometric process [THERMO] A constant-volume, frictionless thermodynamic process in which the system is confined by mechanically rigid boundaries.

isostatics [MECH] In photoelasticity studies of stress analyses, those curves, the tangents to which represent the progressive change in principal-plane directions. Also known as stress trajectories. Also known as stress lines.

isostatic surface [MECH] A surface in a three-dimensional elastic body such that at each point of the surface one of the principal planes of stress at that point is tangent to the surface.

isotherm [THERMO] A curve or formula showing the relationship between two variables, such as pressure and volume, when the temperature is held constant. Also known as isothermal.

isothermal [THERMO] 1. Having constant temperature; at constant temperature. 2. *See* isotherm.

isothermal calorimeter [THERMO] A calorimeter in which the heat received by a reservoir, containing a liquid in equilibrium with its solid at the melting point or with its vapor at the boiling point, is determined by the change in volume of the liquid.

isothermal compression [THERMO] Compression at constant temperature.

isothermal equilibrium [THERMO] The condition in which two or more systems are at the same temperature, so that no heat flows between them.

isothermal expansion [THERMO] Expansion of a substance while its temperature is held constant.

isothermal flow [THERMO] Flow of a gas in which its temperature does not change.

isothermal layer [THERMO] A layer of fluid, all points of which have the same temperature.

isothermal magnetization [THERMO] Magnetization of a substance held at constant temperature; used in combination with adiabatic demagnetization to produce temperatures close to absolute zero.

isothermal process [THERMO] Any constant-temperature process, such as expansion or compression of a gas, accompanied by heat addition or removal from the system at a rate just adequate to maintain the constant temperature.

isothermal transformation [THERMO] Any transformation of a substance which takes place at a constant temperature.

isotropic fluid [FL MECH] A fluid whose properties are not dependent on the direction along which they are measured.

isotropic turbulence [FL MECH] Turbulence whose properties, especially statistical correlations, do not depend on direction.

J

J *See* joule.

jack [MECH ENG] A portable device for lifting heavy loads through a short distance, operated by a lever, a screw, or a hydraulic press.

jackbit [DES ENG] A drilling bit used to provide the cutting end in rock drilling; the bit is detachable and either screws on or is taper-fitted to a length of drill steel. Also known as ripbit.

jack chain [DES ENG] **1.** A chain made of light wire, with links arranged in figure-eights with loops at right angles. **2.** A toothed endless chain for moving logs.

jacket [MECH ENG] The space around an engine cylinder through which a cooling liquid circulates.

jackhammer [MECH ENG] A hand-held rock drill operated by compressed air.

jack plane [DES ENG] A general-purpose bench plane measuring over 1 foot (30 centimeters) in length.

jackscrew [MECH ENG] **1.** A jack operated by a screw mechanism. Also known as screw jack. **2.** The screw of such a jack.

jackshaft [MECH ENG] A countershaft, especially when used as an auxiliary shaft between two other shafts.

Jacobs taper [DES ENG] A machine tool used for mounting drill chucks in drilling machines.

Jaeger method [FL MECH] A method of determining surface tension of a liquid in which one measures the pressure required to cause air to flow from a capillary tube immersed in the liquid.

Jaeger-Steinwehr method [THERMO] A refinement of the Griffiths method for determining the mechanical equivalent of heat, in which a large mass of water, efficiently stirred, is used, the temperature rise of the water is small, and the temperature of the surroundings is carefully controlled.

jag bolt [DES ENG] An anchor bolt with barbs on a flaring shank.

Jamin effect [FL MECH] Resistance to flow of a column of liquid divided by air bubbles in a capillary tube, even when subjected to a substantial pressure difference between the ends of the tube.

jam nut *See* locknut.

JATO engine [AERO ENG] Derived from jet-assisted-takeoff engine. **1.** An auxiliary jet-producing unit or units, usually rockets, for additional thrust. **2.** A JATO bottle or unit; the complete auxiliary power system used for assisted takeoff.

jawbreaker *See* jaw crusher.

jaw clutch [MECH ENG] A clutch that provides positive connection of one shaft with another by means of interlocking faces; may be square or spiral; the most common type of positive clutch.

jaw crusher [MECH ENG] A machine for breaking rock between two steel jaws, one fixed and the other swinging. Also known as jawbreaker.

J bolt [DES ENG] A J-shaped bolt, threaded on the long leg of the J.

Jeans viscosity equation [THERMO] An equation which states that the viscosity of a gas is proportional to the temperature raised to a constant power, which is different for different gases.

jeep [MECH ENG] A one-quarter-ton, four-wheel-drive utility vehicle in wide use in all United States military services.

jerk [MECH] 1. The rate of change of acceleration; it is the third derivative of position with respect to time. 2. A unit of rate of change of acceleration, equal to 1 foot (30.48 centimeters) per second squared per second.

jerk pump [MECH ENG] A pump that supplies a precise amount of fuel to the fuel injection valve of an internal combustion engine at the time the valve opens; used for fuel injection.

jet [FL MECH] A strong, well-defined stream of compressible fluid, either gas or liquid, issuing from an orifice or nozzle or moving in a contracted duct.

jet aircraft [AERO ENG] An aircraft with a jet engine or engines.

jet bit [DES ENG] A modification of a drag bit or a roller bit that utilizes the hydraulic jet principle to increase drilling rate.

jet compressor [MECH ENG] A device, utilizing an actuating nozzle and a combining tube, for the pumping of a compressible fluid.

jet condenser [MECH ENG] A direct-contact steam condenser utilizing the aspirating effect of a jet for the removal of noncondensables.

jet drilling [MECH ENG] A drilling method that utilizes a chopping bit, with a water jet run on a string of hollow drill rods, to chop through soils and wash the cuttings to the surface. Also known as wash boring.

jet engine [AERO ENG] An aircraft engine that derives all or most of its thrust by reaction to its ejection of combustion products (or heated air) in a jet and that obtains oxygen from the atmosphere for the combustion of its fuel.   [MECH ENG] Any engine that ejects a jet or stream of gas or fluid, obtaining all or most of its thrust by reaction to the ejection.

jet flap [AERO ENG] A sheet of fluid discharged at high speed close to the trailing edge of a wing so as to induce lift over the whole wing.

jet mixer [MECH ENG] A type of flow mixer or line mixer, depending on impingement of one liquid on the other to produce mixing.

jet nozzle [DES ENG] A nozzle, usually specially shaped, for producing a jet, such as the exhaust nozzle on a jet or rocket engine.

jet propulsion [AERO ENG] The propulsion of a rocket or other craft by means of a jet engine.

jet pump [MECH ENG] A pump in which an accelerating jet entrains a second fluid to deliver it at elevated pressure.

jet stream [AERO ENG] The stream of gas or fluid expelled by any reaction device, in particular the stream of combustion products expelled from a jet engine, rocket engine, or rocket motor.

J factor [THERMO] A dimensionless equation used for the calculation of free convection heat transmission through fluid films.

jib boom [MECH ENG] An extension that is hinged to the upper end of a crane boom.

jib crane [MECH ENG] Any of various cranes having a projecting arm (jib).

jig [MECH ENG] A device used to position and hold parts for machining operations and to guide the cutting tool.

jig back [MECH ENG] An aerial ropeway with a pair of containers that move in opposite directions and are loaded or stopped alternately at opposite stations but do not pass around the terminals. Also known as reversible tramway; to-and-fro ropeway.

jig borer [MECH ENG] A machine tool resembling a vertical milling machine designed for locating and drilling holes in jigs.

jig grinder [MECH ENG] A precision grinding machine used to locate and grind holes to size, especially in hardened steels and carbides.

jigsaw [MECH ENG] A tool with a narrow blade suitable for cutting intricate curves and lines.

jim crow [DES ENG] A device with a heavy buttress screw thread used for bending rails by hand.

jobber's reamer [DES ENG] A machine reamer that is solid with straight or helical flutes and taper shanks.

joggle [DES ENG] **1.** A flangelike offset on a flat piece of metal. **2.** A projection or notch on a sheet of building material to prevent protrusion. **3.** A dowel for joining blocks of masonry.

Johansson block [DES ENG] A type of gage block ground to an accuracy of at least 1/100,000 inch (0.25 micrometer). Also known as Jo block.

jointer gage [DES ENG] An attachment to a bench vise that holds a board at any angle desired for planing.

jordan [MECH ENG] A machine or engine used to refine paper pulp, consisting of a rotating cone, with cutters, that fits inside another cone, also with cutters.

joule [MECH] The unit of energy or work in the meter-kilogram-second system of units, equal to the work done by a force of 1 newton magnitude when the point at which the force is applied is displaced 1 meter in the direction of the force. Symbolized J. Also known as newton-meter of energy.

Joule cycle *See* Brayton cycle.

Joule equivalent [THERMO] The numerical relation between quantities of mechanical energy and heat; the present accepted value is 1 fifteen-degrees calorie equals 4.1855 ± 0.0005 joules. Also known as mechanical equivalent of heat.

Joule-Kelvin effect *See* Joule-Thomson effect.

Joule's law [THERMO] The law that at constant temperature the internal energy of a gas tends to a finite limit, independent of volume, as the pressure tends to zero.

Joule-Thomson effect [THERMO] A change of temperature in a gas undergoing Joule-Thomson expansion. Also known as Joule-Kelvin effect.

Joule-Thomson expansion [THERMO] The adiabatic, irreversible expansion of a fluid flowing through a porous plug or partially opened valve. Also known as Joule-Thomson process.

Joule-Thomson process *See* Joule-Thomsom expansion.

journal [MECH ENG] That part of a shaft or crank which is supported by and turns in a bearing.

journal bearing [MECH ENG] A cylindrical bearing which supports a rotating cylindrical shaft.

journal friction [MECH ENG] Friction of the axle in a journal bearing arising mainly from viscous sliding friction between journal and lubricant.

joystick [AERO ENG] A lever used to control the motion of an aircraft; fore-and-aft motion operates the elevators while lateral motion operates the ailerons.

jumbo *See* drill carriage.

jumper tube [MECH ENG] A short tube used to bypass the flow of fluid in a boiler or tubular heater.

jump phenomenon [CONT SYS] A phenomenon occurring in a nonlinear system subjected to a sinusoidal input at constant frequency, in which the value of the amplitude of the forced oscillation can jump upward or downward as the input amplitude is varied through either of two fixed values, and the graph of the forced amplitude versus the input amplitude follows a hysteresis loop.

jump resonance [CONT SYS] A jump discontinuity occurring in the frequency response of a nonlinear closed-loop control system with saturation in the loop.

Junkers engine [MECH ENG] A double-opposed-piston, two-cycle internal combustion engine with intake and exhaust ports at opposite ends of the cylinder.

Jurin rule [FL MECH] The rule that a height to which a liquid rises in a capillary tube is twice the liquid's surface tension times the cosine of its contact angle with the capillary, divided by the product of the liquid's weight density and the internal radius of the tube.

just ton *See* ton.

K

K$_E$ *See* elasticity number 2.

Kalman filter [CONT SYS] A linear system in which the mean squared error between the desired output and the actual output is minimized when the input is a random signal generated by white noise.

Kaplan turbine [MECH ENG] A propeller-type hydraulic turbine in which the positions of the runner blades and the wicket gates are adjustable for load change with sustained efficiency.

Kármán constant [FL MECH] A dimensionless number formed from the velocity of turbulent flow parallel to a plane wall, the distance from the wall, the shear stress, and the density of the fluid; for a wide range of flow patterns it has a constant value.

Kármán vortex street [FL MECH] A double row of line vortices in a fluid which, under certain conditions, is shed in the wake of cylindrical bodies when the relative fluid velocity is perpendicular to the axis of the cylinder.

Kater's reversible pendulum [MECH] A gravity pendulum designed to measure the acceleration of gravity and consisting of a body with two knife-edge supports on opposite sides of the center of mass.

Kauertz engine [MECH ENG] A type of cat-and-mouse rotary engine in which the pistons are vanes which are sections of a right circular cylinder; two pistons are attached to one rotor so that they rotate with constant angular velocity, while the other two pistons are controlled by a gear-and-crank mechanism, so that angular velocity varies.

kb *See* kilobar.

KC-97 *See* Stratofreighter.

KC-135 *See* Stratotanker.

kcal *See* kilocalorie.

kellering [MECH ENG] Three-dimensional machining of a contoured surface by tracer-milling the die block or punch; the cutter path is controlled by a tracer that follows the contours on a die model.

Kelvin [THERMO] A unit of absolute temperature equal to 1/273.16 of the absolute temperature of the triple point of water. Symbolized K. Formerly known as degree Kelvin.

Kelvin body [MECH] An ideal body whose shearing (tangential) stress is the sum of a term proportional to its deformation and a term proportional to the rate of change of its deformation with time. Also known as Voigt body.

Kelvin's circulation theorem [FL MECH] The theorem that, if the external forces acting on an inviscid fluid are conservative and if the fluid density is a function of the pressure only, then the circulation along a closed curve which moves with the fluid does not change with time.

Kelvin's minimum-energy theorem [FL MECH] The theorem that the irrotational motion of an incompressible, inviscid fluid occupying a simply connected region has less kinetic energy than any other fluid motion consistent with the boundary condition of zero relative velocity normal to the boundaries of the region.

Kennedy key [DES ENG] A square taper key fitted into a keyway of square section and driven from opposite ends of the hub.

key [DES ENG] 1. An instrument that is inserted into a lock to operate the bolt. 2. A device used to move in some manner in order to secure or tighten. 3. One of the levers of a keyboard. 4. *See* machine key.

keyhole [DES ENG] A hole or a slot for receiving a key.

keyhole saw [DES ENG] A fine compass saw with a blade 11–16 inches (28–41 centimeters) long.

key seat *See* keyway.

keyseater [MECH ENG] A machine for milling beds or grooves in mechanical parts which receive keys.

keyway [DES ENG] 1. An opening in a lock for passage of a flat metal key. 2. The pocket in the driven element to provide a driving surface for the key. 3. A groove or channel for a key in any mechanical part. Also known as key seat.

kg *See* kilogram; kilogram force.

kgf *See* kilogram force.

kickback [MECH ENG] A backward thrust, such as the backward starting of an internal combustion engine as it is cranked, or the reverse push of a piece of work as it is fed to a rotary saw.

kickdown [MECH ENG] 1. Shifting to lower gear in an automotive vehicle. 2. The device for shifting.

kick over [MECH ENG] To start firing; applied to internal combustion engines.

kilobar [MECH] A unit of pressure equal to 1000 bars (100 megapascals). Abbreviated kb.

kilocalorie [THERMO] A unit of heat energy equal to 1000 calories. Abbreviated kcal. Also known as kilogram-calorie (kg-cal); large calorie (Cal).

kilogram [MECH] 1. The unit of mass in the meter-kilogram-second system, equal to the mass of the international prototype kilogram stored at Sèvres, France. Abbreviated kg. 2. *See* kilogram force.

kilogram-calorie *See* kilocalorie.

kilogram force [MECH] A unit of force equal to the weight of a 1-kilogram mass at a point on the earth's surface where the acceleration of gravity is 9.80665 meters/sec^2. Abbreviated kgf. Also known as kilogram (kg); kilogram weight (kg-wt).

kilogram-meter *See* meter-kilogram.

kilogram weight *See* kilogram force.

kiloliter [MECH] A unit of volume equal to 1000 liters or to 1 cubic meter. Abbreviated kl.

kilometer [MECH] A unit of length equal to 1000 meters. Abbreviated km.

kinematic boundary condition [FL MECH] The condition that the component of fluid velocity perpendicular to a solid boundary must vanish on the boundary itself; when the boundary is a fluid surface, the condition applies to the vector difference of velocities across the interface.

kinematic fluidity [FL MECH] The reciprocal of the kinematic viscosity.

kinematics [MECH] The study of the motion of a system of material particles without reference to the forces which act on the system.

kinematic similarity [FL MECH] A relationship between fluid-flow systems in which corresponding fluid velocities and velocity gradients are in the same ratios at corresponding locations.

kinematic viscosity [FL MECH] The absolute viscosity of a fluid divided by its density. Also known as coefficient of kinematic viscosity.

kinetic energy [MECH] The energy which a body possesses because of its motion; in classical mechanics, equal to one-half of the body's mass times the square of its speed.

kinetic friction [MECH] The friction between two surfaces which are sliding over each other.

kinetic momentum [MECH] The momentum which a particle possesses because of its motion; in classical mechanics, equal to the particle's mass times its velocity.

kinetic potential *See* Lagrangian.

kinetic pressure [FL MECH] The kinetic energy per unit volume of a fluid, equal to one-half the product of its density and the square of its velocity.

kinetic reaction [MECH] The negative of the mass of a body multiplied by its acceleration.

kinetics [MECH] The dynamics of material bodies.

kingpin [MECH ENG] The pin for articulation between an automobile stub axle and an axle-beam or steering head. Also known as swivel pin.

kip [MECH] A 1000-pound (453.6-kilogram) load.

Kirchhoff's law [THERMO] The law that the ratio of the emissivity of a heat radiator to the absorptivity of the same radiator is the same for all bodies, depending on frequency and temperature alone, and is equal to the emissivity of a blackbody. Also known as Kirchhoff's principle.

Kirchhoff's principle *See* Kirchhoff's law.

Kirkwood-Brinkely's theory [MECH] In terminal ballistics, a theory formulating the scaling laws from which the effect of blast at high altitudes may be inferred, based upon observed results at ground level.

kl *See* kiloliter.

km *See* kilometer.

knee [MECH ENG] In a knee-and-column type of milling machine, the part which supports the saddle and table and which can move vertically on the column.

knee frequency *See* break frequency.

knee tool [MECH ENG] A tool holder with a shape resembling a knee, such as the holder for simultaneous cutting and interval operations on a screw machine or turret lathe.

knife [DES ENG] A sharp-edged blade for cutting.

knife-edge [DES ENG] A sharp narrow edge resembling that of a knife, such as the fulcrum for a lever arm in a measuring instrument.

knife-edge bearing [MECH ENG] A balance beam or lever arm fulcrum in the form of a hardened steel wedge; used to minimize friction.

knife-edge cam follower [DES ENG] A cam follower having a sharp narrow edge or point like that of a knife; useful in developing cam profile relationships.

knife file [DES ENG] A tapered file with a thin triangular cross section resembling that of a knife.

knob [DES ENG] A component that is placed on a control shaft to facilitate manual rotation of the shaft; sometimes has a pointer or markings to indicate shaft position.

knock-off [MECH ENG] 1. The automatic stopping of a machine when it is operating improperly. 2. The device that causes automatic stopping.

knuckle joint [DES ENG] A hinge joint between two rods in which an eye on one piece fits between two flat projections with eyes on the other piece and is retained by a round pin.

knuckle joint press [MECH ENG] A short-stroke press in which the slide is actuated by a crank attached to a knuckle joint hinge.

knuckle pin [DES ENG] The pin of a knuckle joint.

knuckle post [MECH ENG] A post which acts as the pivot for the steering knuckle in an automobile.

Knudsen number [FL MECH] The ratio of the mean free path length of the molecules of a fluid to a characteristic length; used to describe the flow of low-density gases.

Kollsman window [AERO ENG] A small window on the dial face of an aircraft pressure altimeter in which the altimeter setting in inches of mercury is indicated.

Kozeny-Carmen equation [FL MECH] Equation for streamline flow of fluids through a powdered bed.

K ratio [AERO ENG] The ratio of propellant surface to nozzle throat area.

Kullenberg piston corer [MECH ENG] A piston-operated coring device used to obtain 2-inch-diameter (5-centimeter-diameter) core samples.

Kutta-Joukowski airfoil [FL MECH] A class of airfoils that may be produced by mapping circles with the complex variable transform $w = z + (c^2/z)$.

Kutta-Joukowski equation [FL MECH] An equation which states that the lift force exerted on a body by an ideal fluid, per unit length of body perpendicular to the flow, is equal to the product of the mass density of the fluid, the linear velocity of the fluid relative to the body, and the fluid circulation. Also known as Kutta-Joukowski theorem.

Kutta-Joukowski theorem *See* Kutta-Joukowski equation.

kytoon [AERO ENG] A captive balloon used to maintain meteorological equipment aloft at approximately a constant height; it is streamlined, and combines the aerodynamic properties of a balloon and a kite.

L

l *See* liter.

laboratory coordinate system [MECH] A reference frame attached to the laboratory of the observer, in contrast to the center-of-mass system.

ladder-bucket dredge *See* bucket-ladder dredge.

ladder dredge *See* bucket-ladder dredge.

ladder drilling [MECH ENG] An arrangement of retractable drills with pneumatic powered legs mounted on banks of steel ladders connected to a holding frame; used in large-scale rock tunneling, with the advantage that many drills can be worked at the same time by a small labor force.

ladder trencher [MECH ENG] A machine that digs trenches by means of a bucket-ladder excavator. Also known as ladder ditcher.

ladle [DES ENG] A deep-bowled spoon with a long handle for dipping up, transporting, and pouring liquids.

lag bolt *See* coach screw.

lagging network *See* integral network.

lag-lead network *See* lead-lag network.

lag network *See* integral network.

Lagrange bracket [MECH] Given two functions of coordinates and momenta in a system, their Lagrange bracket is an expression measuring how coordinates and momenta change jointly with respect to the two functions.

Lagrange function *See* Lagrangian.

Lagrange-Hamilton theory [MECH] The formalized study of continuous systems in terms of field variables where a Lagrangian density function and Hamiltonian density function are introduced to produce equations of motion.

Lagrange's equations [MECH] Equations of motion of a mechanical system for which a classical (non-quantum-mechanical) description is suitable, and which relate the kinetic energy of the system to the generalized coordinates, the generalized forces, and the time. Also known as Lagrangian equations of motion.

Lagrange stream function [FL MECH] A scalar function of position used to describe steady, incompressible two-dimensional flow; constant values of this function give the streamlines, and the rate of flow between a pair of streamlines is equal to the difference between the values of this function on the streamlines. Also known as current function; stream function.

Lagrangian [MECH] **1.** The difference between the kinetic energy and the potential energy of a system of particles, expressed as a function of generalized coordinates and velocities from which Lagrange's equations can be derived. Also known as kinetic potential; Lagrange function. **2.** For a dynamical system of fields, a function which plays the same role as the Lagrangian of a system of particles; its integral over a time interval is a maximum or a minimum with respect to infinitesimal variations of the fields, provided the initial and final fields are held fixed.

Lagrangian coordinates   *See* generalized coordinates.

Lagrangian density   [MECH] For a dynamical system of fields or continuous media, a function of the fields, of their time and space derivatives, and the coordinates and time, whose integral over space is the Lagrangian.

Lagrangian equations of motion   *See* Lagrange's equations.

Lagrangian function   [MECH] The function which measures the difference between the kinetic and potential energy of a dynamical system.

Lagrangian generalized velocity   *See* generalized velocity.

Lagrangian method   [FL MECH] A method of studying fluid motion and the mechanics of deformable bodies in which one considers volume elements which are carried along with the fluid or body, and across whose boundaries material does not flow; in contrast to Euler method.

lag screw   *See* coach screw.

lambda   [MECH] A unit of volume equal to 10^{-6} liter or 10^{-9} cubic meter.

lambda point   [THERMO] A temperature at which the specific heat of a substance has a sharply peaked maximum, observed in many second-order transitions.

Lambert surface   [THERMO] An ideal, perfectly diffusing surface for which the intensity of reflected radiation is independent of direction.

Lamé constants   [MECH] Two constants which relate stress to strain in an isotropic, elastic material.

laminar boundary layer   [FL MECH] A thin layer over the surface of a body immersed in a fluid, in which the fluid velocity relative to the surface increases rapidly with distance from the surface and the flow is laminar.

laminar flow   [FL MECH] Streamline flow of an incompressible, viscous Newtonian fluid; all particles of the fluid move in distinct and separate lines.

laminar flow control   [AERO ENG] The removal of a small amount of boundary-layer air from the surface of an aircraft wing with the result that the airflow is laminar rather than turbulent; frictional drag is greatly reduced.

laminar sublayer   [FL MECH] The laminar boundary layer underlying a turbulent boundary layer.

laminar wing   [AERO ENG] A low-drag wing in which the distribution of thickness along the chord is so selected as to maintain laminar flow over as much of the wing surface as possible.

laminated spring   [DES ENG] A flat or curved spring made of thin superimposed plates and forming a cantilever or beam of uniform strength.

Lami's theorem   [MECH] When three forces act on a particle in equilibrium, the magnitude of each is proportional to the sine of the angle between the other two.

Lancashire boiler   [MECH ENG] A cylindrical steam boiler consisting of two longitudinal furnace tubes which have internal grates at the front.

lance door   [MECH ENG] The door to a boiler furnace through which a hand lance is inserted.

Lanchester balancer   [MECH ENG] A device for balancing four-cylinder engines; consists of two meshed gears with eccentric masses, driven by the crankshaft.

land   [AERO ENG] Of an aircraft, to alight on land or a ship deck.   [DES ENG] The top surface of the tooth of a cutting tool, behind the cutting edge.

landing area   [AERO ENG] An area intended primarily for landing and takeoff of aircraft.

landing circle   [AERO ENG] The approximately circular path flown by an airplane to get into the landing pattern; used particularly with naval aircraft landing on an aircraft carrier.

landing flap   [AERO ENG] A movable airfoil-shaped structure located aft of the rear beam or spar of the wing; extends about two-thirds of the span of the wing and functions to substantially increase the lift, permitting lower takeoff and landing speeds.

landing gear   [AERO ENG] Those components of an aircraft or spacecraft that support and provide mobility for the craft on land, water, or other surface.

landing light   [AERO ENG] One of the floodlights mounted on the leading edge of the wing and below the nose of the fuselage to enable an airplane to land at night.

landing load   [AERO ENG] The load on an aircraft's wings produced during landing; depends on descent velocity and landing attitude.

landing strip   [AERO ENG] A portion of the landing area prepared for the landing and takeoff of aircraft in a particular direction; it may include one or more runways.   Also known as air strip.

land measure   [MECH] 1. Units of area used in measuring land. 2. Any system for measuring land.

land mile   *See* mile.

lang lay   [DES ENG] A wire rope lay in which the wires of each strand are twisted in the same direction as the strands.

lantern pinion   [DES ENG] A pinion with bars (between parallel disks) instead of teeth.

lantern ring   [DES ENG] A ring or sleeve around a rotating shaft; an opening in the ring provides for forced feeding of oil or grease to bearing surfaces; particularly effective for pumps handling liquids.

Laplace irrotational motion   [FL MECH] Irrotational flow of an inviscid, incompressible fluid.

Laplacian speed of sound   [FL MECH] The phase speed of a sound wave in a compressible fluid under the assumption that the expansions and compressions are adiabatic.

large calorie   *See* kilocalorie.

large dyne   *See* newton.

large-systems control theory   [CONT SYS] A branch of the theory of control systems concerned with the special problems that arise in the design of control algorithms (that is, control policies and strategies) for complex systems.

Larson-Miller parameter   [MECH] The effects of time and temperature on creep, defined empirically as $P = T (C + \log t) \times 10^{-3}$, where T = test temperature in degrees Rankine (degrees Fahrenheit + 460) and t = test time in hours; the constant C depends upon the material but is frequently taken to be 20.

latch bolt   [DES ENG] A self-acting spring bolt with a beveled head.

latent heat   [THERMO] The amount of heat absorbed or evolved by 1 mole, or a unit mass, of a substance during a change of state (such as fusion, sublimation or vaporization) at constant temperature and pressure.

latent heat of sublimation   *See* heat of sublimation.

latent heat of vaporization   *See* heat of vaporization.

latent load [MECH ENG] Cooling required to remove unwanted moisture from an air-conditioned space.

lateral acceleration [AERO ENG] The component of the linear acceleration of an aircraft or missile along its lateral, or Y, axis.

lateral controller [AERO ENG] A primary flight control mechanism, generally a part of the longitudinal controller, which controls the ailerons; often resembles an automobile steering wheel but may be a control column.

lathe [MECH ENG] A machine for shaping a workpiece by gripping it in a holding device and rotating it under power against a suitable cutting tool for turning, boring, facing, or threading.

launch [AERO ENG] **1.** To send off a rocket vehicle under its own rocket power, as in the case of guided aircraft rockets, artillery rockets, and space vehicles. **2.** To send off a missile or aircraft by means of a catapult or by means of inertial force, as in the release of a bomb from a flying aircraft. **3.** To give a space probe an added boost for flight into space just before separation from its launch vehicle.

launch complex [AERO ENG] The composite of facilities and support equipment needed to assemble, check out, and launch a rocket vehicle.

launching angle [AERO ENG] The angle between the horizontal plane and the longitudinal axis of a rocket or missile at the time of launching.

launching ramp [AERO ENG] A ramp used for launching an aircraft or missile into the air.

launching site [AERO ENG] **1.** A site from which launching is done. **2.** The platform, ramp, rack, or other installation at such a site.

launch pad [AERO ENG] The load-bearing base or platform from which a rocket vehicle is launched. Also known as pad.

launch vehicle [AERO ENG] A rocket or other vehicle used to launch a probe, satellite, or the like. Also known as booster.

launch window [AERO ENG] The time period during which a spacecraft or missile must be launched in order to achieve a desired encounter, rendezvous, or impact.

Laval nozzle *See* de Laval nozzle.

lawnmower [MECH ENG] A machine for cutting grass on lawns.

law of action and reaction *See* Newton's third law.

law of gravitation *See* Newton's law of gravitation.

lay [DES ENG] The direction, length, or angle of twist of the strands in a rope or cable.

lb *See* pound.

lb ap *See* pound.

lb apoth *See* pound.

lbf *See* pound.

lbf-ft *See* foot-pound.

lbf in.$^{-2}$ abs *See* pounds per square inch absolute.

lb t *See* pound.

lb tr *See* pound.

lb UK *See* pound.

lead [DES ENG] The distance that a screw will advance or move into a nut in one complete turn.

lead angle [DES ENG] The angle that the tangent to a helix makes with the plane normal to the axis of the helix.

lead compensation [CONT SYS] A type of feedback compensation primarily employed for stabilization or for improving a system's transient response; it is generally characterized by a series compensation transfer function of the type

$$G_c(s) = K \frac{(s - z)}{(s - p)}$$

where $z < p$ and K is a constant.

leading edge [AERO ENG] The front edge of an airfoil or wing. [DES ENG] The surfaces or inset cutting points on a bit that face in the same direction as the rotation of the bit.

leading edge slat [AERO ENG] A small airfoil attached to the leading edge of a wing of an aircraft that automatically improves airflow at large angles of attack.

lead-in groove [DES ENG] A blank spiral groove at the outside edge of a disk recording, generally of a pitch much greater than that of the recorded grooves, provided to bring the pickup stylus quickly to the first recorded groove. Also known as lead-in spiral.

leading truck [MECH ENG] A swiveling frame with wheels under the front end of a locomotive.

lead-in spiral *See* lead-in groove.

lead-lag network [CONT SYS] Compensating network which combines the characteristics of the lag and lead networks, and in which the phase of a sinusoidal response lags a sinusoidal input at low frequencies and leads it at high frequencies. Also known as lag-lead network.

lead network *See* derivative network.

lead-out groove [DES ENG] A blank spiral groove at the end of a disk recording, generally of a pitch much greater than that of the recorded grooves, connected to either the locked or eccentric groove. Also known as throw-out spiral.

lead-over groove [DES ENG] A groove cut between separate selections or sections on a disk recording to transfer the pickup stylus from one cut to the next. Also known as cross-over spiral.

lead screw [MECH ENG] A threaded shaft used to convert rotation to longitudinal motion; in a lathe it moves the tool carriage when cutting threads; in a disk recorder it guides the cutter at a desired rate across the surface of an ungrooved disk.

leaf spring [DES ENG] A beam of cantilever design, firmly anchored at one end and with a large deflection under a load. Also known as flat spring.

league [MECH] A unit of length equal to 3 miles or 4828.032 meters.

leak test pressure [MECH ENG] The inlet pressure used for a standard quantitative seat leakage test.

lean fuel mixture *See* lean mixture.

lean mixture [MECH ENG] A fuel-air mixture containing a low percentage of fuel and a high percentage of air, as compared with a normal or rich mixture. Also known as lean fuel mixture.

least-action principle *See* principle of least action.

least-energy principle [MECH] The principle that the potential energy of a system in stable equilibrium is a minimum relative to that of nearby configurations.

least-work theory [MECH] A theory of statically indeterminate structures based on the fact that when a stress is applied to such a structure the individual parts of it are deflected so that the energy stored in the elastic members is minimized.

lee eddies [FL MECH] The small, irregular motions or eddies produced immediately in the rear of an obstacle in a turbulent fluid.

lee wave [FL MECH] Any wave disturbance which is caused by, and is therefore stationary with respect to, some barrier in the fluid flow.

left-hand [DES ENG] Of drilling and cutting tools, screw threads, and other threaded devices, designed to rotate clockwise or cut to the left.

left-handed *See* left-laid.

left-hand screw [DES ENG] A screw that advances when turned counterclockwise.

left-laid [DES ENG] The lay of a wire or fiber rope or cable in which the individual wires or fibers in the strands are twisted to the right and the strands to the left. Also known as left-handed; regular-lay left twist.

leg [MECH ENG] The case that encloses the vertical part of the belt carrying the buckets within a grain elevator.

Leidenfrost point [THERMO] The lowest temperature at which a hot body submerged in a pool of boiling water is completely blanketed by a vapor film; there is a minimum in the heat flux from the body to the water at this temperature.

Leidenfrost's phenomenon [THERMO] A phenomenon in which a liquid dropped on a surface that is above a critical temperature becomes insulated from the surface by a layer of vapor, and does not wet the surface as a result.

LEM *See* lunar excursion module.

length [MECH] Extension in space.

length of lay [DES ENG] The distance measured along a line parallel to the axis of the rope in which the strand makes one complete turn about the axis of the rope, or the wires make a complete turn about the axis of the strand.

lentor *See* stoke.

leo [MECH] A unit of acceleration, equal to 10 meters per second per second; it has rarely been employed.

LES *See* Lincoln experimental satellite.

letdown [AERO ENG] Gradual and orderly reduction in altitude, particularly in preparation for landing.

level [DES ENG] A device consisting of a bubble tube that is used to find a horizontal line or plane. Also known as spirit level.

level measurement [MECH] The determination of the linear vertical distance between a reference point or datum plane and the surface of a liquid or the top of a pile of divided solid.

level off [AERO ENG] To bring an aircraft to level flight after an ascent or descent.

level-off position [AERO ENG] That position over which a craft ends an ascent or descent and begins relatively horizontal motion.

level point *See* point of fall.

level valve [MECH ENG] A valve operated by a lever which travels through a maximum arc of 180°.

leverage [MECH] The multiplication of force or motion achieved by a lever.

Leverett function [FL MECH] A dimensionless number used in studying two-phase flow in porous mediums, written as $(\xi/e)^{1/2}(p/\sigma)$, where ξ is the permeability of a medium (as defined by Darcy's law), e is the medium's porosity, σ is the surface tension between two liquids flowing through it, and p is the capillary pressure.

lever shears [DES ENG] A shears in which the input force at the handles is related to the output force at the cutting edges by the principle of the lever. Also known as alligator shears; crocodile shears.

levitated vehicle [MECH ENG] A train or other vehicle which travels at high speed at some distance above an electrically conducting track by means of levitation.

lewis [DES ENG] A device for hoisting heavy stones; employs a dovetailed tenon that fits into a mortise in the stone.

lewis bolt [DES ENG] A bolt with an enlarged, tapered head that is inserted into masonry or stone and fixed with lead; used as a foundation bolt.

L-head engine [MECH ENG] A type of four-stroke cycle internal combustion engine having both inlet and exhaust valves on one side of the engine block which are operated by pushrods actuated by a single camshaft.

lift *See* aerodynamic lift; elevator.

lift coefficient [AERO ENG] The quantity $C_L = 2L/\rho V^2 S$, where L is the lift of a whole airplane wing, ρ is the mass density of the air, V is the free-stream velocity, and S is the wing area; this is also applicable to other airfoils.

lift-drag ratio [AERO ENG] The lift of an aerodynamic form, such as an airplane wing, divided by the drag.

lifter flight [DES ENG] Spaced plates or projections on the inside surfaces of cylindrical rotating equipment (such as rotary dryers) to lift and shower the solid particles through the gas-drying stream during their passage through the dryer cylinder.

lift fan [AERO ENG] A special turbofan engine used primarily for lift in VTOL/STOL aircraft and often mounted in a wing with vertical thrust axis.

lifting block [MECH ENG] A combination of pulleys and ropes which allows heavy weights to be lifted with least effort.

lifting reentry [AERO ENG] A reentry into the atmosphere by a space vehicle where aerodynamic lift is used, allowing a more gradual descent, greater accuracy in landing at a predetermined spot; it can accommodate greater errors in the guidance system and greater temperature control.

lifting reentry vehicle [AERO ENG] A space vehicle designed to utilize aerodynamic lift upon entering the atmosphere.

lift-off [AERO ENG] The action of a rocket vehicle as it leaves its launch pad in a vertical ascent.

lift pump [MECH ENG] A pump for lifting fluid to the pump's own level.

lift truck [MECH ENG] A small hand- or power-operated dolly equipped with a platform or forklift.

lift valve [MECH ENG] A valve that moves perpendicularly to the plane of the valve seat.

light bomber [AERO ENG] Any bomber with a gross weight of less than 100,000 pounds (45,000 kilograms), including bombs; for example, the A-20 and A-26 bombers in World War II.

lighter-than-air craft [AERO ENG] An aircraft, such as a dirigible, that weighs less than the air it displaces.

light-inspection car [MECH ENG] A railway motorcar weighing 400–600 pounds (180–270 kilograms) and having a capacity of 650–800 pounds (295–360 kilograms).

light section car [MECH ENG] A railway motorcar weighing 750–900 pounds (340–408 kilograms) and propelled by 4–6-horsepower (3000–4500-watt) engines.

Lilly controller [MECH ENG] A device on steam and electric winding engines that protects against overspeed, overwind, and other incidents injurious to workers and the engine.

limb [DES ENG] 1. The graduated margin of an arc or circle in an instrument for measuring angles, as that part of a marine sextant carrying the altitude scale. 2. The graduated staff of a leveling rod.

limit control [MECH ENG] 1. In boiler operation, usually a device, electrically controlled, that shuts down a burner at a prescribed operating point. 2. In machine-tool operation, a sensing device which terminates motion of the workpiece or tool at prescribed points.

limit dimensioning method [DES ENG] Method of dimensioning and tolerancing wherein the maximum and minimum permissible values for a dimension are stated specifically to indicate the size or location of the element in question.

limited-pressure cycle *See* mixed cycle.

limited-rotation hydraulic actuator [MECH ENG] A type of hydraulic actuator that produces limited reciprocating rotary force and motion; used for lifting, lowering, opening, closing, indexing, and transferring movements; examples are the piston-rack actuator, single-vane actuator, and double-vane actuator.

limit governor [MECH ENG] A mechanical governor that takes over control from the main governor to shut the machine down when speed reaches a predetermined excess above the allowable rated. Also known as topping governor.

limit-load design *See* ultimate-load design.

limits [DES ENG] In dimensioning, the maximum and minimum values prescribed for a specific dimension; the limits may be of size if the dimension concerned is a size dimension, or they may be of location if the dimension concerned is a location dimension.

limit velocity [MECH] In armor and projectile testing, the lowest possible velocity at which any one of the complete penetrations is obtained; since the limit velocity is difficult to obtain, a more easily obtainable value, designated as the ballistic limit, is usually employed.

Lincoln experimental satellite [AERO ENG] A series of military communication satellites initiated in 1965; carried out successful X band experiments. Abbreviated LES.

linear [CONT SYS] Having an output that varies in direct proportion to the input.

linear actuator [MECH ENG] A device that converts some kind of power, such as hydraulic or electric power, into linear motion.

linear control system [CONT SYS] A linear system whose inputs are forced to change in a desired manner as time progresses.

linear feedback control [CONT SYS] Feedback control in a linear system.

linearization [CONT SYS] 1. The modification of a system so that its outputs are approximately linear functions of its inputs, in order to facilitate analysis of the system. 2. The mathematical approximation of a nonlinear system, whose departures from linearity are small, by a linear system corresponding to small changes in the variables about their average values.

linear motion *See* rectilinear motion.

linear-quadratic-Gaussian problem [CONT SYS] An optimal-state regulator problem, containing Gaussian noise in both the state and measurement equations, in which the expected value of the quadratic performance index is to be minimized. Abbreviated LQG problem.

linear regulator problem [CONT SYS] A type of optimal control problem in which the system to be controlled is described by linear differential equations and the performance index to be minimized is the integral of a quadratic function of the system state and control functions. Also known as optimal regulator problem; regulator problem.

linear strain [MECH] The ratio of the change in the length of a body to its initial length. Also known as longitudinal strain.

linear system [CONT SYS] A system in which the outputs are components of a vector which is equal to the value of a linear operator applied to a vector whose components are the inputs.

linear velocity See velocity.

line lubricator See line oiler.

line of action [MECH ENG] The locus of contact points as gear teeth profiles go through mesh.

line of fall [MECH] The line tangent to the ballistic trajectory at the level point.

line of flight [MECH] The line of movement, or the intended line of movement, of an aircraft, guided missile, or projectile in the air.

line of impact [MECH] A line tangent to the trajectory of a missile at the point of impact.

line-of-sight velocity See radial velocity.

line of thrust [MECH] Locus of the points through which the resultant forces pass in an arch or retaining wall.

line oiler [MECH ENG] An apparatus inserted in a line conducting air or steam to an air- or steam-activated machine that feeds small controllable amounts of lubricating oil into the air or steam. Also known as air-line lubricator; line lubricator.

liner [DES ENG] A replaceable tubular sleeve inside a hydraulic or pump-pressure cylinder in which the piston travels.

liner bushing [DES ENG] A bushing, provided with or without a head, that is permanently installed in a jig to receive the renewable wearing bushings. Also known as master bushing.

line shafting [MECH ENG] One or more pieces of assembled shafting to transmit power from a central source to individual machines.

line space lever [MECH ENG] A lever on a typewriter used to move the carriage to a new line.

line vortex [FL MECH] A type of fluid motion in which fluid flows approximately in circles about a line, at speeds inversely proportional to the distance from the line, so that there is an infinite concentration of vorticity on the line, and vorticity vanishes elsewhere.

lining bar [DES ENG] A crowbar with a pinch, wedge, or diamond point at its working end.

link [DES ENG] **1.** One of the rings of a chain. **2.** A connecting piece in the moving parts of a machine.

linkage [MECH ENG] A mechanism that transfers motion in a desired manner by using some combination of bar links, slides, pivots, and rotating members.

link V belt　[DES ENG] A V belt composed of a large number of rubberized-fabric links joined by metal fasteners.

linter　[MECH ENG] A machine for removing fuzz linters from ginned cottonseed.

lip　[DES ENG] Cutting edge of a fluted drill formed by the intersection of the flute and the lip clearance angle, and extending from the chisel edge at the web to the circumference.

liquid-bubble tracer　[FL MECH] A method of observing the motion of a liquid by following tiny particles of an immiscible liquid of the same density as the moving liquid.

liquid-cooled dissipator　*See* cold plate.

liquid-cooled engine　[MECH ENG] An internal combustion engine with a jacket cooling system in which liquid, usually water, is circulated to maintain acceptable operating temperatures of machine parts.

liquid flow　[FL MECH] The flow or movement of materials in the liquid phase.

liquid holdup　[FL MECH] A condition in two-phase flow through a vertical pipe; when gas flows at a greater linear velocity than the liquid, slippage takes place and liquid holdup occurs.

liquid measure　[MECH] A system of units used to measure the volumes of liquid substances in the United States; the units are the fluid dram, fluid ounce, gill, pint, quart, and gallon.

liquid pint　*See* pint.

liquid piston rotary compressor　[MECH ENG] A rotary compressor in which a multiblade rotor revolves in a casing partly filled with liquid, for example, water.

liquid-sorbent dehumidifier　[MECH ENG] A sorbent type of dehumidifier consisting of a main circulating fan, sorbent-air contactor, sorbent pump, and reactivator; dehumidification and reactivation are continuous operations, with a small part of the sorbent constantly bled off from the main circulating system and reactivated to the concentration required for the desired effluent dew point.

liquidus line　[THERMO] For a two-component system, a curve on a graph of temperature versus concentration which connects temperatures at which fusion is completed as the temperature is raised.

liter　[MECH] A unit of volume or capacity, equal to 1 decimeter cubed, or 0.001 cubic meter, or 1000 cubic centimeters. Abbreviated l.

live axle　[MECH ENG] An axle to which wheels are rigidly fixed.

live center　[MECH ENG] A lathe center that fits into the headstock spindle.

live load　[MECH] A moving load or a load of variable force acting upon a structure, in addition to its own weight.

live-roller conveyor　[MECH ENG] Conveying machine which moves objects over a series of rollers by the application of power to all or some of the rollers.

livre　[MECH] A unit of mass, used in France, equal to 0.5 kilogram.

Ljungström heater　[MECH ENG] Continuous, regenerative, heat-transfer air heater (recuperator) made of slow-moving rotors packed with closely spaced metal plates or wires with a housing to confine the hot and cold gases to opposite sides.

Ljungström steam turbine　[MECH ENG] A radial outward-flow turbine having two opposed rotation rotors.

load　[MECH] 1. The weight that is supported by a structure. 2. Mechanical force that is applied to a body. 3. The burden placed on any machine, measured by units such as horsepower, kilowatts, or tons.

load-and-carry equipment [MECH ENG] Earthmoving equipment designed to load and transport material.

load compensation [CONT SYS] Compensation in which the compensator acts on the output signal after it has generated feedback signals. Also known as load stabilization.

loader [MECH ENG] A machine such as a mechanical shovel used for loading bulk materials.

load factor [MECH] The ratio of load to the maximum rated load.

loading head [MECH ENG] The part of a loader which gathers the bulk materials.

loading station [MECH ENG] A device which receives material and puts it on a conveyor; may be one or more plates or a hopper.

load stabilization *See* load compensation.

load stress [MECH] Stress that results from a pressure or gravitational load.

lobe [DES ENG] A projection on a cam wheel or a noncircular gear wheel.

local buckling [MECH] Buckling of thin elements of a column section in a series of waves or wrinkles.

local coefficient of heat transfer [THERMO] The heat transfer coefficient at a particular point on a surface, equal to the amount of heat transferred to an infinitesimal area of the surface at the point by a fluid passing over it, divided by the product of this area and the difference between the temperatures of the surface and the fluid.

local controller *See* first-level controller.

local derivative [FL MECH] The rate of change of a quantity with respect to time at a fixed point of a fluid, $\partial f/\partial t$; it is related to the individual derivative df/dt through the expression $\partial f/\partial t = df/dt - V \cdot \nabla f$, where f is a thermodynamic property $f(x,y,z,t)$ of the fluid, V the vector velocity of the fluid, and ∇ the del operator.

localized vector [MECH] A vector whose line of application or point of application is prescribed, in addition to its direction.

local Mach number [AERO ENG] The Mach number of an isolated section of an airplane or its airframe.

local structural discontinuity [MECH] The effect of intensified stress on a small portion of a structure.

locating [MECH ENG] A function of tooling operations accomplished by designing and constructing the tooling device so as to bring together the proper contact points or surfaces between the workpiece and the tooling.

locating hole [MECH ENG] A hole used to position the part in relation to a cutting tool or to other parts and gage points.

location dimension [DES ENG] A dimension which specifies the position or distance relationship of one feature of an object with respect to another.

location fit [DES ENG] The characteristic wherein mechanical sizes of mating parts are such that, when assembled, the parts are accurately positioned in relation to each other.

lock [DES ENG] A fastening device in which a releasable bolt is secured.

locked-coil rope [DES ENG] A completely smooth wire rope that resists wear, made of specially formed wires arranged in concentric layers about a central wire core. Also known as locked-wire rope.

locked groove [DES ENG] A blank and continuous groove placed at the end of the modulated grooves on a disk recording to prevent further travel of the pickup. Also known as concentric groove.

locked-wire rope *See* locked-coil rope.

lock front [DES ENG] On a door lock or latch, the plate through which the latching or locking bolt (or bolts) projects.

locking fastener [DES ENG] A fastening used to prevent loosening of a threaded fastener in service, for example, a seating lock, spring stop nut, interference wedge, blind, or quick release.

lock joint [DES ENG] A joint made by interlocking the joined elements, with or without other fastening.

locknut [DES ENG] 1. A nut screwed down firmly against another or against a washer to prevent loosening. Also known as jam nut. 2. A nut that is self-locking when tightened. 3. A nut fitted to the end of a pipe to secure it and prevent leakage.

lock washer [DES ENG] A solid or split washer placed underneath a nut or screw to prevent loosening by exerting pressure.

locomotive [MECH ENG] A self-propelling machine with flanged wheels, for moving loads on railroad tracks; utilizes fuel (for steam or internal combustion engines), compressed air, or electric energy.

locomotive boiler [MECH ENG] An internally fixed horizontal fire-tube boiler with integral furnace; the doubled furnace walls contain water which mixes with water in the boiler shell.

locomotive crane [MECH ENG] A crane mounted on a railroad flatcar or a special chassis with flanged wheels. Also known as rail crane.

logarithmic profile of velocity [FL MECH] The mean velocity parallel to a boundary of a fluid in turbulent motion as a function of distance from the boundary, on the assumption that the shearing stress is independent of distance from the boundary, and the mixing length is proportional either to the distance from the boundary or to the ratio of the first derivative of the profile of velocity itself to the second derivative.

log-mean temperature difference [THERMO] The log-mean temperature difference T_{LM} = $(T_2 - T_1)/\ln T_2/T_1$, where T_2 and T_1 are the absolute (K or °R) temperatures of the two extremes being averaged; used in heat transfer calculations in which one fluid is cooled or heated by a second held separate by pipes or process vessel walls.

longeron [AERO ENG] A principal longitudinal member of the structural framework of a fuselage, nacelle, or empennage boom.

longitudinal acceleration [MECH] The component of the linear acceleration of an aircraft, missile, or particle parallel to its longitudinal, or X, axis.

longitudinal controller [AERO ENG] A primary flight control mechanism which controls pitch attitude; located in the cockpit, this may be a control column or a side stick.

longitudinal drum boiler [MECH ENG] A boiler in which the axis of the horizontal drum is parallel to the tubes, both lying in the same plane.

longitudinal strain *See* linear strain.

longitudinal vibration [MECH] A continuing periodic change in the displacement of elements of a rod-shaped object in the direction of the long axis of the rod.

long-nose pliers [DES ENG] Small pincer with long, tapered jaws.

long ton *See* ton.

look angle [AERO ENG] The elevation and azimuth at which a particular satellite is predicted to be found at a specified time.

loop [AERO ENG] A flight maneuver in which an airplane flies a circular path in an approximately vertical plane, with the lateral axis of the airplane remaining horizontal, that is, an inside loop.

loop gain [CONT SYS] The ratio of the magnitude of the primary feedback signal in a feedback control system to the magnitude of the actuating signal.

loop ratio *See* loop transfer function.

loop transfer function [CONT SYS] For a feedback control system, the ratio of the Laplace transform of the primary feedback signal to the Laplace transform of the actuating signal. Also known as loop ratio.

loop transmittance [CONT SYS] **1.** The transmittance between the source and sink created by the splitting of a specified node in a signal flow graph. **2.** The transmittance between the source and sink created by the splitting of a node which has been inserted in a specified branch of a signal flow graph in such a way that the transmittance of the branch is unchanged.

loose fit [DES ENG] A fit with enough clearance to allow free play of the joined members.

loose-joint butt [DES ENG] A knuckle hinge in which the pin on one half slides easily into a slot on the other half.

loose pulley [MECH ENG] In belt-driven machinery, a pulley which turns freely on a shaft so that the belt can be shifted from the driving pulley to the loose pulley, thereby causing the machine to stop.

lopping shears [DES ENG] Long-handled shears used for pruning branches.

loss-in-weight feeder [MECH ENG] A device to apportion the output of granulated or powdered solids at a constant rate from a feed hopper; weight-measured decrease in hopper content actuates further opening of the discharge chute to compensate for flow loss as the hopper overburden decreases; used in the chemical, fertilizer, and plastics industries.

loss of head [FL MECH] Energy decrease between two points in a hydraulic system due to such causes as friction, bends, obstructions, or expansions.

lost motion [MECH ENG] The delay between the movement of a driver and the movement of a follower.

Lowenhertz thread [DES ENG] A screw thread that differs from U.S. Standard form in that the angle between the flanks measured on an axial plane is 53°8′; height equals 0.75 times the pitch, and width of flats at top and bottom equals 0.125 times the pitch.

lower pair [MECH ENG] A link in a mechanism in which the mating parts have surface (instead of line or point) contact.

lowest safe waterline [MECH ENG] The lowest water level in a boiler drum at which the burner may safely operate.

low heat value [THERMO] The heat value of a combustion process assuming that none of the water vapor resulting from the process is condensed out, so that its latent heat is not available. Also known as lower heating value; net heating value.

low-helix drill [DES ENG] A two-flute twist drill with a lower helix angle than a conventional drill. Also known as slow-spiral drill.

low-intensity atomizer [MECH ENG] A type of electrostatic atomizer operating on the principle that atomization is the result of Rayleigh instability, in which the presence of charge in the surface counteracts surface tension.

low-level condenser [MECH ENG] A direct-contact water-cooled steam condenser that uses a pump to remove liquid from a vacuum space.

low-lift truck [MECH ENG] A hand or powered lift truck that raises the load sufficiently to make it mobile.

low-pressure area [MECH ENG] The point in a bearing where the pressure is the least and the area or space for a lubricant is the greatest.

low-pressure fluid flow [FL MECH] Flow of fluids below atmospheric pressures, particularly gases and vapors following ideal gas laws, in pipes, fittings, and other common configurations.

low velocity [MECH] Muzzle velocity of an artillery projectile of 2499 feet (762 meters) per second or less.

low-water fuel cutoff [MECH ENG] A float device which shuts off fuel supply and burner when boiler water level drops below the lowest safe waterline.

lozenge file [DES ENG] A small file with four sides and a lozenge-shaped cross section; used in forming dies.

LQG problem See linear-quadratic-Gaussian problem.

Luenberger observer [CONT SYS] A compensator driven by both the inputs and measurable outputs of a control system.

lug [DES ENG] A projection or head on a metal part to serve as a cap, handle, support, or fitting connection.

lug bolt [DES ENG] 1. A bolt with a flat extension or hook instead of a head. 2. A bolt designed for securing a lug.

Luna program [AERO ENG] A series of Soviet space probes launched for flight missions to the moon.

lunar excursion module [AERO ENG] A manned spacecraft designed to be carried on top of the Apollo service module and having its own power plant for making a manned landing on the moon and a return from the moon to the orbiting Apollo spacecraft. Abbreviated LEM. Also known as lunar module (LM).

lunar flight [AERO ENG] Flight by a spacecraft to the moon.

lunar module See lunar excursion module.

lunar orbit [AERO ENG] Orbit of a spacecraft around the moon.

lunar probe [AERO ENG] Any space probe launched for flight missions to the moon.

lunar satellite [AERO ENG] A satellite making one or more revolutions about the moon.

lunar spacecraft [AERO ENG] A spacecraft designed for flight to the moon.

Lyapunov stability criterion [CONT SYS] A method of determining the stability of systems (usually nonlinear) by examining the sign-definitive properties of an associated Lyapunov function.

m *See* meter; milli-.

mμ *See* millimicron.

Mach angle [FL MECH] The vertex half angle of the Mach cone generated by a body in supersonic flight.

Mach cone [FL MECH] **1.** The cone-shaped shock wave theoretically emanating from an infinitesimally small particle moving at supersonic speed through a fluid medium; it is the locus of the Mach lines. **2.** The cone-shaped shock wave generated by a sharp-pointed body, as at the nose of a high-speed aircraft.

machete [DES ENG] A knife with a broad blade 2 to 3 feet (60 to 90 centimeters) long.

Mach front *See* Mach stem.

machine [MECH ENG] A combination of rigid or resistant bodies having definite motions and capable of performing useful work.

machine bolt [DES ENG] A heavy-weight bolt with a square, hexagonal, or flat head used in the automotive, aircraft, and machinery fields.

machine design [DES ENG] Application of science and invention to the development, specification, and construction of machines.

machine drill [MECH ENG] Any mechanically driven diamond, rotary, or percussive drill.

machine element [DES ENG] Any of the elementary mechanical parts, such as gears, bearings, fasteners, screws, pipes, springs, and bolts used as essentially standardized components for most devices, apparatus, and machinery.

machine file [DES ENG] A file that can be clamped in the chuck of a power-driven machine.

machine key [DES ENG] A piece inserted between a shaft and a hub to prevent relative rotation. Also known as key.

machinery [MECH ENG] A group of parts or machines arranged to perform a useful function.

machine screw [DES ENG] A blunt-ended screw with a standardized thread and a head that may be flat, round, fillister, or oval, and may be slotted, or constructed for wrenching; used to fasten machine parts together.

machine setting *See* mechanical setting.

machine taper [MECH ENG] A taper that provides a connection between a tool, arbor, or center and its mating part to ensure and maintain accurate alignment between the parts; permits easy separation of parts.

machine tool [MECH ENG] A stationary power-driven machine for the shaping, cutting, turning, boring, drilling, grinding, or polishing of solid parts, especially metals.

machinist's file [DES ENG] A type of double-cut file that removes metal fast and is used for rough metal filing.

Mach line [FL MECH] 1. A line representing a Mach wave. 2. *See* Mach wave.

Mach number [FL MECH] The ratio of the speed of a body or of a point on a body with respect to the surrounding air or other fluid, or the ratio of the speed of a fluid, to the speed of sound in the medium. Symbolized N_{Ma}. Also known as relative Mach number.

Mach reflection [FL MECH] The reflection of a shock wave from a rigid wall in which the shock strength of the reflected wave and the angle of reflection both have the smaller of the two values which are theoretically possible.

Mach stem [FL MECH] A shock wave or front formed above the surface of the earth by the fusion of direct and reflected shock waves resulting from an airburst bomb. Also known as Mach front.

Mach wave [FL MECH] Also known as Mach line. 1. A shock wave theoretically occurring along a common line of intersection of all the pressure disturbances emanating from an infinitesimally small particle moving at supersonic speed through a fluid medium, with such a wave considered to exert no changes in the condition of the fluid passing through it. 2. A very weak shock wave appearing, for example, at the nose of a very sharp body, where the fluid undergoes no substantial change in direction.

Macleod equation [FL MECH] An equation which states that the fourth root of the surface tension of a liquid is proportional to the difference between the densities of the liquid and of its vapor.

MacMichael degree [FL MECH] An arbitrary unit used in measuring viscosity with a type of Couette viscometer; its size depends on the stiffness of the suspension of the inner cylinder of the viscometer.

macrorheology [MECH] A branch of rheology in which materials are treated as homogeneous or quasi-homogeneous, and processes are treated as isothermal.

macroscopic property [THERMO] *See* thermodynamic property.

magnetic brake [MECH ENG] A friction brake under the control of an electromagnet.

magnetic chuck [MECH ENG] A chuck in which the workpiece is held by magnetic force.

magnetic clutch *See* magnetic fluid clutch; magnetic friction clutch.

magnetic fluid clutch [MECH ENG] A friction clutch that is engaged by magnetizing a liquid suspension of powdered iron located between pole pieces mounted on the input and output shafts. Also known as magnetic clutch.

magnetic friction clutch [MECH ENG] A friction clutch in which the pressure between the friction surfaces is produced by magnetic attraction. Also known as magnetic clutch.

magnetocaloric effect [THERMO] The reversible change of temperature accompanying the change of magnetization of a ferromagnetic material.

magnetohydrodynamic arcjet [AERO ENG] An electromagnetic propulsion system utilizing a plasma that is heated in an electric arc and then adiabatically expanded through a nozzle and further accelerated by a crossed electric and magnetic field.

Magnus effect [FL MECH] A force on a rotating cylinder in a fluid flowing perpendicular to the axis of the cylinder; the force is perpendicular to both flow direction and cylinder axis. Also known as Magnus force.

Magnus force *See* Magnus effect.

Magnus moment [FL MECH] A torque associated with the Magnus effect, such as moments about the pitch and yaw axes of a missile or aircraft due to rotation about the roll axis.

main shaft [MECH ENG] The line of shafting receiving its power from the engine or motor and transmitting power to other parts.

major diameter [DES ENG] The largest diameter of a screw thread, measured at the crest for an external (male) thread and at the root for an internal (female) thread.

mallet [DES ENG] An implement with a barrel-shaped head made of wood, rubber, or other soft material; used for driving another tool, such as a chisel, or for striking a surface without causing damage.

mandrel [MECH ENG] A shaft inserted through a hole in a component to support the work during machining.

mandrel press [MECH ENG] A press for driving mandrels into holes.

mangle gearing [MECH ENG] Gearing for producing reciprocating motion; a pinion rotating in a single direction drives a rack with teeth at the ends and on both sides.

manifold pressure [MECH ENG] The pressure in the intake manifold of an internal combustion engine.

manned orbiting laboratory [AERO ENG] An earth-orbiting satellite containing instrumentation and personnel for continuous measurement and surveillance of the earth, its atmosphere, and space. Abbreviated MOL.

manned spacecraft [AERO ENG] A vehicle capable of sustaining a person above the terrestrial atmosphere.

Manning equation [FL MECH] An equation used to compute the velocity of uniform flow in an open channel.

manocryometer [THERMO] An instrument for measuring the change of a substance's melting point with change in pressure; the height of a mercury column in a U-shaped capillary supported by an equilibrium between liquid and solid in an adjoining bulb is measured, and the whole apparatus is in a thermostat.

M-A-N scavenging system [MECH ENG] A system for removing used oil and waste gases from a cylinder of an internal combustion engine in which the exhaust ports are located above the intake ports on the same side of the cylinder, so that gases circulate in a loop, leaving a dead spot in the center of the loop.

many-body problem [MECH] The problem of predicting the motions of three or more objects obeying Newton's laws of motion and attracting each other according to Newton's law of gravitation. Also known as *n*-body problem.

margin of safety [DES ENG] A design criterion, usually the ratio between the load that would cause failure of a member or structure and the load that is imposed upon it in service.

Margoulis number *See* Stanton number.

Mariner program [AERO ENG] A United States program, begun in 1962, to send a series of uncrewed, solar-powered spacecraft to the vicinity of Venus and Mars, to carry out observations of cosmic rays and solar wind in interplanetary space, to investigate interactions of the solar wind with planets, and to make atmospheric and surface measurements of planets, including photographic scanning of the surface of Mars.

MARISAT [AERO ENG] A geostationary communication satellite equipped with a repeater operating at microwave frequencies, for ship-to-shore communication by satellite. Derived from maritime satellite.

maritime satellite *See* MARISAT.

Mars probe [AERO ENG] A United States uncrewed spacecraft intended to be sent to the vicinity of the planet Mars, such as in the Mariner or Viking programs.

mask [DES ENG] A frame used in front of a television picture tube to conceal the rounded edges of the screen.

masonry drill [DES ENG] A drill tipped with cemented carbide for drilling in concrete or masonry.

masonry nail [DES ENG] Spiral-fluted nail designed to be driven into mortar joints in masonry.

Mason's theorem [CONT SYS] A formula for the overall transmittance of a signal flow graph in terms of transmittances of various paths in the graph.

mass [MECH] A quantitative measure of a body's resistance to being accelerated; equal to the inverse of the ratio of the body's acceleration to the acceleration of a standard mass under otherwise identical conditions.

mass divergence [FL MECH] The divergence of the momentum field, a measure of the rate of net flux of mass out of a unit volume of a system; in symbols, $\nabla \cdot \rho V$, where ρ is the fluid density, V the velocity vector, and ∇ the del operator.

mass flow [FL MECH] The mass of a fluid in motion which crosses a given area in a unit time.

Massieu function [THERMO] The negative of the Helmholtz free energy divided by the temperature.

mass ratio [AERO ENG] The ratio of the mass of the propellant charge of the rocket to the total mass of the rocket when charged with the propellant.

mass transport [FL MECH] 1. Carrying of loose materials in a moving medium such as water or air. 2. The movement of fluid, especially water, from one place to another.

mass units [MECH] Units of measurement having to do with masses of materials, such as pounds or grams.

mass velocity [FL MECH] The weight flow rate of a fluid divided by the cross-sectional area of the enclosing chamber or conduit; for example, lb/hr ft^2.

master bushing *See* liner bushing.

master cylinder [MECH ENG] The container for the fluid and the piston, forming part of a device such as a hydraulic brake or clutch.

master gage [DES ENG] A locating device with fixed hole locations or part positions; locates in three dimensions and generally occupies the same space as the part it represents.

master layout [DES ENG] A permanent template record laid out in reference planes and used as a standard of reference in the development and coordination of other templates.

material particle [MECH] An object which has rest-mass and an observable position in space, but has no geometrical extension, being confined to a single point. Also known as particle.

Matheson joint [DES ENG] A wrought-pipe joint made by enlarging the end of one pipe length to receive the male end of the next length.

mattock [DES ENG] A tool with the combined features of an adz, an ax, and a pick.

Maupertius' principle [MECH] The principle of least action is sufficient to determine the motion of a mechanical system.

maximum allowable working pressure [MECH ENG] The maximum gage pressure in a pressure vessel at a designated temperature, used for the determination of the set pressure for relief valves.

maximum angle of inclination   [MECH ENG] The maximum angle at which a conveyor may be inclined and still deliver an amount of bulk material within a given time.

maximum belt slope   [MECH ENG] A slope beyond which the material on the belt of a conveyor tends to roll downhill.

maximum belt tension   [MECH ENG] The total of the starting and operating tensions in a conveyor.

maximum continuous load   [MECH ENG] The maximum load that a boiler can maintain for a designated length of time.

maximum gradability   [MECH ENG] Steepest slope a vehicle can negotiate in low gear; usually expressed in precentage of slope, namely, the ratio between the vertical rise and the horizontal distance traveled; sometimes expressed by the angle between the slope and the horizontal.

maximum ordinate   [MECH] Difference in altitude between the origin and highest point of the trajectory of a projectile.

maximum production life   [MECH ENG] The length of time that a cutting tool performs at cutting conditions of maximum tool efficiency.

Maxwell body   *See* Maxwell liquid.

Maxwell equal area rule   [THERMO] At temperatures for which the theoretical isothermal of a substance, on a graph of pressure against volume, has a portion with positive slope (as occurs in a substance with liquid and gas phases obeying the van der Waals equation), a horizontal line drawn at the equilibrium vapor pressure and connecting two parts of the isothermal with negative slope has the property that the area between the horizontal and the part of the isothermal above it is equal to the area between the horizontal and the part of the isothermal below it.

Maxwell liquid   [FL MECH] A liquid whose rate of deformation is the sum of a term proportional to the shearing stress acting on it and a term proportional to the rate of change of this stress. Also known as Maxwell body.

Maxwell relation   [THERMO] One of four equations for a system in thermal equilibrium, each of which equates two partial derivatives, involving the pressure, volume, temperature, and entropy of the system.

Maxwell's coefficient of diffusion   [FL MECH] A number in an equation for the difference between mean velocities of two gases which are allowed to mix, which determines the contribution to this quantity of the concentration gradient.

Maxwell's demon   *See* demon of Maxwell.

Maxwell's stress functions   [MECH] Three functions of position, ϕ_1, ϕ_2, and ϕ_3, in terms of which the elements of the stress tensor σ of a body may be expressed, if the body is in equilibrium and is not subjected to body forces; the elements of the stress tensor are given by $\sigma_{11} = \partial^2\phi_2/\partial x_3^2 + \partial^2\phi_3/\partial x_2^2$, $\sigma_{23} = -\partial^2\phi_1/\partial x_2\partial x_3$, and cyclic permutations of these equations.

Maxwell's theorem   [MECH] If a load applied at one point A of an elastic structure results in a given deflection at another point B, then the same load applied at B will result in the same deflection at A.

mayer   [THERMO] A unit of heat capacity equal to the heat capacity of a substance whose temperature is raised $1°$ centigrade by 1 joule.

mb   *See* millibar.

McLeod gage   [FL MECH] A type of instrument used to measure vacuum by measuring the height of a column of mercury supported by the gas whose pressure is to be measured, when this gas is trapped and compressed into a capillary tube.

M contour [CONT SYS] A line on a Nyquist diagram connecting points having the same magnitude of the primary feedback ratio.

M-design bit [DES ENG] A long-shank, box-threaded core bit made to fit M-design core barrels.

M-design core barrel [DES ENG] A double-tube core barrel in which a 2½°-taper core lifter is carried inside a short tubular sleeve coupled to the bottom end of the inner tube, and the sleeve extends downward inside the bit shank to within a very short distance behind the face of the core bit.

mean British thermal unit *See* British thermal unit.

mean calorie [THERMO] One-hundredth of the heat needed to raise 1 gram of water from 0 to 100°C.

mean camber line [AERO ENG] A line on a cross section of a wing of an aircraft which is equidistant from the upper and lower surfaces of the wing.

mean chord [AERO ENG] That chord of an airfoil that is equal to the sum of all the airfoil's chord lengths divided by the number of chord lengths; equivalently, that chord whose length is equal to the area of the airfoil section divided by the span.

mean effective pressure [MECH ENG] A term commonly used in the evaluation for positive displacement machinery performance which expresses the average net pressure difference in pounds per square inch on the two sides of the piston in engines, pumps, and compressors. Abbreviated mep; mp. Also known as mean pressure.

mean normal stress [MECH] In a system stressed multiaxially, the algebraic mean of the three principal stresses.

mean pressure *See* mean effective pressure.

mean specific heat [THERMO] The average over a specified range of temperature of the specific heat of a substance.

mean-square-error criterion [CONT SYS] Evaluation of the performance of a control system by calculating the square root of the average over time of the square of the difference between the actual output and the output that is desired.

mean stress [MECH] 1. The algebraic mean of the maximum and minimum values of a periodically varying stress. 2. *See* octahedral normal stress.

mean trajectory [MECH] The trajectory of a missile that passes through the center of impact or center of burst.

measured relieving capacity [DES ENG] The measured amounts of fluid which can be exhausted through a relief device at its rated operating pressure.

mechanical advantage [MECH ENG] The ratio of the force produced by a machine such as a lever or pulley to the force applied to it.

mechanical analysis [MECH ENG] Mechanical separation of soil, sediment, or rock by sieving, screening, or other means to determine particle-size distribution.

mechanical classification [MECH ENG] A sorting operation in which mixtures of particles of mixed sizes, and often of different specific gravities, are separated into fractions by the action of a stream of fluid, usually water.

mechanical classifier [MECH ENG] Any of various machines that are commonly used to classify mixtures of particles of different sizes, and sometimes of different specific gravities; the Dorr classifier is an example.

mechanical draft [MECH ENG] A draft that depends upon the use of fans or other mechanical devices; may be induced or forced.

mechanical-draft cooling tower [MECH ENG] Cooling tower that depends upon fans for introduction and circulation of its air supply.

mechanical efficiency [MECH ENG] In an engine, the ratio of brake horsepower to indicated horsepower.

mechanical equivalent of heat [THERMO] The amount of mechanical energy equivalent to a unit of heat.

mechanical hysteresis [MECH] The dependence of the strain of a material not only on the instantaneous value of the stress but also on the previous history of the stress; for example, the elongation is less at a given value of tension when the tension is increasing than when it is decreasing.

mechanical impedance [MECH] The complex ratio of a phasor representing a sinusoidally varying force applied to a system to a phasor representing the velocity of a point in the system.

mechanical linkage [MECH ENG] A set of rigid bodies, called links, joined together at pivots by means of pins or equivalent devices.

mechanical loader [MECH ENG] A power machine for loading mineral, coal, or dirt.

mechanical ohm [MECH] A unit of mechanical resistance, reactance, and impedance, equal to a force of 1 dyne divided by a velocity of 1 centimeter per second.

mechanical press [MECH ENG] A press whose slide is operated by mechanical means.

mechanical property [MECH] A property that involves a relationship between stress and strain or a reaction to an applied force.

mechanical pulping [MECH ENG] Mechanical, rather than chemical, recovery of cellulose fibers from wood; unpurified, finely ground wood is made into newsprint, cheap Manila papers, and tissues.

mechanical pump [MECH ENG] A pump through which fluid is conveyed by direct contact with a moving part of the pumping machinery.

mechanical reactance [MECH] The imaginary part of mechanical impedance.

mechanical refrigeration [MECH ENG] The removal of heat by utilizing a refrigerant subjected to cycles of refrigerating thermodynamics and employing a mechanical compressor.

mechanical resistance *See* resistance.

mechanical rotational impedance *See* rotational impedance.

mechanical rotational reactance *See* rotational reactance.

mechanical rotational resistance *See* rotational resistance.

mechanical seal [MECH ENG] Mechanical assembly that forms a leakproof seal between flat, rotating surfaces to prevent high-pressure leakage.

mechanical separation [MECH ENG] A group of industrial operations by means of which particles of solid or drops of liquid are removed from a gas or liquid, or are separated into individual fractions, or both, by gravity separation (settling), centrifugal action, and filtration.

mechanical setting [MECH ENG] Producing bits by setting diamonds in a bit mold into which a cast or powder metal is placed, thus embedding the diamonds and forming the bit crown; opposed to hand setting. Also known as cast setting; machine setting; sinter setting.

mechanical shovel [MECH ENG] A loader limited to level or slightly graded drivages; when full, the shovel is swung over the machine, and the load is discharged into containers or vehicles behind.

mechanical spring *See* spring.

mechanical stoker *See* automatic stoker.

mechanical torque converter [MECH ENG] A torque converter, such as a pair of gears, that transmits power with only incidental losses.

mechanical units [MECH] Units of length, time, and mass, and of physical quantities derivable from them.

mechanical vibration [MECH] A motion, usually unintentional and often undesirable, of parts of machines and structures.

mechanism [MECH ENG] That part of a machine which contains two or more pieces so arranged that the motion of one compels the motion of the others.

mechanize [MECH ENG] 1. To substitute machinery for human or animal labor. 2. To produce or reproduce by machine.

mechanomotive force [MECH] The root-mean-square value of a periodically varying force.

megasecond [MECH] A unit of time, equal to 1,000,000 seconds. Abbreviated Ms.

megawatt [MECH] A unit of power, equal to 1,000,000 watts. Abbreviated MW.

Meinzer unit *See* permeability coefficient.

melt fracture [MECH] Melt flow instability through a die during plastics molding, leading to helicular, rippled surface irregularities on the finished product.

melting point [THERMO] 1. The temperature at which a solid of a pure substance changes to a liquid. Abbreviated mp. 2. For a solution of two or more components, the temperature at which the first trace of liquid appears as the solution is heated.

melt instability [MECH] Instability of the plastic melt flow through a die.

melt strength [MECH] Strength of a molten plastic.

membrane stress [MECH] Stress which is equivalent to the average stress across the cross section involved and normal to the reference plane.

meniscus [FL MECH] The free surface of a liquid which is near the walls of a vessel and which is curved because of surface tension.

mep *See* mean effective pressure.

Mercer engine [MECH ENG] A revolving-block engine in which two opposing pistons operate in a single cylinder with two rollers attached to each piston; intake ports are uncovered when the pistons are closest together, and exhaust ports are uncovered when they are farthest apart.

Mercury program [AERO ENG] First United States program to use crewed spacecraft, carried out in 1961–1963; the craft carried one man.

Merrick Weightometer [MECH ENG] An instrument which measures the weight of material being carried by a belt conveyor, in which the weight of the load on a portion of the conveyor is balanced, using a system of levers, by the buoyancy of a cylindrical steel float partly immersed in a bath of mercury.

Mersenne's law [MECH] The fundamental frequency of a vibrating string is proportional to the square root of the tension and inversely proportional both to the length and the square root of the mass per unit length.

Merton nut [DES ENG] A nut whose threads are made of an elastic material such as cork, and are formed by compressing the material into a screw.

mesh [DES ENG] A size of screen or of particles passed by it in terms of the number of openings occurring per linear inch. Also known as mesh size. [MECH ENG] Engagement or working contact of teeth of gears or of a gear and a rack.

metacenter [FL MECH] The intersection of a vertical line through the center of buoyancy of a floating body, slightly displaced from its equilibrium position, with a line connecting the center of gravity and the equilibrium center of buoyancy; the floating body is stable if the metacenter lies above the center of gravity.

metal-slitting saw [MECH ENG] A milling cutter similar to a circular saw blade but sometimes with side teeth as well as teeth around the circumference; used for deep slotting and sinking in cuts.

metal spinning *See* spinning.

metarheology [MECH] A branch of rheology whose approach is intermediate between those of macrorheology and microrheology; certain processes that are not isothermal are taken into consideration, such as kinetic elasticity, surface tension, and rate processes.

meteorological satellite [AERO ENG] Earth-orbiting spacecraft carrying a variety of instruments for measuring visible and invisible radiations from the earth and its atmosphere.

meter [MECH] The international standard unit of length, equal to 1,650,763.73 times the wavelength of the orange light emitted when a gas consisting of the pure krypton isotope of mass number 86 is excited in an electrical discharge. Abbreviated m.

metering screw [MECH ENG] An extrusion-type screw feeder or conveyor section used to feed pulverized or doughy material at a constant rate.

meter-kilogram [MECH] **1.** A unit of energy or work in a meter-kilogram-second gravitational system, equal to the work done by a kilogram-force when the point at which the force is applied is displaced 1 meter in the direction of the force; equal to 9.80665 joules. Abbreviated m-kgf. Also known as meter kilogram-force. **2.** A unit of torque, equal to the torque produced by a kilogram-force acting at a perpendicular distance of 1 meter from the axis of rotation. Also known as kilogram-meter (kgf-m).

meter kilogram-force *See* meter-kilogram.

meter-kilogram-second system [MECH] A metric system of units in which length, mass, and time are fundamental quantities, and the units of these quantities are the meter, the kilogram, and the second respectively. Abbreviated mks system.

meter sizing factor [FL MECH] A dimensionless number used in calculating the rate of flow of fluid through a pipe from the readings of a flowmeter that measures the drop in pressure when the fluid is forced to flow through a circular orifice; it is equal to $K(d/D)^2$, where K is the flow coefficient, d is the orifice bore diameter, and D is the internal diameter of the pipe.

meter-ton-second system [MECH] A modification of the meter-kilogram-second system in which the metric ton (1000 kilograms) replaces the kilogram as the unit of mass.

method of mixtures [THERMO] A method of determining the heat of fusion of a substance whose specific heat is known, in which a known amount of the solid is combined with a known amount of the liquid in a calorimeter, and the decrease in the liquid temperature during melting of the solid is measured.

metric centner [MECH] **1.** A unit of mass equal to 50 kilograms. **2.** A unit of mass equal to 100 kilograms. Also known as quintal.

metric grain [MECH] A unit of mass, equal to 50 milligrams; used in commercial transactions in precious stones.

metric line *See* millimeter.

metric ounce *See* mounce.

metric slug *See* metric-technical unit of mass.

metric system [MECH] A system of units used in scientific work throughout the world and employed in general commercial transactions and engineering applications; its units of length, time, and mass are the meter, second, and kilogram respectively, or decimal multiples and submultiples thereof.

metric-technical unit of mass [MECH] A unit of mass, equal to the mass which is accelerated by 1 meter per second per second by a force of 1 kilogram-force; it is equal to 9.80665 kilograms. Abbreviated TME. Also known as hyl; metric slug.

metric thread gearing [DES ENG] Gears that may be interchanged in change-gear systems to provide feeds suitable for cutting metric and module threads.

metric ton *See* tonne.

mi *See* mile.

microangstrom [MECH] A unit of length equal to one-millionth of an angstrom, or 10^{-16} meter. Abbreviated μA.

microbar *See* barye.

microfluid [FL MECH] A fluid in which the effects of local motion of contained material particles on properties and behavior of the fluid are not disregarded.

microgram [MECH] A unit of mass equal to one-millionth of a gram. Abbreviated μg.

micrometeorite penetration [AERO ENG] Penetration of the thin outer shell (skin) of space vehicles by small particles traveling in space at high velocities.

micrometer [MECH] A unit of length equal to one-millionth of a meter. Abbreviated μm. Formerly known as micron (μ).

micrometer of mercury *See* micron.

micromicrosecond *See* picosecond.

micromicrowatt *See* picowatt.

micron [MECH] 1. A unit of pressure equal to the pressure exerted by a column of mercury 1 micrometer high, having a density of 13.5951 grams per cubic centimeter, under the standard acceleration of gravity; equal to 0.133322387415 pascal; it differs from the millitorr by less than one part in seven million. Also known as micrometer of mercury. 2. *See* micrometer.

micro-reciprocal-degree *See* mired.

microrheology [MECH] A branch of rheology in which the heterogeneous nature of dispersed systems is taken into account.

microsecond [MECH] A unit of time equal to one-millionth of a second. Abbreviated μs.

microwatt [MECH] A unit of power equal to one-millionth of a watt. Abbreviated μW.

Midas [AERO ENG] A two-object trajectory-measuring system whereby two complete cotar antenna systems and two sets of receivers at each station, with the multiplexing done after phase comparison, are utilized in tracking more than one object at a time.

midcourse correction [AERO ENG] A change in the course of a spacecraft some time between the end of the launching phase and some arbitrary point when terminal guidance begins.

Mie-Grüneisen equation [THERMO] An equation of state particularly useful at high pressure, which states that the volume of a system times the difference between the pressure and the pressure at absolute zero equals the product of a number which depends only on the volume times the difference between the internal energy and the internal energy at absolute zero.

mil [MECH] 1. A unit of length, equal to 0.001 inch, or to 2.54×10^{-5} meter. Also known as milli-inch; thou. 2. *See* milliliter.

mile [MECH] A unit of length in common use in the United States, equal to 5280 feet, or 1609.344 meters. Abbreviated mi. Also known as land mile; statute mile.

military aircraft [AERO ENG] Aircraft that are designed or modified for highly specialized use by the armed services of a nation.

military satellite [AERO ENG] An artificial earth satellite used for military purposes; the six mission categories are communication, navigation, geodesy, nuclear test detection, surveillance, and research and technology.

miller *See* milling machine.

millibar [MECH] A unit of pressure equal to one-thousandth of a bar. Abbreviated mb. Also known as vac.

millier *See* tonne.

milligal [MECH] A unit of acceleration commonly used in geodetic measurements, equal to 10^{-3} galileo, or 10^{-5} meter per second per second. Abbreviated mGal.

milligram [MECH] A unit of mass equal to one-thousandth of a gram. Abbreviated mg.

millihg *See* millimeter of mercury.

milli-inch *See* mil.

milliliter [MECH] A unit of volume equal to 10^{-3} liter or 10^{-6} cubic meter. Abbreviated ml. Also known as mil.

millimeter [MECH] A unit of length equal to one-thousandth of a meter. Abbreviated mm. Also known as metric line; strich.

millimeter of mercury [MECH] A unit of pressure, equal to the pressure exerted by a column of mercury 1 millimeter high with a density of 13.5951 grams per cubic centimeter under the standard acceleration of gravity; equal to 133.322387415 pascals; it differs from the torr by less than 1 part in 7,000,000. Abbreviated mmHg. Also known as millihg.

millimeter of water [MECH] A unit of pressure, equal to the pressure exerted by a column of water 1 millimeter high with a density of 1 gram per cubic centimeter under the standard acceleration of gravity; equal to 9.80665 pascals. Abbreviated mmH_2O.

millimicron [MECH] A unit of length equal to one-thousandth of a micron (or micrometer) or one-billionth of a meter. Abbreviated mμ.

milling [MECH ENG] Mechanical treatment of materials to produce a powder, to change the size or shape of metal powder particles, or to coat one powder mixture with another.

milling cutter [DES ENG] A rotary tool-steel cutting tool with peripheral teeth, used in a milling machine to remove material from the workpiece through the relative motion of workpiece and cutter.

milling machine [MECH ENG] A machine for the removal of metal by feeding a workpiece through the periphery of a rotating circular cutter. Also known as miller.

milling planer [MECH ENG] A planer that uses a rotary cutter rather than single-point tools.

millisecond [MECH] A unit of time equal to one-thousandth of a second. Abbreviated ms.

milliwatt [MECH] A unit of power equal to one-thousandth of a watt. Abbreviated mW.

min *See* minim.

mine car [MECH ENG] An industrial car, usually of the four-wheel type, with a low body; the door is at one end, pivoted at the top with a latch at the bottom used for hauling bulk materials.

minim [MECH] A unit of volume in the apothecaries' measure; equals $\frac{1}{60}$ fluidram (approximately 0.061612 cubic centimeter) or about 1 drop (of water). Abbreviated min.

minimal realization [CONT SYS] In linear system theory, a set of differential equations, of the smallest possible dimension, which have an input/output transfer function matrix equal to a given matrix function $G(s)$.

mini-maxi regret [CONT SYS] In decision theory, a criterion which selects that strategy which has the smallest maximum difference between its payoff and that of the best hindsight choice.

minimum flight altitude [AERO ENG] The lowest altitude at which aircraft may safely operate.

minimum metal condition [DES ENG] The condition corresponding to the removal of the greatest amount of material permissible in a machined part.

minimum-phase system [CONT SYS] A linear system for which the poles and zeros of the transfer function all have negative or zero real parts.

minimum resolvable temperature difference [THERMO] The change in equivalent blackbody temperature that corresponds to a change in radiance which will produce a just barely resolvable change in the output of an infrared imaging device, taking into account the characteristics of the device, the display, and the observer. Abbreviated MRTD.

minitrack [AERO ENG] A satellite tracking system consisting of a field of separate antennas and associated receiving equipment interconnected so as to form interferometers which track a transmitting beacon in the payload itself.

minor diameter [DES ENG] The diameter of a cylinder bounding the root of an external thread or the crest of an internal thread.

minor loop [CONT SYS] A portion of a feedback control system that consists of a continuous network containing both forward elements and feedback elements.

minute [MECH] A unit of time, equal to 60 seconds.

mired [THERMO] A unit used to measure the reciprocal of color temperature, equal to the reciprocal of a color temperature of 10^6 kelvins. Derived from micro-reciprocal-degree.

missile attitude [MECH] The position of a missile as determined by the inclination of its axes (roll, pitch, and yaw) in relation to another object, as to the earth.

mist [FL MECH] Fine liquid droplets suspended in or falling through a moving or stationary gas atmosphere.

miter bend [DES ENG] A pipe bend made by mitering (angle cutting) and joining pipe ends.

miter gear [DES ENG] A bevel gear whose bevels are in 1:1 ratio.

miter joint [DES ENG] A joint, usually perpendicular, in which the mating ends are beveled.

miter saw [DES ENG] A hollow-ground saw in diameters from 6 to 16 inches (15.24 to 40.64 centimeters), used for cutting off and mitering on light stock such as moldings and cabinet work.

miter valve [DES ENG] A valve in which a disk fits in a seat making a 45° angle with the axis of the valve.

mixed cycle [MECH ENG] An internal combustion engine cycle which combines the Otto cycle constant-volume combustion and the Diesel cycle constant-pressure combustion in high-speed compression-ignition engines. Also known as combination cycle; commercial Diesel cycle; limited-pressure cycle.

mixed-flow impeller [MECH ENG] An impeller for a pump or compressor which combines radial- and axial-flow principles.

mixture ratio [AERO ENG] The ratio of the weight of oxidizer used per unit of time to the weight of fuel used per unit of time.

m-kgf *See* meter-kilogram.

mks system *See* meter-kilogram-second system.

ml *See* milliliter.

mm *See* millimeter.

mmHg *See* millimeter of mercury.

mmH₂O *See* millimeter of water.

mobile crane [MECH ENG] 1. A cable-controlled crane mounted on crawlers or rubber-tired carriers. 2. A hydraulic-powered crane with a telescoping boom mounted on truck-type carriers or as self-propelled models.

mobile hoist [MECH ENG] A platform hoist mounted on a pair of pneumatic-tired road wheels, so it can be towed from one site to another.

mobile loader [MECH ENG] A self-propelling power machine for loading coal, mineral, or dirt.

model reduction [CONT SYS] The process of discarding certain modes of motion while retaining others in the model used by an active control system, in order that the control system can compute control commands with sufficient rapidity.

model reference system [CONT SYS] An ideal system whose response is agreed to be optimum; computer simulation in which both the model system and the actual system are subjected to the same stimulus is carried out, and parameters of the actual system are adjusted to minimize the difference in the outputs of the model and the actual system.

mode of oscillation *See* mode of vibration.

mode of vibration [MECH] A characteristic manner in which a system which does not dissipate energy and whose motions are restricted by boundary conditions can oscillate, having a characteristic pattern of motion and one of a discrete set of frequencies. Also known as mode of oscillation.

modulation [MECH ENG] Regulation of the fuel-air mixture to a burner in response to fluctuations of load on a boiler.

module [AERO ENG] A self-contained unit which serves as a building block for the overall structure in space technology; usually designated by its primary function, such as command module or lunar landing module.

modulus of compression *See* bulk modulus of elasticity.

modulus of deformation [MECH] The modulus of elasticity of a material that deforms other than according to Hooke's law.

modulus of elasticity [MECH] The ratio of the increment of some specified form of stress to the increment of some specified form of strain, such as Young's modulus, the bulk modulus, or the shear modulus. Also known as coefficient of elasticity; elasticity modulus; elastic modulus.

modulus of elasticity in shear [MECH] A measure of a material's resistance to shearing stress, equal to the shearing stress divided by the resultant angle of deformation expressed in radians. Also known as coefficient of rigidity; modulus of rigidity; shear modulus.

modulus of resilience [MECH] The maximum mechanical energy stored per unit volume of material when it is stressed to its elastic limit.

modulus of rigidity *See* modulus of elasticity in shear.

modulus of rupture in bending [MECH] The maximum stress per unit area that a specimen can withstand without breaking when it is bent, as calculated from the breaking load under the assumption that the specimen is elastic until rupture takes place.

modulus of rupture in torsion [MECH] The maximum stress per unit area that a specimen can withstand without breaking when its ends are twisted, as calculated from the breaking load under the assumption that the specimen is elastic until rupture takes place.

modulus of torsion *See* torsional modulus.

modulus of volume elasticity *See* bulk modulus of elasticity.

mohm [MECH] A unit of mechanical mobility, equal to the reciprocal of 1 mechanical ohm.

Mohr liter [MECH] A unit of volume, equal to 1000 Mohr cubic centimeters.

Mohr's circle [MECH] A graphical construction making it possible to determine the stresses in a cross section if the principal stresses are known.

moisture content [MECH] The quantity of water in a mass of soil, sewage, sludge, or screenings; expressed in percentage by weight of water in the mass.

moisture flux *See* eddy flux.

moisture loss [MECH ENG] The difference in heat content between the moisture in the boiler exit gases and that of moisture at ambient air temperature.

moisture-vapor transmission [FL MECH] The rate at which water vapor permeates a porous film (such as plastic or paper) or a wall.

MOL *See* manned orbiting laboratory.

molded-fabric bearing [DES ENG] A bearing composed of laminations of cotton or other fabric impregnated with a phenolic resin and molded under heat and pressure.

mole [MECH ENG] A mechanical tunnel excavator.

molecular diffusion [FL MECH] The transfer of mass between adjacent layers of fluid in laminar flow.

molecular effusion [FL MECH] Mass-transfer flow mechanism of free-molecule transfer through pores or orifices.

molecular flow [FL MECH] Gas-flow phenomenon at low pressures or in small channels when the mean free path is of the same order of magnitude as the channel diameter; a gas molecule thus migrates along the channel independent of other gas molecules present.

molecular heat [THERMO] The heat capacity per mole of a substance.

molecular heat diffusion [THERMO] Transfer of heat through the motion of molecules.

molecular pump [MECH ENG] A vacuum pump in which the molecules of the gas to be exhausted are carried away by the friction between them and a rapidly revolving disk or drum.

Mollier diagram [THERMO] Graph of enthalpy versus entropy of a vapor on which isobars, isothermals, and lines of equal dryness are plotted.

moment [MECH] Static moment of some quantity, except in the term "moment of inertia."

momental ellipsoid *See* inertia ellipsoid.

moment coefficient [AERO ENG] The coefficients used for moment are similar to coefficients of lift, drag, and thrust, and are likewise dimensionless; however, these must include a characteristic length, in addition to the area; the span is used for rolling or yawing moment, and the chord is used for pitching moment.

moment diagram [MECH] A graph of the bending moment at a section of a beam versus the distance of the section along the beam.

moment of force *See* torque.

moment of inertia [MECH] The sum of the products formed by multiplying the mass (or sometimes, the area) of each element of a figure by the square of its distance from a specified line. Also known as rotational inertia.

moment of momentum *See* angular momentum.

momentum [MECH] Also known as linear momentum; vector momentum. 1. For a single nonrelativistic particle, the product of the mass and the velocity of a particle. 2. For a single relativistic particle, $mv/(1 - v^2/c^2)^{1/2}$, where m is the rest-mass, v the velocity, and c the speed of light. 3. For a system of particles, the vector sum of the momenta (as in the first or second definition) of the particles.

momentum conservation *See* conservation of momentum.

momentum-transport hypothesis [FL MECH] The hypothesis that the principle of conservation of momentum is valid in turbulent eddy transfer.

monkey wrench [DES ENG] A wrench having one jaw fixed and the other adjustable, both of which are perpendicular to a straight handle.

monocable [MECH ENG] An aerial ropeway that uses one rope to both support and haul a load.

monochromatic temperature scale [THERMO] A temperature scale based upon the amount of power radiated from a blackbody at a single wavelength.

monocoque [AERO ENG] A type of construction, as of a rocket body, in which all or most of the stresses are carried by the skin.

monofuel propulsion [AERO ENG] Propulsion system which obtains its power from a single fuel; in rocket units, the fuel furnishes both oxygen supply and the hydrocarbon for combustion.

monotropy coefficient [FL MECH] A coefficient ν related to the ratio of velocity coefficients, A_y/A_x, in an equation developed by P. Raethjen for the velocity profile in a fluid.

Moody formula [MECH ENG] A formula giving the efficiency e' of a field turbine, whose runner has diameter D', in terms of the efficiency e of a model turbine, whose runner has diameter D; $e' = 1 - (1-e)(D/D')^{1/5}$.

Moody friction factor [FL MECH] Modification of the friction factor–Reynolds number–fluid flow relationship into which a roughness factor has been incorporated.

moon shot [AERO ENG] The launching of a rocket intended to travel to the vicinity of the moon.

mooring mast [AERO ENG] A mast or pole with fittings at the top to secure any lighter-than-air craft, such as a dirigible or blimp.

Morera's stress functions [MECH] Three functions of position, ψ_1, ψ_2, and ψ_3, in terms of which the elements of the stress tensor σ of a body may be expressed, if the body is in equilibrium and is not subjected to body forces; the elements of the stress tensor are given by $\sigma_{11} = -2\partial^2\psi_1/\partial x_2\partial x_3$, $\sigma_{23} = \partial^2\psi_2/\partial x_1\partial x_2 + \partial^2\psi_3/\partial x_1\partial x_3$, and cyclic permutations of these equations.

Morgan equation [THERMO] A modification of the Ramsey-Shields equation, in which the expression for the molar surface energy is set equal to a quadratic function of the temperature rather than to a linear one.

Morse taper reamer [DES ENG] A machine reamer with a taper shank.

mortise and tenon [DES ENG] A type of joint, principally used for wood, in which a hole, slot, or groove (mortise) in one member is fitted with a projection (tenon) from the second member.

mortise lock [DES ENG] A lock designed to be installed in a mortise rather than applied to a door's surface.

mortising machine [MECH ENG] A machine employing an auger and a chisel to produce a square or rectangular mortise in wood.

motion [MECH] A continuous change of position of a body.

motorcycle [MECH ENG] An automotive vehicle, essentially a motorized bicycle, with two tandem and sometimes three rubber wheels.

motor reducer [MECH ENG] Speed-reduction power transmission equipment in which the reducing gears are integral with drive motors.

motortruck [MECH ENG] An automotive vehicle which is used to transport freight.

motor vehicle [MECH ENG] Any automotive vehicle that does not run on rails, and generally having rubber tires.

mounce [MECH] A unit of mass, equal to 25 grams. Also known as metric ounce.

moving constraint [MECH] A constraint that changes with time, as in the case of a system on a moving platform.

moving load [MECH] A load that can move, such as vehicles or pedestrians.

mp *See* mean effective pressure; melting point.

MRTD *See* minimum resolvable temperature difference.

msec *See* millisecond.

mud auger [DES ENG] A diamond-point bit with the wings of the point twisted in a shallow augerlike spiral. Also known as clay bit; diamond-point bit; mud bit.

mud bit *See* mud auger.

multilayer bit [DES ENG] A bit set with diamonds arranged in successive layers beneath the surface of the crown.

multilevel control theory [CONT SYS] An approach to the control of large-scale systems based on decomposition of the complex overall control problem into simpler and more easily managed subproblems, and coordination of the subproblems so that overall system objectives and constraints are satisfied.

multiple-loop system [CONT SYS] A system whose block diagram has at least two closed paths, along each of which all arrows point in the same direction.

multiple-slide press [MECH ENG] A press with individual adjustable slides built into the main slide or connected independently to the main shaft.

multiple-stage rocket *See* multistage rocket.

multiple-strand conveyor [MECH ENG] A conveyor with two or more spaced strands of chain, belts, or cords as the supporting or propelling medium.

multirope friction winder   [MECH ENG] A winding system in which the drive to the winding ropes is the frictional resistance between the ropes and the driving sheaves.

multistage compressor   [MECH ENG] A machine for compressing a gaseous fluid in a sequence of stages, with or without intercooling between stages.

multistage pump   [MECH ENG] A pump in which the head is developed by multiple impellers operating in series.

multistage rocket   [AERO ENG] A vehicle having two or more rocket units, each unit firing after the one in back of it has exhausted its propellant; normally, each unit, or stage, is jettisoned after completing its firing. Also known as multiple-stage rocket; step rocket.

multivariable system   [CONT SYS] A dynamical system in which the number of either inputs or outputs is greater than 1.

Muskhelishvili's method   [MECH] A method of solving problems concerning the elastic deformation of a planar body that involves using methods from the theory of functions of a complex variable to calculate analytic functions which determine the plane strain of the body.

mW   *See* milliwatt.

MW   *See* megawatt.

Myklestad method   [AERO ENG] A method of determining the mode shapes and frequencies of the lateral bending modes of space vehicles, taking into account secondary effects of shear and rotary inertia, in which one imagines masses to be concentrated at a finite number of points along the beam, with elastic properties remaining constant between consecutive mass points.

N

nacelle [AERO ENG] A separate streamlined enclosure on an airplane for sheltering or housing something, as the crew or an engine.

nail [DES ENG] A slender, usually pointed fastener with a head, designed for insertion by impact.

nailhead [DES ENG] Flat protuberance at the end of a nail opposite the point.

nanogram [MECH] One-billionth (10^{-9}) of a gram. Abbreviated ng.

nanometer [MECH] A unit of length equal to one-billionth of a meter, or 10^{-9} meter. Also known as nanon.

nanosecond [MECH] A unit of time equal to one-billionth of a second, or 10^{-9} second.

natural convection [THERMO] Convection in which fluid motion results entirely from the presence of a hot body in the fluid, causing temperature and hence density gradients to develop, so that the fluid moves under the influence of gravity. Also known as free convection.

natural coordinates [FL MECH] An orthogonal, or mutually perpendicular, system of curvilinear coordinates for the description of fluid motion, consisting of an axis t tangent to the instantaneous velocity vector and an axis n normal to this velocity vector to the left in the horizontal plane, to which a vertically directed axis z may be added for the description of three-dimensional flow; such a coordinate system often permits a concise formulation of atmospheric dynamical problems, especially in the Lagrangian system of hydrodynamics.

natural draft [FL MECH] Unforced gas flow through a chimney or vertical duct, directly related to chimney height and the temperature difference between the ascending gases and the atmosphere, and not dependent upon the use of fans or other mechanical devices.

natural-draft cooling tower [MECH ENG] A cooling tower that depends upon natural convection of air flowing upward and in contact with the water to be cooled.

natural laminar flow [AERO ENG] Airflow over a portion of the wing such that local pressure decreases in the direction of flow and flow in the boundary layer is laminar rather than turbulent; frictional drag on the aircraft is greatly reduced.

nautical chain [MECH] A unit of length equal to 15 feet or 4.572 meters.

Navier's equation [MECH] A vector partial differential equation for the displacement vector of an elastic solid in equilibrium and subjected to a body force.

Navier-Stokes equations [FL MECH] The equations of motion for a viscous fluid which may be written $d\mathbf{V}/dt = -(1/\rho)\nabla p + \mathbf{F} + \nu\nabla^2\mathbf{V} + (1/3)\nu\nabla(\nabla\cdot\mathbf{V})$, where p is the pressure, ρ the density, F the total external force per unit mass, $\mathbf{V}$ the fluid velocity, and ν the kinematic viscosity; for an incompressible fluid, the term in $\nabla\cdot\mathbf{V}$ (divergence) vanishes, and the effects of viscosity then play a role analogous to that of temperature in thermal conduction and to that of density in simple diffusion.

navigational satellite [AERO ENG] An artificial earth-orbiting satellite designed for use in at least four widely different navigational systems.

navigation dome *See* astrodome.

n-body problem *See* many-body problem.

NC *See* numerical control.

N/C *See* numerical control.

needle [DES ENG] 1. A device made of steel pointed at one end with a hole at the other; used for sewing. 2. A device made of steel with a hook at one end; used for knitting.

needle bearing [DES ENG] A roller-type bearing with long rollers of small diameter; the rollers are retained in a flanged cup, have no retainer, and bear directly on the shaft.

needle file [DES ENG] A small file with an extended tang that serves as a needle.

needle nozzle [MECH ENG] A streamlined hydraulic turbine nozzle with a movable element for converting the pressure and kinetic energy in the pipe leading from the reservoir to the turbine into a smooth jet of variable diameter and discharge but practically constant velocity.

needle valve [DES ENG] A slender, pointed rod fitting in a hole or circular or conoidal seat; used in hydraulic turbines and hydroelectric systems.

negative acceleration [MECH] Acceleration in a direction opposite to the velocity, or in the direction of the negative axis of a coordinate system.

negative feedback [CONT SYS] Feedback in which a portion of the output of a circuit, device, or machine is fed back 180° out of phase with the input signal, resulting in a decrease of amplification so as to stabilize the amplification with respect to time or frequency, and a reduction in distortion and noise. Also known as inverse feedback; reverse feedback; stabilized feedback.

negative g [MECH] In designating the direction of acceleration on a body, the opposite of positive g; for example, the effect of flying an outside loop in the upright seated position.

negative rake [MECH ENG] The orientation of a cutting tool whose cutting edge lags the surface of the tooth face.

Nernst approximation formula [THERMO] An equation for the equilibrium constant of a gas reaction based on the Nernst heat theorem and certain simplifying assumptions.

Nernst heat theorem [THERMO] The theorem expressing that the rate of change of free energy of a homogeneous system with temperature, and also the rate of change of enthalpy with temperature, approaches zero as the temperature approaches absolute zero.

Nernst-Simon statement of the third law of thermodynamics [THERMO] The statement that the change in entropy which occurs when a homogeneous system undergoes an isothermal reversible process approaches zero as the temperature approaches absolute zero.

NETD *See* noise equivalent temperature difference.

net flow area [DES ENG] The calculated net area which determines the flow after the complete bursting of a rupture disk.

net head [FL MECH] The difference in elevation between the last free water surface in a power conduit above the waterwheel and the first free water surface in the conduit below the waterwheel, less the friction losses in the conduit.

net positive suction head   [MECH ENG] The minimum suction head required for a pump to operate; depends on liquid characteristics, total liquid head, pump speed and capacity, and impeller design. Abbreviated NPSH.

net thrust   [AERO ENG] The gross thrust of a jet engine minus the drag due to the momentum of the incoming air.

net ton   *See* ton.

Neumann-Kopp rule   [THERMO] The rule that the heat capacity of 1 mole of a solid substance is approximately equal to the sum over the elements forming the substance of the heat capacity of a gram atom of the element times the number of atoms of the element in a molecule of the substance.

neutral axis   [MECH] In a beam bent downward, the line of zero stress below which all fibers are in tension and above which they are in compression.

neutral equilibrium   [FL MECH] A property of the steady state of a system which exhibits neither instability nor stability according to the particular criterion under consideration; a disturbance introduced into such an equilibrium will thus be neither amplified nor damped. Also known as indifferent equilibrium.

neutral fiber   [MECH] A line of zero stress in cross section of a bent beam, separating the region of compressive stress from that of tensile stress.

neutral point   [FL MECH] *See* hyperbolic point.

neutral stability   [CONT SYS] Condition in which the natural motion of a system neither grows nor decays, but remains at its initial amplitude.

neutral surface   [MECH] A surface in a bent beam along which material is neither compressed nor extended.

newton   [MECH] The unit of force in the meter-kilogram-second system, equal to the force which will impart an acceleration of 1 meter per second squared to the International Prototype Kilogram mass. Symbolized N. Formerly known as large dyne.

Newton formula for the stress   *See* Newtonian friction law.

Newtonian attraction   [MECH] The mutual attraction of any two particles in the universe, as given by Newton's law of gravitation.

Newtonian flow   [FL MECH] Flow system in which the fluid performs as a Newtonian fluid, that is, shear stress is proportional to shear rate.

Newtonian fluid   [FL MECH] A simple fluid in which the state of stress at any point is proportional to the time rate of strain at that point; the proportionality factor is the viscosity coefficient.

Newtonian friction law   [FL MECH] The law that shear stress in a fluid is proportional to the shear rate; it holds only for some fluids, which are then called Newtonian. Also known as Newton formula for the stress.

Newtonian mechanics   [MECH] The system of mechanics based upon Newton's laws of motion in which mass and energy are considered as separate, conservative, mechanical properties, in contrast to their treatment in relativistic mechanics.

Newtonian reference frame   [MECH] One of a set of reference frames with constant relative velocity and within which Newton's laws hold; the frames have a common time, and coordinates are related by the Galilean transformation rule.

Newtonian velocity   [MECH] The velocity of an object in a Newtonian reference frame, S, which can be determined from the velocity of the object in any other such frame, S', by taking the vector sum of the velocity of the object in S' and the velocity of the frame S' relative to S.

Newtonian viscosity   [FL MECH] The viscosity of a Newtonian fluid.

newton-meter of energy   *See* joule.

newton-meter of torque   [MECH] The unit of torque in the meter-kilogram-second system, equal to the torque produced by 1 newton of force acting at a perpendicular distance of 1 meter from an axis of rotation. Abbreviated N-m.

Newton's equations of motion   [MECH] Newton's laws of motion expressed in the form of mathematical equations.

Newton's first law   [MECH] The law that a particle not subjected to external forces remains at rest or moves with constant speed in a straight line. Also known as first law of motion; Galileo's law of inertia.

Newton's law of cooling   [THERMO] The law that the rate of heat flow out of an object by both natural convection and radiation is proportional to the temperature difference between the object and its environment, and to the surface area of the object.

Newton's law of gravitation   [MECH] The law that every two particles of matter in the universe attract each other with a force that acts along the line joining them, and has a magnitude proportional to the product of their masses and inversely proportional to the square of the distance between them. Also known as law of gravitation.

Newton's law of resistance   [FL MECH] The law that the force opposing the motion of an object through a fluid at moderate velocities is proportional to the square of the velocity.

Newton's laws of motion   [MECH] Three fundamental principles (called Newton's first, second, and third laws) which form the basis of classical, or Newtonian, mechanics, and have proved valid for all mechanical problems not involving speeds comparable with the speed of light and not involving atomic or subatomic particles.

Newton's second law   [MECH] The law that the acceleration of a particle is directly proportional to the resultant external force acting on the particle and is inversely proportional to the mass of the particle. Also known as second law of motion.

Newton's theory of lift   [FL MECH] A theory of the forces acting on an airfoil in a fluid current in which these forces are assumed to result from the impact of particles of the fluid on the body.

Newton's third law   [MECH] The law that, if two particles interact, the force exerted by the first particle on the second particle (called the action force) is equal in magnitude and opposite in direction to the force exerted by the second particle on the first particle (called the reaction force). Also known as law of action and reaction; third law of motion.

nibbling   [MECH ENG] Contour cutting of material by the action of a reciprocating punch that takes repeated small bites as the work is passed beneath it.

Nichol's chart   [CONT SYS] A plot of curves along which the magnitude M or argument α of the frequency control ratio is constant on a graph whose ordinate is the logarithm of the magnitude of the open-loop transfer function, and whose abscissa is the open-loop phase angle.

nip   *See* angle of nip.

nippers   [DES ENG] Small pincers or pliers for cutting or gripping.

nipple   [DES ENG] A short piece of tubing, usually with an internal or external thread at each end, used to couple pipes.

N-m   *See* newton-meter of torque.

no-atmospheric control   [AERO ENG] Any device or system designed or set up to control a guided rocket missile, rocket craft, or the like outside the atmosphere or in regions where the atmosphere is of such tenuity that it will not affect aerodynamic controls.

noise equivalent temperature difference [THERMO] The change in equivalent black-body temperature that corresponds to a change in radiance which will produce a signal-to-noise ratio of 1 in an infrared imaging device. Abbreviated NETD.

nominal size [DES ENG] Size used for purposes of general identification; the actual size of a part will be approximately the same as the nominal size but need not be exactly the same; for example, a rod may be referred to as ¼ inch, although the actual dimension on the drawing is 0.2495 inch, and in this case ¼ inch is the nominal size.

nonanticipatory system *See* causal system.

nonblackbody [THERMO] A body that reflects some fraction of the radiation incident upon it; all real bodies are of this nature.

nonequilibrium thermodynamics [THERMO] A quantitative treatment of irreversible processes and of rates at which they occur. Also known as irreversible thermodynamics.

nonfeasible method *See* goal coordination method.

nonholonomic system [MECH] A system of particles which is subjected to restraints of such a nature that the system cannot be described by independent coordinates; examples are a rolling hoop, or an ice skate which must point along its path.

nonhoming [CONT SYS] Not returning to the starting or home position, as when the wipers of a stepping relay remain at the last-used set of contacts instead of returning to their home position.

nonimpinging injector [AERO ENG] An injector used in rocket engines which employs parallel streams of propellant usually emerging normal to the face of the injector.

noninteracting control [CONT SYS] A feedback control in a system with more than one input and more than one output, in which feedback transfer functions are selected so that each input influences only one output.

nonlinear control [CONT SYS] An on-line process control which can stabilize operations by means of nonlinear compensatory functions.

nonlinear feedback control system [CONT SYS] Feedback control system in which the relationships between the pertinent measures of the system input and output signals cannot be adequately described by linear means.

nonlinear vibration [MECH] A vibration whose amplitude is large enough so that the elastic restoring force on the vibrating object is not proportional to its displacement.

nonlinear viscoelasticity [FL MECH] The behavior of a fluid which does not obey a first-order differential equation in stress and strain.

non-minimum-phase system [CONT SYS] A linear system whose transfer function has one or more poles or zeros with positive, nonzero real parts.

non-Newtonian fluid [FL MECH] A fluid whose flow behavior departs from that of a Newtonian fluid, so that the rate of shear is not proportional to the corresponding stress. Also known as non-Newtonian system.

non-Newtonian fluid flow [FL MECH] The flow behavior of non-Newtonian fluids, whose study has applications in many important problems of practical significance such as flow in tubes, extrusion, flow through dies, coating operations, rolling operations, and mixing of fluids.

non-Newtonian viscosity [FL MECH] The behavior of a fluid which, when subjected to a constant rate of shear, develops a stress which is not proportional to the shear. Also known as anomalous viscosity.

nonquantum mechanics [MECH] The classical mechanics of Newton and Einstein as opposed to the quantum mechanics of Heisenberg, Schrödinger, and Dirac; particles have definite position and velocity, and they move according to Newton's laws.

nonreclosing pressure relief device [MECH ENG] A device which remains open after relieving pressure and must be reset before it can operate again.

nonrelativistic kinematics [MECH] The study of motions of systems of objects at speeds which are small compared to the speed of light, without reference to the forces which act on the system.

nonrelativistic mechanics [MECH] The study of the dynamics of systems in which all speeds are small compared to the speed of light.

nonspinning rope *See* nonstranded rope.

nonstranded rope [DES ENG] A wire rope with the wires in concentric sheaths instead of in strands, and in opposite directions in the different sheaths, giving the rope nonspinning properties. Also known as nonspinning rope.

nonuniform flow [FL MECH] Fluid flow which does not have the same velocity at all points in a medium, at a given instant.

nonviscous flow *See* inviscid flow.

nonviscous fluid *See* inviscid fluid.

normal acceleration [MECH] 1. The component of the linear acceleration of an aircraft or missile along its normal, or Z, axis. 2. The usual or typical acceleration.

normal axis [MECH] The vertical axis of an aircraft or missile.

normal coordinates [MECH] A set of coordinates for a coupled system such that the equations of motion each involve only one of these coordinates.

normal depth [FL MECH] The depth in an open channel at which a given flow has uniform velocity.

normal frequencies [MECH] The frequencies of the normal modes of vibration of a system.

normal impact [MECH] 1. Impact on a plane perpendicular to the trajectory. 2. Striking of a projectile against a surface that is perpendicular to the line of flight of the projectile.

normal mode of vibration [MECH] Vibration of a coupled system in which the value of one of the normal coordinates oscillates and the values of all the other coordinates remain stationary.

normal operation [MECH ENG] The operation of a boiler or pressure vessel at or below the conditions of coincident pressure and temperature for which the vessel has been designed.

normal pitch [MECH ENG] The distance between working faces of two adjacent gear teeth, measured between the intersections of the line of action with the faces.

normal reaction [MECH] The force exerted by a surface on an object in contact with it which prevents the object from passing through the surface; the force is perpendicular to the surface, and is the only force that the surface exerts on the object in the absence of frictional forces.

normal stress [MECH] The stress component at a point in a structure which is perpendicular to the reference plane.

northerly turning error [AERO ENG] An acceleration error in the magnetic compass of an aircraft in a banked attitude during a turn, so called because it was first noted and is most pronounced during turns made from initial north-south courses; during a turn the magnetic needle is tilted from the horizontal, due to acceleration and the banking of the aircraft; in this position the compass needle will be acted upon by the vertical as well as the horizontal component of the earth's magnetic field; in addition, the compass needle is mechanically restricted in movement, due to tilt. Also known as turning error.

nose   [FL MECH]   The dense, forward part of a turbidity current.

nose cone   [AERO ENG]   A protective cone-shaped case for the nose section of a missile or rocket; may include the warhead, fusing system, stabilization system, heat shield, and supporting structure and equipment.

nose-heavy   [AERO ENG]   Pertaining to an airframe in which the nose tends to sink when the longitudinal control is released in any attitude of normal flight.

nose radius   [MECH ENG]   The radius measured in the back rake or top rake plane of a cutting tool.

notching   [MECH ENG]   Cutting out various shapes from the ends or edges of a work-piece.

notching press   [MECH ENG]   A mechanical press for notching straight or rounded edges.

nozzle   [DES ENG]   A tubelike device, usually streamlined, for accelerating and directing a fluid, whose pressure decreases as it leaves the nozzle.

nozzle blade   [AERO ENG]   Any one of the blades or vanes in a nozzle diaphragm. Also known as nozzle vane.

nozzle-contraction-area ratio   [DES ENG]   Ratio of the cross-sectional area for gas flow at the nozzle inlet to that at the throat.

nozzle-divergence loss factor   [FL MECH]   The ratio between the momentum of the gases in a nozzle and the momentum of an ideal nozzle.

nozzle efficiency   [MECH ENG]   The efficiency with which a nozzle converts potential energy into kinetic energy, commonly expressed as the ratio of the actual change in kinetic energy to the ideal change at the given pressure ratio.

nozzle exit area   [DES ENG]   The cross-sectional area of a nozzle available for gas flow measured at the nozzle exit.

nozzle-expansion ratio   [DES ENG]   Ratio of the cross-sectional area for gas flow at the exit of a nozzle to the cross-sectional area available for gas flow at the throat.

nozzle throat   [DES ENG]   The portion of a nozzle with the smallest cross section.

nozzle throat area   [DES ENG]   The area of the minimum cross section of a nozzle.

nozzle thrust coefficient   [AERO ENG]   A measure of the amplification of thrust due to gas expansion in a particular nozzle as compared with the thrust that would be exerted if the chamber pressure acted only over the throat area. Also known as thrust coefficient.

nozzle vane   *See* nozzle blade.

NPSH   *See* net positive suction head.

N rod bit   [DES ENG]   A Canadian standard noncoring bit having a set diameter of 2.940 inches (74.676 millimeters).

nuclear-electric propulsion   [AERO ENG]   A system of propulsion utilizing a nuclear reactor to generate electricity which is then used in an electric propulsion system or as a heat source for the working fluid.

nuclear-electric rocket engine   [AERO ENG]   A rocket engine in which a nuclear reactor is used to generate electricity that is used in an electric propulsion system or as a heat source for the working fluid.

nuclear power plant   [MECH ENG]   A power plant in which nuclear energy is converted into heat for use in producing steam for turbines, which in turn drive generators that produce electric power.

nuclear rocket   *See* atomic rocket.

numerical control   [CONT SYS] A control system for machine tools and some industrial processes, in which numerical values corresponding to desired positions of tools or controls are recorded on punched paper tapes, punched cards, or magnetic tapes so that they can be used to control the operation automatically. Abbreviated NC; N/C.

Nusselt equation   [THERMO] Dimensionless equation used to calculate convection heat transfer for heating or cooling of fluids outside a bank of 10 or more rows of tubes to which the fluid flow is normal.

Nusselt number   [THERMO] A dimensionless number used in the study of forced convection which gives a measure of the ratio of the total heat transfer to conductive heat transfer, and is equal to the heat-transfer coefficient times a characteristic length divided by the thermal conductivity. Symbolized N_{Nu}.

nut   [DES ENG] An internally threaded fastener for bolts and screws.

nutation   [MECH] A bobbing or nodding up-and-down motion of a spinning rigid body, such as a top, as it precesses about its vertical axis.

Nyquist contour   [CONT SYS] A directed closed path in the complex frequency plane used in constructing a Nyquist diagram, which runs upward, parallel to the whole length of the imaginary axis at an infinitesimal distance to the right of it, and returns from $+j\infty$ to $-j\infty$ along a semicircle of infinite radius in the right half-plane.

Nyquist diagram   [CONT SYS] A plot in the complex plane of the open-loop transfer function as the complex frequency is varied along the Nyquist contour; used to determine stability of a control system.

Nyquist stability criterion   *See* Nyquist stability theorem.

Nyquist stability theorem   [CONT SYS] The theorem that the net number of counterclockwise rotations about the origin of the complex plane carried out by the value of an analytic function of a complex variable, as its argument is varied around the Nyquist contour, is equal to the number of poles of the variable in the right half-plane minus the number of zeros in the right half-plane. Also known as Nyquist stability criterion.

0

oblique shock   *See* oblique shock wave.

oblique shock wave   [FL MECH] A shock wave inclined at an oblique angle to the direction of flow in a supersonic flow field. Also known as oblique shock.

observability   [CONT SYS] Property of a system for which observation of the output variables at all times is sufficient to determine the initial values of all the state variables.

observation spillover   [CONT SYS] The part of the sensor output of an active control system caused by modes that have been omitted from the control algorithm in the process of model reduction.

observer   [CONT SYS] A linear system *B* driven by the inputs and outputs of another linear system *A* which produces an output that converges to some linear function of the state of system *A*. Also known as state estimator; state observer.

ocean thermal energy conversion   [MECH ENG] The conversion of energy arising from the temperature difference between warm surface water of oceans and cold deep-ocean current into electrical energy or other useful forms of energy.

octahedral normal stress   [MECH] The normal component of stress across the faces of a regular octahedron whose vertices lie on the principal axes of stress; it is equal in magnitude to the spherical stress across any surface. Also known as mean stress.

octahedral shear stress   [MECH] The tangential component of stress across the faces of a regular octahedron whose vertices lie on the principal axes of stress; it is a measure of the strength of the deviatoric stress.

octane requirement   [MECH ENG] The fuel octane number needed for efficient operation (without knocking or spark retardation) of an internal combustion engine.

octoid   [DES ENG] Pertaining to a gear tooth form used to generate the teeth in bevel gears; the octoid form closely resembles the involute form.

OD   *See* outside diameter.

odd-leg caliper   [DES ENG] A caliper in which the legs bend in the same direction instead of opposite directions.

off-airways   [AERO ENG] Pertaining to any aircraft course or track that does not lie within the bounds of prescribed airways.

offset   [CONT SYS] The steady-state difference between the desired control point and that actually obtained in a process control system.   [MECH] The value of strain between the initial linear portion of the stress-strain curve and a parallel line that intersects the stress-strain curve of an arbitrary value of strain; used as an index of yield stress; a value of 0.2% is common.

offset cylinder   [MECH ENG] A reciprocating part in which the crank rotates about a center off the centerline.

offset screwdriver [DES ENG] A screwdriver with the blade set perpendicular to the shank for access to screws in otherwise awkward places.

offset yield strength [MECH] That stress at which the strain surpasses by a specific amount (called the offset) an extension of the initial proportional portion of the stress-strain curve; usually expressed in pounds per square inch.

OHV engine *See* overhead-valve engine.

oil dilution valve [MECH ENG] A valve used to mix gasoline with engine oil to permit easier starting of the gasoline engine in cold weather.

oil furnace [MECH ENG] A combustion chamber in which oil is the heat-producing fuel.

oil groove [DES ENG] One of the grooves in a bearing which distribute and collect lubricating oil.

oil-hole drill [DES ENG] A twist drill containing holes through which oil can be fed to the cutting edges.

oilless bearing [MECH ENG] A self-lubricating bearing containing solid or liquid lubricants in its material.

oil lift [MECH ENG] Hydrostatic lubrication of a journal bearing by using oil at high pressure in the area between the bottom of the journal and the bearing itself so that the shaft is raised and supported by an oil film whether it is rotating or not.

oil pump [MECH ENG] A pump of the gear, vane, or plunger type, usually an integral part of the automotive engine; it lifts oil from the sump to the upper level in the splash and circulating systems, and in forced-feed lubrication it pumps the oil to the tubes leading to the bearings and other parts.

oil ring [MECH ENG] 1. A ring located at the lower part of a piston to prevent an excess amount of oil from being drawn up onto the piston during the suction stroke. 2. A ring on a journal, dipping into an oil bath for lubrication.

oleo strut [MECH ENG] A shock absorber consisting of a telescoping cylinder that forces oil into an air chamber, thereby compressing the air; used on aircraft landing gear.

once-through boiler [MECH ENG] A boiler in which water flows, without recirculation, sequentially through the economizer, furnace wall, and evaporating and superheating tubes.

one-dimensional flow [FL MECH] Fluid flow in which all flow is parallel to some straight line, and characteristics of flow do not change in moving perpendicular to this line.

on-off control [CONT SYS] A simple control system in which the device being controlled is either full on or full off, with no intermediate operating positions. Also known as on-off system.

on-off system *See* on-off control.

Onsager reciprocal relations [THERMO] A set of conditions which state that the matrix, whose elements express various fluxes of a system (such as diffusion and heat conduction) as linear functions of the various conjugate affinities (such as mass and temperature gradients) for systems close to equilibrium, is symmetric when certain definitions are chosen for these fluxes and affinities.

on-top flight [AERO ENG] Flight above an overcast.

open-belt drive [DES ENG] A belt drive having both shafts parallel and rotating in the same direction.

open-circuit grinding [MECH ENG] Grinding system in which material passes through the grinder without classification of product and without recycle of oversize lumps; in contrast to closed-circuit grinding.

open cycle [THERMO] A thermodynamic cycle in which new mass enters the boundaries of the system and spent exhaust leaves it; the automotive engine and the gas turbine illustrate this process.

open-cycle engine [MECH ENG] An engine in which the working fluid is discharged after one pass through boiler and engine.

open-cycle gas turbine [MECH ENG] A gas turbine prime mover in which air is compressed in the compressor element, fuel is injected and burned in the combustor, and the hot products are expanded in the turbine element and exhausted to the atmosphere.

open-end wrench [DES ENG] A wrench consisting of fixed jaws at one or both ends of a handle.

opening die [MECH ENG] A die head for cutting screws that opens automatically to release the cut thread.

opening pressure [MECH ENG] The static inlet pressure at which discharge is initiated.

open-loop control system [CONT SYS] A control system in which the system outputs are controlled by system inputs only, and no account is taken of actual system output.

open-side planer [DES ENG] A planer constructed with one upright or housing to support the crossrail and tools.

open-side tool block [DES ENG] A toolholder on a cutting machine consisting of a T-slot clamp, a C-shaped block, and two or more tool clamping screws. Also known as heavy-duty tool block.

open system [THERMO] A system across whose boundaries both matter and energy may pass.

operating stress [MECH] The stress to which a structural unit is subjected in service.

operating water level [MECH ENG] The water level in a boiler drum which is normally maintained above the lowest safe level.

opposed engine [MECH ENG] A reciprocating engine having the pistons on opposite sides of the crankshaft, with the piston strokes on each side working in a direction opposite to the direction of the strokes on the other side.

optimal smoother [CONT SYS] An optimal filer algorithm which generates the best estimate of a dynamical variable at a certain time based on all available data, both past and future.

optimizing control function [CONT SYS] That level in the functional decomposition of a large-scale control system which determines the necessary relationships among the variables of the system to achieve an optimal, or suboptimal, performance based on a given approximate model of the plant and its environment.

optimum flight [AERO ENG] An aircraft flight so planned and navigated that it is completed under the optimum conditions of minimum time and minimum exposure to dangerous flying weather.

orange-peel bucket [DES ENG] A type of grab bucket that is multileaved and generally round in configuration.

orbital angular momentum [MECH] The angular momentum associated with the motion of a particle about an origin, equal to the cross product of the position vector with the linear momentum. Also known as orbital momentum.

orbital curve [AERO ENG] One of the tracks on a primary body's surface traced by a satellite that orbits about it several times in a direction other than normal to the primary body's axis of rotation; each track is displaced in a direction opposite and by an amount equal to the degrees of rotation between each satellite orbit.

orbital decay [AERO ENG] The lessening of the eccentricity of the elliptical orbit of an artificial satellite.

orbital direction [AERO ENG] The direction that the path of an orbiting body takes; in the case of an earth satellite, this path may be defined by the angle of inclination of the path to the equator.

orbital moment *See* orbital angular momentum.

orbital momentum *See* orbital angular momentum.

orbital plane [MECH] The plane which contains the orbit of a body or particle in a central force field; it passes through the center of force.

orbital rendezvous [AERO ENG] 1. The meeting of two or more orbiting objects with zero relative velocity at a preconceived time and place. 2. The point in space at which such an event occurs.

orbit point [AERO ENG] A geographically defined reference point over land or water, used in stationing airborne aircraft.

order of phase transition [THERMO] A phase transition in which there is a latent heat and an abrupt change in properties, such as in density, is a first-order transition; if there is not such a change, the order of the transition is one greater than the lowest derivative of such properties with respect to temperature which has a discontinuity.

ordinary gear train [MECH ENG] A gear train in which all axes remain stationary relative to the frame.

organic bonded wheel [DES ENG] A grinding wheel in which organic bonds are used to hold the abrasive grains.

orifice mixer [MECH ENG] Arrangement in which two or more liquids are pumped through an orifice constriction to cause turbulence and consequent mixing action.

orifice plate [DES ENG] A disk, with a hole, placed in a pipeline to measure flow.

O ring [DES ENG] A flat ring made from synthetic rubber, used as an airtight seal or a seal against high pressures.

orthotropic [MECH] Having elastic properties such as those of timber, that is, with considerable variations of strength in two or more directions perpendicular to one another.

oscillating conveyor [MECH ENG] A conveyor on which pulverized solids are moved by a pan or trough bed attached to a vibrator or oscillating mechanism. Also known as vibrating conveyor.

oscillating granulator [MECH ENG] Solids size-reducer in which particles are broken by a set of oscillating bars arranged in cylindrical form over a screen of suitable mesh.

oscillating screen [MECH ENG] Solids separator in which the sifting screen oscillates at 300 to 400 revolutions per minute in a plane parallel to the screen.

oscillation *See* cycling.

oscillatory shear [FL MECH] Application of small-amplitude oscillations to produce shear in viscoelastic fluids for the study of dynamic viscosity.

Oseen's flow [FL MECH] Fluid flow in which the velocity of flow is very small but the Reynold's number is greater than 1.

Ostwald's adsorption isotherm [THERMO] An equation stating that at a constant temperature the weight of material adsorbed on an adsorbent dispersed through a gas or solution, per unit weight of adsorbent, is proportional to the concentration of the adsorbent raised to some constant power.

Otto cycle [THERMO] A thermodynamic cycle for the conversion of heat into work, consisting of two isentropic phases interspersed between two constant-volume phases. Also known as spark-ignition combustion cycle.

Otto engine [MECH ENG] An internal combustion engine that operates on the Otto cycle, where the phases of suction, compression, combustion, expansion, and exhaust occur sequentially in a four-stroke-cycle or two-stroke-cycle reciprocating mechanism.

Otto-Lardillon method [MECH] A method of computing trajectories of missiles with low velocities (so that drag is proportional to the velocity squared) and quadrant angles of departure that may be high, in which exact solutions of the equations of motion are arrived at by numerical integration and are then tabulated.

ounce [MECH] **1.** A unit of mass in avoirdupois measure equal to 1/16 pound or to approximately 0.0283495 kilogram. Abbreviated oz. **2.** A unit of mass in either troy or apothecaries' measure equal to 480 grains or exactly 0.0311034768 kilogram. Also known as apothecaries' ounce or troy ounce (abbreviations are oz ap and oz t in the United States, and oz apoth and oz tr in the United Kingdom).

ouncedal [MECH] A unit of force equal to the force which will impart an acceleration of 1 foot per second per second to a mass of 1 ounce; equal to 0.0086409346485 newton.

output shaft [MECH ENG] The shaft that transfers motion from the prime mover to the driven machines.

outside caliper [DES ENG] A caliper having two curved legs which point toward each other; used for measuring outside dimensions of a workpiece.

outside diameter [DES ENG] The outer diameter of a pipe, including the wall thickness; usually measured with calipers. Abbreviated OD.

overall efficiency [AERO ENG] The efficiency of a jet engine, rocket engine, or rocket motor in converting the total heat energy of its fuel first into available energy for the engine, then into effective driving energy.

overarm [MECH ENG] One of the adjustable supports for the end of a milling-cutter arbor farthest from the machine spindle.

overdrive [MECH ENG] An automobile engine device that lowers the gear ratio, thereby reducing fuel consumption.

overfire draft [MECH ENG] The air pressure in a boiler furnace during occurrence of the main flame.

overgear [MECH ENG] A gear train in which the angular velocity ratio of the driven shaft to driving shaft is greater than unity, as when the propelling shaft of an automobile revolves faster than the engine shaft.

overhead camshaft [MECH ENG] A camshaft mounted above the cylinder head.

overhead shovel [MECH ENG] A tractor which digs with a shovel at its front end, swings the shovel rearward overhead, and dumps the shovel at its rear end.

overhead-traveling crane [MECH ENG] A hoisting machine with a bridgelike structure moved on wheels along overhead trackage which is usually fixed to the building structure.

overhead-valve engine [MECH ENG] A four-stroke-cycle internal combustion engine having its valves located in the cylinder head, operated by pushrods that actuate rocker arms. Abbreviated OHV engine. Also known as valve-in-head engine.

overpressure [FL MECH] The transient pressure, usually expressed in pounds per square inch, exceeding existing atmospheric pressure and manifested in the blast wave from an explosion.

override [CONT SYS] To cancel the influence of an automatic control by means of a manual control.

overriding process control [CONT SYS] Process control in which any one of several controllers associated with one control valve can be made to override another in accordance with a priority requirement of the process.

overrunning clutch [MECH ENG] A clutch that allows the driven shaft to turn freely only under certain conditions; for example, a clutch in an engine starter that allows the crank to turn freely when the engine attempts to run.

overshot wheel [MECH ENG] A horizontal-shaft waterwheel with buckets around the circumference; the weight of water pouring into the buckets from the top rotates the wheel.

overspeed governor [MECH ENG] A governor that stops the prime mover when speed is excessive.

overspin [MECH] In a spin-stabilized projectile, the overstability that results when the rate of spin is too great for the particular design of projectile, so that its nose does not turn downward as it passes the summit of the trajectory and follows the descending branch. Also known as overstabilization.

oversquare engine [MECH ENG] An engine with bore diameter greater than the stroke length.

overstabilization *See* overspin.

overtone [MECH] One of the normal modes of vibration of a vibrating system whose frequency is greater than that of the fundamental mode.

oxygen point [THERMO] The temperature at which liquid oxygen and its vapor are in equilibrium, that is, the boiling point of oxygen, at standard atmospheric pressure; it is taken as a fixed point on the International Practical Temperature Scale of 1968, at $-182.962°C$.

oz *See* ounce.

oz ap *See* ounce.

oz apoth *See* ounce.

oz t *See* ounce.

oz tr *See* ounce.

P

P *See* poise.

Pa *See* pascal.

packing ring *See* piston ring.

pad *See* launch pad.

padded bit *See* castellated bit.

paddle [AERO ENG] A large, flat, paddle-shaped support for solar cells, used on some satellites. [DES ENG] Any of various implements consisting of a shaft with a broad, flat blade or bladelike part at one or both ends.

paddle wheel [MECH ENG] **1.** A device used to propel shallow-draft vessels, consisting of a wheel with paddles or floats on its circumference, the wheel rotating in a plane parallel to the ship's length. **2.** A wheel with paddles used to move leather in a processing vat.

padlock [DES ENG] An unmounted lock with a shackle that can be opened and closed; the shackle is usually passed through an eye, then closed to secure a hasp.

pail [DES ENG] A cylindrical or slightly tapered container.

pair [MECH ENG] Two parts in a kinematic mechanism that mutually constrain relative motion; for example, a sliding pair composed of a piston and cylinder.

pairing element [MECH ENG] Either of two machine parts connected to permit motion.

pallet [MECH ENG] One of the disks or pistons in a chain pump.

palpable coordinate [MECH] A generalized coordinate that appears explicitly in the Lagrangian of a system.

pan bolt [DES ENG] A bolt with a head resembling an upside-down pan.

pancake auger [DES ENG] An auger having one spiral web, 12 to 15 inches (30 to 38 centimeters) in diameter, attached to the bottom end of a slender central shaft; used as removable deadman to which a drill rig or guy line is anchored.

pancake engine [MECH ENG] A compact engine with cylinders arranged radially.

pancake landing [AERO ENG] Landing of an aircraft at a low forward speed and at a very high rate of descent.

pan conveyor [MECH ENG] A conveyor consisting of a series of pans.

pan crusher [MECH ENG] Solids-reduction device in which one or more grinding wheels or mullers revolve in a pan containing the material to be pulverized.

pane [DES ENG] One of the sides on a nut or on the head of a bolt.

panel coil *See* plate coil.

pan head [DES ENG] The head of a screw or rivet in the shape of a truncated cone.

paper cutter [DES ENG] A hand-operated paper cutter and trimmer, consisting of a cutting blade bolted at one end to a ruled board; when the blade is drawn flush with the board, which has a metal strip at the cutting edge, a shearing action takes place which cuts the paper cleanly and evenly.

paper machine [MECH ENG] A synchronized series of mechanical devices for transforming a dilute suspension of cellulose fibers into a dry sheet of paper.

parabolic flight [AERO ENG] A space flight occurring in a parabolic orbit.

paracentric [DES ENG] Pertaining to a key and keyway with longitudinal ribs and grooves that project beyond the center, as used in pin-tumbler cylinder locks to deter lockpicking.

parachute [AERO ENG] 1. A contrivance that opens out somewhat like an umbrella and catches the air so as to retard the movement of a body attached to it. 2. The canopy of this contrivance.

parachute-opening shock [AERO ENG] The shock or jolt exerted on a suspended parachute load when the parachute fully catches the air.

parallel axis theorem [MECH] A theorem which states that the moment of inertia of a body about any given axis is the moment of inertia about a parallel axis through the center of mass, plus the moment of inertia that the body would have about the given axis if all the mass of the body were located at the center of mass. Also known as Steiner's theorem.

parallel baffle muffler [DES ENG] A muffler constructed of a series of ducts placed side by side in which the duct cross section is a narrow but long rectangle.

parallel compensation See feedback compensation.

parallel drum [DES ENG] A cylindrical form of drum on which the haulage or winding rope is coiled.

parallel linkage [MECH ENG] A linkage system in which reciprocating motion is amplified.

parasheet [AERO ENG] A simple form of parachute in which the canopy is a single piece of material or two or more pieces sewed together; it may have any geometrical form, such as square or hexagonal, and the hem may be gathered to assist in the development of a crown when the parasheet is opened.

parasite drag [FL MECH] The portion of the total drag of an aircraft exclusive of the induced drag of the wings.

parking orbit [AERO ENG] A temporary earth orbit during which the space vehicle is checked out and its trajectory carefully measured to determine the amount and time of increase in velocity required to send it into a final orbit or into space in the desired direction.

Parsons-stage steam turbine [MECH ENG] A reaction-type steam-turbine stage in which the pressure drop occurs partially across the stationary nozzles and partly across the rotating blades.

particle See material particle.

particle dynamics [MECH] The study of the dependence of the motion of a single material particle on the external forces acting upon it, particularly electromagnetic and gravitational forces.

particle energy [MECH] For a particle in a potential, the sum of the particle's kinetic energy and potential energy.

particle mechanics [MECH] The study of the motion of a single material particle.

pascal [MECH] A unit of pressure equal to the pressure resulting from a force of 1 newton acting uniformly over an area of 1 square meter. Symbolized Pa.

Pascal's law [FL MECH] The law that a confined fluid transmits externally applied pressure uniformly in all directions, without change in magnitude.

pass [AERO ENG] 1. A single circuit of the earth made by a satellite; it starts at the time the satellite crosses the equator from the Southern Hemisphere into the Northern Hemisphere. 2. The period of time in which a satellite is within telemetry range of a data acquisition station. [MECH ENG] 1. The number of times that combustion gases are exposed to heat transfer surfaces in boilers (that is, single-pass, double-pass, and so on). 2. In metal rolling, the passage in one direction of metal deformed between rolls. 3. In metal cutting, transit of a metal cutting tool past the workpiece with a fixed tool setting.

passive communications satellite [AERO ENG] A satellite that reflects communications signals between stations, without providing amplification; an example is the Echo satellite. Also known as passive satellite.

patch bolt [DES ENG] A bolt with a countersunk head having a square knob that twists off when the bolt is screwed in tightly; used to repair boilers and steel ship hulls.

pattern [AERO ENG] The flight path flown by an aircraft, or prescribed to be flown, as in making an approach to a landing.

paver [MECH ENG] Any of several machines which, moving along the road, carry and lay paving material.

pawl [MECH ENG] The driving link or holding link of a ratchet mechanism, permits motion in one direction only.

payload [AERO ENG] That which an aircraft, rocket, or the like carries over and above what is necessary for the operation of the vehicle in its flight.

payload–mass ratio [AERO ENG] Of a rocket, the ratio of the effective propellant mass to the initial vehicle mass.

pdl-ft *See* foot-poundal.

Peaucellier linkage [MECH ENG] A mechanical linkage to convert circular motion exactly into straight-line motion.

pebble mill [MECH ENG] A solids size-reduction device with a cylindrical or conical shell rotating on a horizontal axis, and with a grinding medium such as balls of flint, steel, or porcelain.

peck [MECH] Abbreviated pk. 1. A unit of volume used in the United States for measurement of solid substances, equal to 8 dry quarts, or ¼ bushel, or 537.605 cubic inches, or 0.00880976754172 cubic meter. 2. A unit of volume used in the United Kingdom for measurement of solid and liquid substances, although usually the former, equal to 2 gallons, or approximately 0.00909218 cubic meter.

pedal [DES ENG] A lever operated by foot.

peen [DES ENG] The end of a hammer head with a hemispherical, wedge, or other shape; used to bend, indent, or cut.

peepdoor [MECH ENG] A small door in a furnace with a glass opening through which combustion may be observed.

pellet mill [MECH ENG] Device for injecting particulate, granular or pasty feed into holes of a roller, then compacting the feed into a continuous solid rod to be cut off by a knife at the periphery of the roller.

Pelton turbine *See* Pelton wheel.

Pelton wheel [MECH ENG] An impulse hydraulic turbine in which pressure of the water supply is converted into velocity by a few stationary nozzles, and the water jets then impinge on the buckets mounted on the rim of a wheel; usually limited to high head installations, exceeding 500 feet (150 meters). Also known as Pelton turbine.

pendulous gyroscope [MECH] A gyroscope whose axis of rotation is constrained by a suitable weight to remain horizontal; it is the basis of one type of gyrocompass.

pendulum press [MECH ENG] A punch press actuated by a swinging treadle operated by the foot.

pendulum saw [MECH ENG] A circular saw that swings in a vertical arc for crosscuts.

penetration [AERO ENG] That phase of the letdown from high altitude to a specified approach altitude.

penetration ballistics [MECH] A branch of terminal ballistics concerned with the motion and behavior of a missile during and after penetrating a target.

penetration rate [MECH ENG] The actual rate of penetration of drilling tools.

penetration speed [MECH ENG] The speed at which a drill can cut through rock or other material.

pennyweight [MECH] A unit of mass equal to $\frac{1}{20}$ troy ounce or to 1.55517384 grams; the term is employed in the United States and in England for the valuation of silver, gold, and jewels. Abbreviated dwt; pwt.

perch [MECH] Also known as pole; rod. **1.** A unit of length, equal to 5.5 yards, or 16.5 feet, or 5.0292 meters. **2.** A unit of area, equal to 30.25 square yards, or 272.25 square feet, or 25.29285264 square meters.

percussion bit [MECH ENG] A rock-drilling tool with chisellike cutting edges, which when driven by impacts against a rock surface drills a hole by a chipping action.

percussion drill [MECH ENG] A drilling machine usually using compressed air to drive a piston that delivers a series of impacts to the shank end of a drill rod or steel and attached bit.

percussion drilling [MECH ENG] A drilling method in which hammer blows are transmitted by the drill rods to the drill bit.

perfect fluid *See* inviscid fluid.

perfect gas *See* ideal gas.

periodic motion [MECH] Any motion that repeats itself identically at regular intervals.

peripheral speed *See* cutting speed.

permanent axis [MECH] The axis of the greatest moment of inertia of a rigid body, about which it can rotate in equilibrium.

permanent gas [THERMO] A gas at a pressure and temperature far from its liquid state.

permanent set [MECH] Permanent plastic deformation of a structure or a test piece after removal of the applied load. Also known as set.

permanent wave [FL MECH] A wave (in a fluid) which moves with no change in streamline pattern, and which, therefore, is a stationary wave relative to a coordinate system moving with the wave.

permeability [FL MECH] **1.** The ability of a membrane or other material to permit a substance to pass through it. **2.** Quantitatively, the amount of substance which passes through the material under given conditions.

permeability coefficient [FL MECH] The rate of water flow in gallons per day through a cross section of 1 square foot under a unit hydraulic gradient, at the prevailing temperature or at 60°F (16°C). Also known as coefficient of permeability; hydraulic conductivity; Meinzer unit.

perpendicular axis theorem [MECH] A theorem which states that the sum of the moments of inertia of a plane lamina about any two perpendicular axes in the plane of the lamina is equal to the moment of inertia about an axis through their intersection perpendicular to the lamina.

phase [THERMO] The type of state of a system, such as solid, liquid, or gas.

phase crossover [CONT SYS] A point on the plot of the loop ratio at which it has a phase angle of 180°.

phase diagram [THERMO] 1. A graph showing the pressures at which phase transitions between different states of a pure compound occur, as a function of temperature. 2. A graph showing the temperatures at which transitions between different phases of a binary system occur, as a function of the relative concentrations of its components.

phase integral *See* action.

phase margin [CONT SYS] The difference between 180° and the phase of the loop ratio of a stable system at the gain-crossover frequency.

phase plane analysis [CONT SYS] A method of analyzing systems in which one plots the time derivative of the system's position (or some other quantity characterizing the system) as a function of position for various values of initial conditions.

phase portrait [CONT SYS] A graph showing the time derivative of a system's position (or some other quantity characterizing the system) as a function of position for various values of initial conditions.

Philips hot-air engine [MECH ENG] A compact hot-air engine that is a Philips Research Lab (Holland) design; it uses only one cylinder and piston, and operates at 3000 revolutions per minute, with hot-chamber temperature of 1200°F (650°C), maximum pressure of 50 atmospheres (5.07 megapascals), and mean effective pressure of 14 atmospheres (1.42 megapascals).

Phillips screw [DES ENG] A screw having in its head a recess in the shape of a cross; it is inserted or removed with a Phillips screwdriver that automatically centers itself in the screw.

photodraft [DES ENG] A photographic reproduction of a master layout or design on a specially prepared emulsion-coated piece of sheet metal; used as a master in a tool-construction department.

photoelectric door opener [CONT SYS] A control system that employs a photocell or other photo device, used to open and close a power-operated door.

photoelectric flame-failure detector [CONT SYS] A photoelectric control that cuts off fuel flow when the fuel-consuming flame is extinguished.

photoelectric loop control [CONT SYS] A photoelectric control system used as a position regulator for a loop of material passing from one strip-processing line to another that may travel at a different speed. Also known as loop control.

photoelectric register control [CONT SYS] A register control using a light source, one or more phototubes, a suitable optical system, an amplifier, and a relay to actuate control equipment when a change occurs in the amount of light reflected from a moving surface due to register marks, dark areas of a design, or surface defects. Also known as photoelectric scanner.

photoelectric scanner *See* photoelectric register control.

photoelectric smoke-density control [CONT SYS] A photoelectric control system used to measure, indicate, and control the density of smoke in a flue or stack.

photoelectric sorter [CONT SYS] A photoelectric control system used to sort objects according to color, size, shape, or other light-changing characteristics.

photon sail *See* solar sail.

phugoid [AERO ENG] Pertaining to variations in the longitudinal motion or course of the center of mass of an aircraft.

physical realizability [CONT SYS] For a transfer function, the possibility of constructing a network with this transfer function.

physical system *See* causal system.

pick [DES ENG] 1. The steel cutting points used on a coal-cutter chain. 2. A miner's steel or iron digging tool with sharp points at each end.

pickax [DES ENG] A pointed steel or iron tool mounted on a wooden handle and used for breaking earth and stone.

pick hammer [DES ENG] A hammer with a point at one end of the head and a blunt surface at the other end.

pick lacing [DES ENG] The pattern to which the picks are set in a cutter chain.

pickoff [MECH ENG] A mechanical device for automatic removal of the finished part from a press die.

pickup [AERO ENG] A potentiometer used in an automatic pilot to detect the motion of the airplane around the gyro and initiate corrective adjustments.

picosecond [MECH] A unit of time equal to 10^{-12} second, or one-millionth of a microsecond. Abbreviated ps. Formerly known as micromicrosecond.

picowatt [MECH] A unit of power equal to 10^{-12} watt, or one-millionth of a microwatt. Abbreviated pW. Formerly known as micromicrowatt.

piecewise-linear system [CONT SYS] A system for which one can divide the range of values of input quantities into a finite number of intervals such that the output quantity is a linear function of the input quantity within each of these intervals.

pièze [MECH] A unit of pressure equal to 1 sthène per square meter, or to 1000 pascals. Abbreviated pz.

piezotropic [FL MECH] Characterized by piezotropy.

piezotropy [FL MECH] The property of a fluid in which processes are characterized by a functional dependence of the thermodynamic functions of state: $d\rho/dt = b(dp/dt)$, where ρ is the density, p the pressure, and b a function of the thermodynamic variables, called the coefficient of piezotropy.

pile driver [MECH ENG] A hoist and movable steel frame equipped to handle piles and drive them into the ground.

pile extractor [MECH ENG] 1. A pile hammer which strikes the pile upward so as to loosen its grip and remove it from the ground. 2. A vibratory hammer which loosens the pile by high-frequency jarring.

pile formula [MECH] An equation for the forces acting on a pile at equilibrium: $P = pA + tS + Sn \sin \phi$, where P is the load, A is the area of the pile point, p is the force per unit area on the point, S is the embedded surface of the pile, t is the force per unit area parallel to S, n is the force per unit area normal to S, and ϕ is the taper angle of the pile.

pile hammer [MECH ENG] The heavy weight of a pile driver that depends on gravity for its striking power and is used to drive piles into the ground. Also known as drop hammer.

pillar bolt [DES ENG] A bolt projecting from a part so as to support it.

pillar crane [MECH ENG] A crane whose mechanism can be rotated about a fixed pillar.

pillar press [MECH ENG] A punch press framed by two upright columns; the driving shaft passes through the columns, and the slide operates between them.

pilot [AERO ENG] **1.** A person who handles the controls of an aircraft or spacecraft from within the craft, and guides or controls the craft in flight. **2.** A mechanical system designed to exercise control functions in an aircraft or spacecraft. [MECH ENG] A cylindrical steel bar extending through, and about 8 inches (20 centimeters) beyond the face of, a reaming bit; it acts as a guide that follows the original unreamed part of the borehole and hence forces the reaming bit to follow, and be concentric with, the smaller-diameter, unreamed portion of the original borehole.

pilot bit [DES ENG] A noncoring bit with a cylindrical diamond-set plug of somewhat smaller diameter than the bit proper, set in the center and projecting beyond the main face of the bit.

pilot chute [AERO ENG] A small parachute canopy attached to a larger canopy to actuate and accelerate the opening of the load-bearing canopy.

pilot drill [MECH ENG] A small drill to start a hole to ensure that a larger drill will run true to center.

pilotless aircraft [AERO ENG] An aircraft adapted to control by or through a preset self-reacting unit or a radio-controlled unit, without the benefit of a human pilot.

pilot wire regulator [CONT SYS] Automatic device for controlling adjustable gains or losses associated with transmission circuits to compensate for transmission changes caused by temperature variations, the control usually depending upon the resistance of a conductor or pilot wire having substantially the same temperature conditions as the conductors of the circuits being regulated.

pin [DES ENG] **1.** A cylindrical fastener made of wood, metal, or other material used to join two members or parts with freedom of angular movement at the joint. **2.** A short, pointed wire with a head used for fastening fabrics, paper, or similar materials.

pinch bar [DES ENG] A pointed lever, used somewhat like a crowbar, to roll heavy wheels.

pinion [MECH ENG] The smaller of a pair of gear wheels or the smallest wheel of a gear train.

pin joint [DES ENG] A joint made with a pin hinge which has a removable pin.

pin rod [DES ENG] A rod designed to connect two parts so they act as one.

pint [MECH] Abbreviated pt. **1.** A unit of volume, used in the United States for measurement of liquid substances, equal to $\frac{1}{8}$ U.S. gallon, or 29⅞ cubic inches, or $4.73176473 \times 10^{-4}$ cubic meters. Also known as liquid pint (liq pt). **2.** A unit of volume used in the United States for measurement of solid substances, equal to $\frac{1}{64}$ U.S. bushel, or 107,521/3200 cubic inches, or approximately 5.50610×10^{-4} cubic meters. Also known as dry pint (dry pt). **3.** A unit of volume, used in the United Kingdom for measurement of liquid and solid substances, although usually the former, equal to $\frac{1}{8}$ imperial gallon, or approximately 5.68261×10^{-4} cubic meters. Also known as imperial pint.

pintle [DES ENG] A vertical pivot pin, as on a rudder or a gun carriage.

pintle chain [DES ENG] A chain with links held together by pivot pins; used with sprocket wheels.

pin-type mill [MECH ENG] Solids pulverizer in which protruding pins on high-speed rotating disk provide the breaking energy.

pipe [DES ENG] A tube made of metal, clay, plastic, wood, or concrete and used to conduct a fluid, gas, or finely divided solid.

pipe bit [DES ENG] A bit designed for attachment to standard coupled pipe for use in socketing the pipe in bedrock.

pipe clamp [DES ENG] A device similar to a casing clamp, but used on a pipe to grasp it and facilitate hoisting or suspension.

pipe cutter [DES ENG] A hand tool consisting of a clamplike device with three cutting wheels which are forced inward by screw pressure to cut into a pipe as the tool is rotated around the pipe circumference.

pipe tee [DES ENG] A T-shaped pipe fitting with two outlets, one at 90° to the connection to the main line.

pipe thread [DES ENG] Most commonly, a 60° thread used on pipes and tubes, characterized by flat crests and roots and cut with 3/4-inch taper per foot (about 1.9 centimeters per 30 centimeters). Also known as taper pipe thread.

pipe wrench [DES ENG] A tool designed to grip and turn a pipe or rod about its axis in one direction only.

piston [MECH ENG] A sliding metal cylinder that reciprocates in a tubular housing, either moving against or moved by fluid pressure.

piston blower [MECH ENG] A piston-operated, positive-displacement air compressor used for stationary, automobile, and marine duty.

piston corer [MECH ENG] A steel tube which is driven into the sediment by a free fall and by lead attached to the upper end, and which is capable of recovering undistorted vertical sections of sediment.

piston displacement [MECH ENG] The volume which a piston in a cylinder displaces in a single stroke, equal to the distance the piston travels times the internal cross section of the cylinder.

piston drill [MECH ENG] A heavy percussion-type rock drill mounted either on a horizontal bar or on a short horizontal arm fastened to a vertical column; drills holes to 6 inches (15 centimeters) in diameter. Also known as reciprocating drill.

piston engine [MECH ENG] A type of engine characterized by reciprocating motion of pistons in a cylinder. Also known as displacement engine; reciprocating engine.

piston flow [FL MECH] Two-phase (vapor-liquid) flow in which the gas flows as large plugs; occurs for gas superficial velocities from about 2 to 30 feet per second (60 to 900 centimeters per second). Also known as plug flow; slug flow.

piston head [MECH ENG] That part of a piston above the top ring.

piston pin [MECH ENG] A cylindrical pin that connects the connecting rod to the piston. Also known as wrist pin.

piston pump [MECH ENG] A pump in which motion and pressure are applied to the fluid by a reciprocating piston in a cylinder. Also known as reciprocating pump.

piston ring [DES ENG] A sealing ring fitted around a piston and extending to the cylinder wall to prevent leakage. Also known as packing ring.

piston rod [MECH ENG] The rod which is connected to the piston, and moves or is moved by the piston.

piston skirt [MECH ENG] That part of a piston below the piston pin bore.

piston speed [MECH ENG] The total distance a piston travels in a given time; usually expressed in feet per minute.

piston valve [MECH ENG] A cylindrical type of steam engine slide valve for admission and exhaust of steam.

pitch [DES ENG] The distance between similar elements arranged in a pattern or between two points of a mechanical part, as the distance between the peaks of two successive grooves on a disk recording or on a screw. [MECH] 1. Of an aerospace vehicle, an angular displacement about an axis parallel to the lateral axis of the vehicle. 2. The rising and falling motion of the bow of a ship or the tail of an airplane as the craft oscillates about a transverse axis.

pitch acceleration [MECH] The angular acceleration of an aircraft or missile about its lateral, or Y, axis.

pitch attitude [MECH] The attitude of an aircraft, rocket, or other flying vehicle, referred to the relationship between the longitudinal body axis and a chosen reference line or plane as seen from the side.

pitch axis [MECH] A lateral axis through an aircraft, missile, or similar body, about which the body pitches. Also known as pitching axis.

pitch circle [DES ENG] In toothed gears, an imaginary circle concentric with the gear axis which is defined at the thickest point on the teeth and along which the tooth pitch is measured.

pitch cone [DES ENG] A cone representing the pitch surface of a bevel gear.

pitch cylinder [DES ENG] A cylinder representing the pitch surface of a spur gear.

pitch diameter [DES ENG] The diameter of the pitch circle of a gear.

pitch indicator [AERO ENG] An instrument for indicating the existence and approximate magnitude of the angular velocity about the lateral axis of an airframe.

pitching axis *See* pitch axis.

pitching moment [MECH] A moment about a lateral axis of an aircraft, rocket, or airfoil.

pitch line *See* cam profile.

pitchover [AERO ENG] The programmed turn from the vertical that a rocket under power takes as it describes an arc and points in a direction other than vertical.

pitot pressure [FL MECH] Pressure at the open end of a pitot tube.

pivot [MECH] A short, pointed shaft forming the center and fulcrum on which something turns, balances, or oscillates.

pivot-bucket conveyor-elevator [MECH ENG] A bucket conveyor having overlapping pivoted buckets on long-pitch roller chains; buckets are always level except when tripped to discharge materials.

pk *See* peck.

Pl *See* poiseuille.

plain-laid [DES ENG] Pertaining to a rope whose strands are twisted together in a direction opposite to that of the twist in the strands.

plain milling cutter [DES ENG] A cylindrical milling cutter with teeth on the periphery only; used for milling plain or flat surfaces. Also known as slab cutter.

plain turning [MECH ENG] Lathe operations involved when machining a workpiece between centers.

plane [DES ENG] A tool consisting of a smooth-soled stock from the face of which extends a wide-edged cutting blade for smoothing and shaping wood.

plane lamina [MECH] A body whose mass is concentrated in a single plane.

plane of departure [MECH] Vertical plane containing the path of a projectile as it leaves the muzzle of the gun.

plane of fire [MECH] Vertical plane containing the gun and the target, or containing a line of site.

plane of maximum shear stress [MECH] Either of two planes that lie on opposite sides of and at angels of 45° to the maximum principal stress axis and that are parallel to the intermediate principal stress axis.

plane of yaw [MECH] The plane determined by the tangent to the trajectory of a projectile in flight and the axis of the projectile.

plane Poiseuille flow [FL MECH] Rheological (viscosity) measurement in which the fluid of interest is propelled through a narrow slot, and the volumetric flow rate and the pressure gradient are measured simultaneously to determine viscosity.

plan equation [MECH ENG] The mathematical statement that horsepower = $plan/33,000$, where p = mean effective pressure (pounds per square inch), l = length of piston stroke (feet), a = net area of piston (square inches), and n = number of cycles completed per minute.

planer [MECH ENG] A machine for the shaping of long, flat, or flat contoured surfaces by reciprocating the workpiece under a stationary single-point tool or tools.

plane strain [MECH] A deformation of a body in which the displacements of all points in the body are parallel to a given plane, and the values of these displacements do not depend on the distance perpendicular to the plane.

plane stress [MECH] A state of stress in which two of the principal stresses are always parallel to a given plane and are constant in the normal direction.

planetary gear train [MECH ENG] An assembly of meshed gears consisting of a central gear, a coaxial internal or ring gear, and one or more intermediate pinions supported on a revolving carrier.

planet gear [MECH ENG] A pinion in a planetary gear train.

planform [AERO ENG] The shape or form of an object, such as an airfoil, as seen from above, as in a plan view.

planishing [MECH ENG] Smoothing the surface of a metal by a rapid series of overlapping, light hammerlike blows or by rolling in a planishing mill.

plant decomposition [CONT SYS] The partitioning of a large-scale control system into subsystems along lines of weak interaction.

plasma engine [AERO ENG] An engine for space travel in which neutral plasma is accelerated and directed by external magnetic fields that interact with the magnetic field produced by current flow through the plasma. Also known as plasma jet.

plasma jet See plasma engine.

plasma propulsion [AERO ENG] Propulsion of spacecraft and other vehicles by using electric or magnetic fields to accelerate both positively and negatively charged particles (plasma) to a very high velocity.

plasma rocket [AERO ENG] A rocket that is accelerated by means of a plasma engine.

plastic [MECH] Displaying, or associated with, plasticity.

plastic deformation [MECH] Permanent change in shape or size of a solid body without fracture resulting from the application of sustained stress beyond the elastic limit.

plastic design See ultimate-load design.

plasticity [MECH] The property of a solid body whereby it undergoes a permanent change in shape or size when subjected to a stress exceeding a particular value, called the yield value.

plasticoviscosity [MECH] Plasticity in which the rate of deformation of a body subjected to stresses greater than the yield stress is a linear function of the stress.

plastic viscosity [FL MECH] A measure of the internal resistance to fluid flow of a Bingham plastic, expressed as the tangential shear stress in excess of the yield stress divided by the resulting rate of shear.

plate [DES ENG] A rolled, flat piece of metal of some arbitrary minimum thickness and width depending on the type of metal.

plate-belt feeder *See* apron feeder.

plate cam [MECH ENG] A flat, open cam that imparts a sliding motion.

plate coil [MECH ENG] Heat-transfer device made from two metal sheets held together, one or both plates embossed to form passages between them for a heating or cooling medium to flow through. Also known as panel coil.

plate conveyor [MECH ENG] A conveyor with a series of steel plates as the carrying medium; each plate is a short trough, all slightly overlapped to form an articulated band, and attached to one center chain or to two side chains; the chains join rollers running on an angle-iron framework and transmit the drive from the driveheads, installed at intermediate points and sometimes also at the head or tail ends.

plate feeder *See* apron feeder.

plate-fin exchanger [MECH ENG] Heat-transfer device made up of a stack or layers, with each layer consisting of a corrugated fin between flat metal sheets sealed off on two sides by channels or bars to form passages for the flow of fluids.

platen [MECH ENG] A flat surface for exchanging heat in a boiler or heat exchanger which may have extended heat transfer surfaces.

plate-type exchanger [MECH ENG] Heat-exchange device similar to a plate-and-frame filter press; fluids flow between the frame-held plates, transferring heat between them.

platform conveyor [MECH ENG] A single- or double-strand conveyor with plates of steel or hardwood forming a continuous platform on which the loads are placed.

pleated cartridge [DES ENG] A filter cartridge made into a convoluted form that resembles the folds of an accordion.

plenum system [MECH ENG] A heating or air conditioning system in which air is forced through a plenum chamber for distribution to ducts.

pli [MECH] A unit of line density (mass per unit length) equal to 1 pound per inch, or approximately 17.8580 kilograms per meter.

pliers [DES ENG] A small instrument with two handles and two grasping jaws, usually long and roughened, working on a pivot; used for holding small objects and cutting, bending, and shaping wire.

plowshare [DES ENG] The pointed part of a plow moldboard, which penetrates and cuts the soil first.

plug cock *See* plug valve.

plug flow *See* piston flow.

plug gage [DES ENG] A steel gage that is used to test the dimension of a hole; may be straight or tapered, plain or threaded, and of any cross-sectional shape.

plug valve [DES ENG] A valve fitted with a plug that has a hole through which fluid flows and that is rotatable through 90° for operation in the open or closed position. Also known as plug cock.

plunge grinding [MECH ENG] Grinding in which the wheel moves radially toward the work.

plunger [DES ENG] A wooden shaft with a large rubber suction cup at the end, used to clear plumbing traps and waste outlets. [MECH ENG] The long rod or piston of a reciprocating pump.

plunger pump [MECH ENG] A reciprocating pump where the packing is on the stationary casing instead of the moving piston.

pneumatic atomizer [MECH ENG] An atomizer that uses compressed air to produce drops in the diameter range of 5–100 micrometers.

pneumatic controller [MECH ENG] A device for the mechanical movement of another device (such as a valve stem) whose action is controlled by variations in pneumatic pressure connected to the controller.

pneumatic control valve [MECH ENG] A valve in which the force of compressed air against a diaphragm is opposed by the force of a spring to control the area of the opening for a fluid stream.

pneumatic conveyor [MECH ENG] A conveyor which transports dry, free-flowing, granular material in suspension, or a cylindrical carrier, within a pipe or duct by means of a high-velocity airstream or by pressure of vacuum generated by an air compressor. Also known as air conveyor.

pneumatic drill [MECH ENG] Compressed-air drill worked by reciprocating piston, hammer action, or turbo drive.

pneumatic drilling [MECH ENG] Drilling a hole when using air or gas in lieu of conventional drilling fluid as the circulating medium; an adaptation of rotary drilling.

pneumatic hammer [MECH ENG] A hammer in which compressed air is utilized for producing the impacting blow.

pneumatic hoist *See* air hoist.

pneumatic riveter [MECH ENG] A riveting machine having a rapidly reciprocating piston driven by compressed air.

pneumatics [FL MECH] Fluid statics and behavior in closed systems when the fluid is a gas.

pneumatic servo *See* valve positioner.

pneumatic servomechanism [CONT SYS] A servomechanism in which power is supplied and transmission of signals is carried out through the medium of compressed air.

pod [AERO ENG] An enclosure, housing, or detachable container of some kind on an airplane or space vehicle, as an engine pod. [DES ENG] **1.** The socket for a bit in a brace. **2.** A straight groove in the barrel of a pod auger.

Pohlé air lift pump [MECH ENG] A pistonless pump in which compressed air fills the annular space surrounding the uptake pipe and is free to enter the rising column at all points of its periphery.

Poinsot ellipsoid *See* inertia ellipsoid.

Poinsot's method [MECH] A method of describing Poinsot motion, by means of a geometrical construction in which the inertia ellipsoid rolls on the invariable plane without slipping.

Poinsot motion [MECH] The motion of a rigid body with a point fixed in space and with zero torque or moment acting on the body about the fixed point.

point angle [DES ENG] The angle at the point or edge of a cutting tool.

point-blank range [MECH] Distance to a target that is so short that the trajectory of a bullet or projectile is practically a straight, rather than a curved, line.

point of contraflexure [MECH] A point at which the direction of bending changes. Also known as point of inflection.

point of fall [MECH] The point in the curved path of a falling projectile that is level with the muzzle of the gun. Also known as level point.

point of inflection *See* point of contraflexure.

poise [FL MECH] A unit of dynamic viscosity equal to the dynamic viscosity of a fluid in which there is a tangential force 1 dyne per square centimeter resisting the flow of two parallel fluid layers past each other when their differential velocity is 1 centimeter per second per centimeter of separation. Abbreviated P.

poiseuille [FL MECH] A unit of dynamic viscosity of a fluid in which there is a tangential force of 1 newton per square meter resisting the flow of two parallel layers past each other when their differential velocity is 1 meter per second per meter of separation; equal to 10 poise; used chiefly in France. Abbreviated Pl.

Poiseuille flow [FL MECH] The steady flow of an incompressible fluid parallel to the axis of a circular pipe of infinite length, produced by a pressure gradient along the pipe.

Poiseuille's law [FL MECH] The law that the volume flow of an incompressible fluid through a circular tube is equal to $\pi/8$ times the pressure differences between the ends of the tube, times the fourth power of the tube's radius divided by the product of the tube's length and the dynamic viscosity of the fluid.

Poisson bracket [MECH] For any two dynamical variables, X and Y, the sum, over all degrees of freedom of the system, of $(\partial X/\partial q)(\partial Y/\partial p) - (\partial X/\partial p)(\partial Y/\partial q)$, where q is a generalized coordinate and p is the corresponding generalized momentum.

Poisson number [MECH] The reciprocal of the Poisson ratio.

Poisson ratio [MECH] The ratio of the transverse contracting strain to the elongation strain when a rod is stretched by forces which are applied at its ends and which are parallel to the rod's axis.

polar orbit [AERO ENG] A satellite orbit running north and south, so the satellite vehicle orbits over both the North Pole and the South Pole.

polar timing diagram [MECH ENG] A diagram of the events of an engine cycle relative to crankshaft position.

pole [MECH] **1.** A point at which an axis of rotation or of symmetry passes through the surface of a body. **2.** See perch.

pole derrick See gin pole.

pole lathe [MECH ENG] A simple lathe in which the work is rotated by a cord attached to a treadle.

pole-positioning [CONT SYS] A design technique used in linear control theory in which many or all of a system's closed-loop poles are positioned as required, by proper choice of a linear state feedback law; if the system is controllable, all of the closed-loop poles can be arbitrarily positioned by this technique.

pole-zero configuration [CONT SYS] A plot of the poles and zeros of a transfer function in the complex plane; used to study the stability of a system, its natural motion, its frequency response, and its transient response.

polhode [MECH] For a rotating rigid body not subject to external torque, the closed curve traced out on the inertia ellipsoid by the intersection with this ellipsoid of an axis parallel to the angular velocity vector and through the center.

Polhode cone See body cone.

polishing [MECH ENG] Smoothing and brightening a surface such as a metal or a rock through the use of abrasive materials.

polishing roll [MECH ENG] A roll or series of rolls on a plastics mold; has highly polished chrome-plated surfaces; used to produce a smooth surface on a plastic sheet as it is extruded.

polishing wheel [DES ENG] An abrasive wheel used for polishing.

polytropic process [THERMO] An expansion or compression of a gas in which the quantity pV^n is held constant, where p and V are the pressure and volume of the gas, and n is some constant.

pond   *See* gram-force.

pontoon   [AERO ENG] A float on an airplane.

pop action   [MECH ENG] The action of a safety valve as it opens under steam pressure when the valve disk is lifted off its seat.

Popov's stability criterion   [CONT SYS] A frequency domain stability test for systems consisting of a linear component described by a transfer function $G(s)$ preceded by a nonlinear component characterized by an input-output function $N(.)$, with a unity gain feedback loop surrounding the series connection.

poppet   [DES ENG] A spring-loaded ball engaging a notch; a ball latch.

poppet valve   [MECH ENG] A cam-operated or spring-loaded reciprocating-engine mushroom-type valve used for control of admission and exhaust of working fluid; the direction of movement is at right angles to the plane of its seat.

popping pressure   [MECH ENG] In compressible fluid service, the inlet pressure at which a safety valve disk opens.

porcupine boiler   [MECH ENG] A boiler having dead end tubes projecting from a vertical shell.

pore diameter   [DES ENG] The average or effective diameter of the openings in a membrane, screen, or other porous material.

pore diffusion   [FL MECH] The movement of fluids (gas or liquid) into the interstices of porous solids or membranes; occurs in membrane separation, zeolite adsorption, dialysis, and reverse osmosis.

porous bearing   [DES ENG] A bearing made from sintered metal powder impregnated with oil by a vacuum treatment.

porous wheel   [DES ENG] A grinding wheel having a porous structure and a vitrified or resinoid bond.

portal crane   [MECH ENG] A jib crane carried on a four-legged portal built to run on rails.

porthole   [DES ENG] The opening or passageway connecting the inside of a bit or core barrel to the outside and through which the circulating medium is discharged.

positional-error constant   [CONT SYS] For a stable unity feedback system, the limit of the transfer function as its argument approaches zero.

positional servomechanism   [CONT SYS] A feedback control system in which the mechanical position (as opposed to velocity) of some object is automatically maintained.

positioning   [MECH ENG] A tooling function concerned with manipulating the workpiece in relationship to the working tools.

positioning action   [CONT SYS] Automatic control action in which there is a predetermined relation between the value of a controlled variable and the position of a final control element.

positive acceleration   [MECH] 1. Accelerating force in an upward sense or direction, such as from bottom to top, or from seat to head; 2. The acceleration in the direction that this force is applied.

positive clutch   [MECH ENG] A clutch designed to transmit torque without slip.

positive-displacement compressor   [MECH ENG] A compressor that confines successive volumes of fluid within a closed space in which the pressure of the fluid is increased as the volume of the closed space is decreased.

positive-displacement pump   [MECH ENG] A pump in which a measured quantity of liquid is entrapped in a space, its pressure is raised, and then it is delivered; for example, a reciprocating piston-cylinder or rotary-vane, gear, or lobe mechanism.

positive draft [MECH ENG] Pressure in the furnace or gas passages of a steam-generating unit which is greater than atmospheric pressure.

positive drive belt *See* timing belt.

positive feedback [CONT SYS] Feedback in which a portion of the output of a circuit or device is fed back in phase with the input so as to increase the total amplification. Also known as reaction (British usage); regeneration; regenerative feedback; retroaction (British usage).

positive motion [MECH ENG] Motion transferred from one machine part to another without slippage.

positive temperature coefficient [THERMO] The condition wherein the resistance, length, or some other characteristic of a substance increases when temperature increases.

post brake [MECH ENG] A brake occasionally fitted on a steam winder or haulage, and consisting of two upright posts mounted on either side of the drum that operate on brake paths bolted to the drum cheeks.

pot die forming [MECH ENG] Forming sheet or plate metal through a hollow die by the application of pressure which causes the workpiece to assume the contour of the die.

potential drop [FL MECH] The difference in pressure head between one equipotential line and another.

potential energy [MECH] The capacity to do work that a body or system has by virtue of its position or configuration.

potential flow [FL MECH] Flow in which the velocity of flow is the gradient of a scalar function, known as the velocity potential.

potential temperature [THERMO] The temperature that would be reached by a compressible fluid if it were adiabatically compressed or expanded to a standard pressure, usually 1 bar.

potential vorticity [FL MECH] The product of the absolute vorticity and the static stability, conservative in adiabatic flow, given by the expression $(\eta/\theta)(\partial\theta/\partial p)$, where η is the absolute vorticity of a fluid parcel, θ the potential temperature, and p the pressure. Also known as absolute potential vorticity.

potentiometric controller [CONT SYS] A controller that operates on the null balance principle, in which an error signal is produced by balancing the sensor signal against a set-point voltage in the input circuit; the error signal is amplified for use in keeping the load at a desired temperature or other parameter.

pound [MECH] **1.** A unit of mass in the English absolute system of units, equal to 0.45359237 kilogram. Abbreviated lb. Also known as avoirdupois pound; pound mass. **2.** A unit of force in the English gravitational system of units, equal to the gravitational force experienced by a pound mass when the acceleration of gravity has its standard value of 9.80665 meters per second per second (approximately 32.1740 ft/sec^2) equal to 4.4482216152605 newtons. Abbreviated lb. Also spelled Pound (Lb). Also known as pound force (lbf). **3.** A unit of mass in the troy and apothecaries' systems, equal to 12 troy or apothecaries' ounces, or 5760 grains, or 5760/7000 avoirdupois pound, or 0.3732417216 kilogram. Also known as apothecaries' pound (abbreviated lb ap in the US or lb apoth in the UK), troy pound (abbreviated lb UK).

poundal [MECH] A unit of force in the British absolute system of units equal to the force which will impart an acceleration of 1 ft/sec^2 to a pound mass, or to 0.138254954376 newton.

poundal-foot *See* foot-poundal.

pound-foot *See* foot-pound.

pound force *See* pound.

pound mass *See* pound.

pound per square foot [MECH] A unit of pressure equal to the pressure resulting from a force of 1 pound applied uniformly over an area of 1 square foot. Abbreviated psf.

pound per square inch [MECH] A unit of pressure equal to the pressure resulting from a force of 1 pound applied uniformly over an area of 1 square inch. Abbreviated psi.

pounds per square inch absolute [MECH] The absolute, thermodynamic pressure, measured by the number of pounds-force exerted on an area of 1 square inch. Abbreviated lbf in^{-2} abs; psia.

pounds per square inch gage [MECH] The gage pressure, measured by the number of pounds-force exerted on an area of 1 square inch. Abbreviated psig.

pour point [FL MECH] Lowest test temperature at which a liquid will flow.

powder clutch [MECH ENG] A type of electromagnetic disk clutch in which the space between the clutch members is filled with dry, finely divided magnetic particles; application of a magnetic field coalesces the particles, creating friction forces between clutch members.

power-actuated pressure relief valve [MECH ENG] A pressure relief valve connected to and controlled by a device which utilizes a separate energy source.

power brake [MECH ENG] An automotive brake with engine-intake-manifold vacuum used to amplify the atmospheric pressure on a piston operated by movement of the brake pedal.

power car [AERO ENG] A suspended structure on an airship that houses an engine. [MECH ENG] 1. A railroad car with equipment for furnishing heat and electric power to a train. 2. A railroad car with controls, which can be operated by itself or as part of a train.

power control valve [MECH ENG] A safety relief device operated by a power-driven mechanism rather than by pressure.

power drill [MECH ENG] A motor-driven drilling machine.

power-driven [MECH ENG] Of a component or piece of equipment, moved, rotated, or operated by electrical or mechanical energy, as in a power-driven fan or power-driven turret.

power-law fluid [FL MECH] A fluid in which the shear stress at any point is proportional to the rate of shear at that point raised to some power.

power package [MECH ENG] A complete engine and its accessories, designed as a single unit for quick installation or removal.

power plant [MECH ENG] Any unit that converts some form of energy into electrical energy, such as a hydroelectric or steam-generating station, a diesel-electric engine in a locomotive, or a nuclear power plant. Also known as electric power plant.

power saw [MECH ENG] A power-operated woodworking saw, such as a bench or circular saw.

power shovel [MECH ENG] A power-operated shovel that carries a short boom on which rides a movable dipper stick carrying an open-topped bucket; used to excavate and remove debris.

power station *See* generating station.

power steering [MECH ENG] A steering control system for a propelled vehicle in which an auxiliary power source assists the driver by providing the major force required to direct the road wheels.

power stroke   [MECH ENG] The stroke in an engine during which pressure is applied to the piston by expanding steam or gases.

power train   [MECH ENG] The part of a vehicle connecting the engine to propeller or driven axle; may include drive shaft, clutch, transmission, and differential gear.

Poynting's law   [THERMO] A special case of the Clapeyron equation, in which the fluid is removed as fast as it forms, so that its volume may be ignored.

Pr$_m$ *See* Prandtl number.

practical entropy   *See* virtual entropy.

Prandtl number   [FL MECH] A dimensionless number used in the study of diffusion in flowing systems, equal to the kinematic viscosity divided by the molecular diffusivity. Symbolized Pr_m. Also known as Schmidt number 1 (N_{Sc}).   [THERMO] A dimensionless number used in the study of forced and free convection, equal to the dynamic viscosity times the specific heat at constant pressure divided by the thermal conductivity. Symbolized N_{Pr}.

prebreaker   [MECH ENG] Device used to break down large masses of solids prior to feeding them to a crushing or grinding device.

precession   [MECH] The angular velocity of the axis of spin of a spinning rigid body, which arises as a result of external torques acting on the body.

precision block   *See* gage block.

precision grinding   [MECH ENG] Machine grinding to specified dimensions and low tolerances.

precombustion chamber   [MECH ENG] A small chamber before the main combustion space of a turbine or reciprocating engine in which combustion is initiated.

precooler   [MECH ENG] A device for reducing the temperature of a working fluid before it is used by a machine.

preheater   [MECH ENG] A device for preliminary heating of a material, substance, or fluid that will undergo further use or treatment by heating.

preignition   [MECH ENG] Ignition of the charge in the cylinder of an internal combustion engine before ignition by the spark.

press   [MECH ENG] Any of various machines by which pressure is applied to a workpiece, by which a material is cut or shaped under pressure, by which a substance is compressed, or by which liquid is expressed.

press slide   [MECH ENG] The reciprocating member of a power press on which the punch and upper die are fastened.

pressure   [MECH] A type of stress which is exerted uniformly in all directions; its measure is the force exerted per unit area.

pressure angle   [MECH ENG] The angle that the line of force makes with a line at right angles to the center line of two gears at the pitch points.

pressure bar   [MECH ENG] A bar that holds the edge of a metal sheet during press operations, such as punching, stamping, or forming, and prevents the sheet from buckling or becoming crimped.

pressure carburetor   *See* injection carburetor.

pressure coefficient   [THERMO] The ratio of the fractional change in pressure to the change in temperature under specified conditions, usually constant volume.

pressure-containing member   [MECH ENG] The part of a pressure-relieving device which is in direct contact with the pressurized medium in the vessel being protected.

pressure drag   *See* pressure resistance.

pressure drop [FL MECH] The difference in pressure between two points in a flow system, usually caused by frictional resistance to a fluid flowing through a conduit, filter media, or other flow-conducting system.

pressure force [FL MECH] The force due to differences of pressure within a fluid mass; the (vector) force per unit volume is equal to the pressure gradient $-\nabla p$, and the force per unit mass (specific force) is equal to the product of the volume force and the specific volume $-\alpha\nabla p$.

pressure gradient [FL MECH] The rate of decrease (that is, the gradient) of pressure in space at a fixed time; sometimes loosely used to denote simply the magnitude of the gradient of the pressure field. Also known as barometric gradient.

pressure head [FL MECH] Also known as head. **1.** The height of a column of fluid necessary to develop a specific pressure. **2.** The pressure of water at a given point in a pipe arising from the pressure in it.

pressure plate [MECH ENG] The part of an automobile disk clutch that presses against the flywheel.

pressure relief device [MECH ENG] **1.** In pressure vessels, a device designed to open in a controlled manner to prevent the internal pressure of a component or system from increasing beyond a specified value, that is, a safety valve. **2.** A spring-loaded machine part which will yield, or deflect, when a predetermined force is exceeded.

pressure relief valve [MECH ENG] A valve which relieves pressure beyond a specified limit and recloses upon return to normal operating conditions.

pressure resistance [FL MECH] In fluid dynamics, a normal stress caused by acceleration of the fluid, which results in a decrease in pressure from the upstream to the downstream side of an object acting perpendicular to the boundary. Also known as pressure drag.

pressure-retaining member [MECH ENG] That part of a pressure-relieving device loaded by the restrained pressurized fluid.

pressure-stabilized [AERO ENG] Referring to membrane-type structures that require internal pressure for maintenance of a stable structure.

pressure thrust [AERO ENG] In rocketry, the product of the cross-sectional area of the exhaust jet leaving the nozzle exit and the difference between the exhaust pressure and the ambient pressure.

pressure-travel curve [MECH] Curve showing pressure plotted against the travel of the projectile within the bore of the weapon.

pressure viscosity [FL MECH] Property of petroleum lubricating oils to increase in viscosity when subjected to pressure.

pretravel [CONT SYST] The distance or angle through which the actuator of a switch moves from the free position to the operating position.

Prevost's theory [THERMO] A theory according to which a body is constantly exchanging heat with its surroundings, radiating an amount of energy which is independent of its surroundings, and increasing or decreasing its temperature depending on whether it absorbs more radiation than it emits, or vice versa.

prick punch [DES ENG] A tool that has a sharp conical point ground to an angle of 30–60°C; used to make a slight indentation on a workpiece to locate the intersection of centerlines.

primary air [MECH ENG] That portion of the combustion air introduced with the fuel in a burner.

primary breaker [MECH ENG] A machine which takes over the work of size reduction from blasting operations, crushing rock to maximum size of about 2-inch (5 centimeter) diameter; may be a gyratory crusher or jaw breaker. Also known as primary crusher.

primary creep [MECH] The initial high strain-rate region in a material subjected to sustained stress.

primary crusher *See* primary breaker.

primary phase [THERMO] The only crystalline phase capable of existing in equilibrium with a given liquid.

primary phase region [THERMO] On a phase diagram, the locus of all compositions having a common primary phase.

primary stress [MECH] A normal or shear stress component in a solid material which results from an imposed loading and which is under a condition of equilibrium and is not self-limiting.

primary structure [AERO ENG] The main framework, of an aircraft including fittings and attachments; any structural member whose failure would seriously impair the safety of the missile is a part of the primary structure.

prime mover [MECH ENG] The component of a power plant that transforms energy from the thermal or the pressure form to the mechanical form.

priming [MECH ENG] In a boiler, the excessive carryover of fine water particles along with the steam because of insufficient steam space, faulty boiler design, or faulty operating conditions.

priming pump [MECH ENG] A device on motor vehicles and tanks, providing a means of injecting a spray of fuel into the engine to facilitate starting.

primitive equations [FL MECH] The Eulerian equations of motion of a fluid in which the primary dependent variables are the fluid's velocity components; these equations govern a wide variety of fluid motions and form the basis of most hydrodynamical analysis; in meteorology, these equations are frequently specialized to apply directly to the cyclonic-scale motions by the introduction of filtering approximations.

principal axis [MECH] One of three perpendicular axes in a rigid body such that the products of inertia about any two of them vanish.

principal axis of strain [MECH] One of the three axes of a body that were mutually perpendicular before deformation. Also known as strain axis.

principal axis of stress [MECH] One of the three mutually perpendicular axes of a body that are perpendicular to the principal planes of stress. Also known as stress axis.

principal function [MECH] The integral of the Lagrangian of a system over time; it is involved in the statement of Hamilton's principle.

principal plane of stress [MECH] For a point in an elastic body, a plane at that point across which the shearing stress vanishes.

principal strain [MECH] The elongation or compression of one of the principal axes of strain relative to its original length.

principal stress [MECH] A stress occurring at right angles to a principal plane of stress.

principle of inaccessibility *See* Carathéodory's principle.

principle of least action [MECH] The principle that, for a system whose total mechanical energy is conserved, the trajectory of the system in configuration space is that path which makes the value of the action stationary relative to nearby paths between the same configurations and for which the energy has the same constant value. Also known as least-action principle.

principle of optimality [CONT SYS] A principle which states that for optimal systems, any portion of the optimal state trajectory is optimal between the states it joins.

principle of virtual work [MECH] The principle that the total work done by all forces acting on a system in static equilibrium is zero for any infinitesimal displacement from equilibrium which is consistent with the constraints of the system. Also known as virtual work principle.

probe [AERO ENG] An instrumented vehicle moving through the upper atmosphere or space or landing upon another celestial body in order to obtain information about the specific environment.

process control system [CONT SYS] The automatic control of a continuous operation.

product design [DES ENG] The determination and specification of the parts of a product and their interrelationship so that they become a unified whole.

product of inertia [MECH] Relative to two rectangular axes, the sum of the products formed by multiplying the mass (or, sometimes, the area) of each element of a figure by the product of the coordinates corresponding to those axes.

profiled keyway [DES ENG] A keyway for a straight key formed by an end-milling cutter. Also known as end-milled keyway.

profile drag [FL MECH] That part of the airfoil drag that results from the skin friction and the shape of the airfoil as indicated by the airfoil profile.

profile thickness [AERO ENG] The maximum distance between the upper and lower contours of an airfoil, measured perpendicular to the mean line of the profile.

profiling machine [MECH ENG] A machine used for milling irregular profiles; the cutting tool is guided by the contour of a model.

program [AERO ENG] In missile guidance, the planned flight path events to be followed by a missile in flight, including all the critical functions, preset in a program device, which control the behavior of the missile.

program device [CONT SYS] In missile guidance, tha automatic device used to control time and sequence of events of a program.

programmable controller [CONT SYS] A control device, normally used in industrial control applications, that employs the hardware architecture of a computer and a relay ladder diagram language. Also known as programmable logic controller.

programmable logic controller *See* programmable controller.

programmed turn [AERO ENG] The automatically controlled turn of a ballistic missile into the curved path that will lead to the correct velocity and vector for the final portion of the trajectory.

programmer [CONT SYS] A device used to control the motion of a missile in accordance with a predetermined plan.

projected planform [AERO ENG] The contour of the planform as viewed from above.

prony brake [MECH ENG] An absorption dynamometer that applies a friction load to the output shaft by means of wood blocks, a flexible band, or other friction surface.

proof resilience [MECH] The tensile strength necessary to stretch an elastomer from zero elongation to the breaking point, expressed in foot-pounds per cubic inch of original dimension.

proof stress [MECH] 1. The stress that causes a specified amount of permanent deformation in a material. 2. A specified stress to be applied to a member or structure in order to assess its ability to support service loads.

propellant injector [AERO ENG] A device for injecting propellants, which include fuel and oxidizer, into the combustion chamber of a rocket engine.

propellant mass ratio [AERO ENG] Of a rocket, the ratio of the effective propellant mass to the initial vehicle mass. Also known as propellant mass fraction.

propellant weight fraction [AERO ENG] The weight of the solid propellant charge divided by weight of the complete solid propellant propulsion unit.

propeller [MECH ENG] A bladed device that rotates on a shaft to produce a useful thrust in the direction of the shaft axis.

propeller blade [DES ENG] One of two or more plates radiating out from the hub of a propeller and normally twisted to form part of a helical surface.

propeller boss [DES ENG] The central portion of the screw propeller which carries the blades, and forms the medium of attachment to the propeller shaft. Also known as propeller hub.

propeller cavitation [FL MECH] Formation of vapor-filled and air-filled bubbles or cavities in water at or on the surface of a rotating propeller, occurring when the pressure falls below the vapor pressure of water.

propeller efficiency [MECH ENG] The ratio of the thrust horsepower delivered by the propeller to the shaft horsepower as delivered by the engine to the propeller.

propeller fan [MECH ENG] An axial-flow blower, with or without a casing, using a propeller-type rotor to accelerate the fluid.

propeller hub *See* propeller boss.

propeller pump *See* axial-flow pump.

propeller shaft [MECH ENG] A shaft, carrying a screw propeller at its end, that transmits power from an engine to the propeller.

propeller slip angle [MECH ENG] The angle between the plane of the blade face and its direction of motion.

propeller tip speed [MECH ENG] The speed in feet per minute swept by the propeller tips.

propeller turbine [MECH ENG] A form of reactive-type hydraulic turbine using an axial-flow propeller rotor.

propeller windmill [MECH ENG] A windmill that extracts wind power from horizontal air movements to rotate the blades of a propeller.

proportional band [CONT SYS] The range of values of the controlled variable that will cause a controller to operate over its full range.

proportional control [CONT SYS] Control in which the amount of corrective action is proportional to the amount of error; used, for example, in chemical engineering to control pressure, flow rate, or temperature in a process system.

proportional controller [CONT SYS] A controller whose output is proportional to the error signal.

proportional dividers [DES ENG] Dividers with two legs, pointed at both ends, and an adjustable pivot; distances measured by the points at one end can be marked off in proportion by the points at the other end.

proportional elastic limit [MECH] The greatest stress intensity for which stress is still proportional to strain.

proportional limit [MECH] The greatest stress a material can sustain without departure from linear proportionality of stress and strain.

proportional-plus-derivative control [CONT SYS] Control in which the control signal is a linear combination of the error signal and its derivative.

proportional-plus-integral control [CONT SYS] Control in which the control signal is a linear combination of the error signal and its integral.

proportional-plus-integral-plus-derivative control [CONT SYS] Control in which the control signal is a linear combination of the error signal, its integral, and its derivative.

propulsion [MECH] The process of causing a body to move by exerting a force against it.

propulsion system [MECH ENG] For a vehicle moving in a fluid medium, such as an airplane or ship, a system that produces a required change in momentum in the vehicle by changing the velocity of the air or water passing through the propulsive device or engine; in the case of a rocket-propelled vehicle operating without a fluid medium, the required momentum change is produced by using up some of the propulsive device's own mass, called the propellant.

Prospector [AERO ENG] A specific uncrewed spacecraft designed to make a soft landing on the moon to take measurements, photographs, and soil samples, and then return to earth.

proving ring [DES ENG] A ring used for calibrating test machines; the diameter of the ring changes when a force is applied along a diameter.

prudent limit of endurance [AERO ENG] The time during which an aircraft can remain airborne and still retain a given safety margin of fuel.

ps *See* picosecond.

pseudoplastic fluid [FL MECH] A fluid whose apparent viscosity or consistency decreases instantaneously with an increase in shear rate.

psf *See* pound per square foot.

psi *See* pound per square inch.

psia *See* pounds per square inch absolute.

psig *See* pounds per square inch gage.

psychromatic ratio [THERMO] Ratio of the heat-transfer coefficient to the product of the mass-transfer coefficient and humid heat for a gas-vapor system; used in calculation of humidity or saturation relationships.

psychrometric chart [THERMO] A graph each point of which represents a specific condition of a gas-vapor system (such as air and water vapor) with regard to temperature (horizontal scale) and absolute humidity (vertical scale); other characteristics of the system, such as relative humidity, wet-bulb temperature, and latent heat of vaporization, are indicated by lines on the chart.

psychrometric formula [THERMO] The semiempirical relation giving the vapor pressure in terms of the barometer and psychrometer readings.

psychrometric tables [THERMO] Tables prepared from the psychrometric formula and used to obtain vapor pressure, relative humidity, and dew point from values of wet-bulb and dry-bulb temperature.

pt *See* pint.

puff [MECH ENG] A small explosion within a furnace due to combustion conditions.

pug mill [MECH ENG] A machine for mixing and tempering a plastic material by the action of blades revolving in a drum or trough.

puller [MECH ENG] A lever-operated chain or wire-rope hoist for lifting or pulling at any angle, which has a reversible ratchet mechanism in the lever permitting short-stroke operation for both tensioning and relaxing, and which holds the loads with a Weston-type friction brake or a releasable ratchet. Also known as come-along.

pulley [DES ENG] A wheel with a flat, round, or grooved rim that rotates on a shaft and carries a flat belt, V-belt, rope, or chain to transmit motion and energy.

pulley lathe [MECH ENG] A lathe for turning pulleys.

pulley top [MECH ENG] A top with a long shank used to tap setscrew holes in pulley hubs.

pull-in torque [MECH ENG] The largest steady torque with which a motor will attain normal speed after accelerating from a standstill.

pull-out torque [MECH ENG] The largest torque under which a motor can operate without sharply losing speed.

pullshovel See backhoe.

pull strength [MECH] A unit in tensile testing; the bond strength in pounds per square inch.

pulper [MECH ENG] A machine that converts materials to pulp, for example, one that reduces paper waste to pulp.

pulsed transfer function [CONT SYS] The ratio of the z-transform of the output of a system to the z-transform of the input, when both input and output are trains of pulses. Also known as discrete transfer function; z-transfer function.

pulsejet engine [AERO ENG] A type of compressorless jet engine in which combustion occurs intermittently so that the engine is characterized by periodic surges of thrust; the inlet end of the engine is provided with a grid to which are attached flap valves; these can be sucked inward by a negative differential pressure to allow a regulated amount of air to flow inward to mix with the fuel. Also known as aeropulse engine.

pulsometer [MECH ENG] A simple, lightweight pump in which steam forces water out of one of two chambers alternately.

pulverizer [MECH ENG] Device for breaking down of solid lumps into a fine material by cleavage along crystal faces.

pump [MECH ENG] A machine that draws a fluid into itself through an entrance port and forces the fluid out through an exhaust port.

pump bob [MECH ENG] A device such as a crank that converts rotary motion into reciprocating motion.

pumping [FL MECH] Unsteadiness of the mercury in the barometer, caused by fluctuations of the air pressure produced by a gusty wind or due to the motion of a vessel.

pumping loss [MECH ENG] Power consumed in purging a cylinder of exhaust gas and sucking in fresh air instead.

punch [MECH ENG] A tool that forces metal into a die for extrusion or similar operations.

punch press [MECH ENG] **1.** A press consisting of a frame in which slides or rams move up and down, of a bed to which the die shoe or bolster plate is attached, and of a source of power to move the slide. Also known as drop press. **2.** Any mechanical press.

punch radius [DES ENG] The radius on the bottom end of the punch over which the metal sheet is bent in drawing.

pure shear [MECH] A particular example of irrotational strain or flattening in which a body is elongated in one direction and shortened at right angles to it as a consequence of differential displacements on two sets of intersecting planes.

purge meter interlock [MECH ENG] A meter to maintain airflow through a boiler furnace at a specific level for a definite time interval; ensures that the proper air-fuel ratio is achieved prior to ignition.

push-bar conveyor [MECH ENG] A type of chain conveyor in which two endless chains are cross-connected at intervals by push bars which propel the load along a stationary bed or trough of the conveyor.

push bench [MECH ENG] A machine used for drawing tubes of moderately heavy gage by cupping metal sheet and applying pressure to the inside bottom of the cup to force it through a die.

push fit [DES ENG] A hand-tight sliding fit between a shaft and a hole.

push nipple [MECH ENG] A short length of pipe used to connect sections of cast iron boilers.

push rod [MECH ENG] A rod, as in an internal combustion engine, which is actuated by the cam to open and close the valves.

putty knife [DES ENG] A knife with a broad flexible blade, used to apply and smooth putty.

pwt *See* pennyweight.

pylon [AERO ENG] A suspension device externally installed under the wing or fuselage of an aircraft; it is aerodynamically designed to fit the configuration of specific aircraft, thereby creating an insignificant amount of drag; it includes means of attaching to accommodate fuel tanks, bombs, rockets, torpedoes, rocket motors, or the like.

pyrometry [THERMO] The science and technology of measuring high temperatures.

pz *See* pièze.

Q

Q [THERMO] A unit of heat energy, equal to 10^{18} British thermal units, or approximately 1.055×10^{21} joules.

qr *See* quarter.

qr tr *See* quarter.

qt *See* quart.

quad [THERMO] A unit of heat energy, equal to 10^{15} British thermal units, or approximately 1.055×10^{18} joules.

quadrant [MECH ENG] A device for converting horizontal reciprocating motion to vertical reciprocating motion.

quadrant angle of fall [MECH] The vertical acute angle at the level point, between the horizontal and the line of fall of a projectile.

quadratic performance index [CONT SYS] A measure of system performance which is, in general, the sum of a quadratic function of the system state at fixed times, and the integral of a quadratic function of the system state and control inputs.

quadruple thread [DES ENG] A multiple thread having four separate helices equally spaced around the circumference of the threaded member; the lead is equal to four times the pitch of the thread.

quarrying machine [MECH ENG] Any machine used to drill holes or cut tunnels in native rock, such as a gang drill or tunneling machine; most commonly, a small locomotive bearing rock-drilling equipment operating on a track.

quart [MECH] Abbreviated qt. **1.** A unit of volume used for measurement of liquid substances in the United States, equal to 2 pints, or ¼ gallon, or 57¾ cubic inches, or $9.46352946 \times 10^{-4}$ cubic meter. **2.** A unit of volume used for measurement of solid substances in the United States, equal to 2 dry pints, or ⅓₂ bushel, or 107,521/ 1,600 cubic inches, or approximately 1.10122×10^{-3} cubic meter. **3.** A unit of volume used for measurement of both liquid and solid substances, although mainly the former, in the United Kingdom, equal to 2 U.K. pints, or ¼ U.K. gallon, or approximately 1.13652×10^{-3} cubic meter.

quarter [MECH] **1.** A unit of mass in use in the United States, equal to ¼ short ton, or 500 pounds, or 226.796185 kilograms. **2.** A unit of mass used in troy measure, equal to ¼ troy hundredweight, or 25 troy pounds, or 9.33104304 kilograms. Abbreviated qr tr. **3.** A unit of mass used in the United Kingdom, equal to ¼ hundredweight, or 28 pounds, or 12.70058636 kilograms. Abbreviated qr. **4.** A unit of volume used in the United Kingdom for measurement of liquid and solid substances, equal to 8 bushels, or 64 gallons, or approximately 0.290950 cubic meter.

quartering machine [MECH ENG] A machine that bores parallel holes simultaneously in such a way that the center lines of adjacent holes are 90° apart.

quarter-turn drive [MECH ENG] A belt drive connecting pulleys whose axes are at right angles.

quasi-linear feedback control system [CONT SYS] Feedback control system in which the relationships between the pertinent measures of the system input and output signals are substantially linear despite the existence of nonlinear elements.

quasi-linear system [CONT SYS] A control system in which the relationships between the input and output signals are substantially linear despite the existence of nonlinear elements.

quasi-static process *See* reversible process.

quick-change gearbox [MECH ENG] A cluster of gears on a machine tool, the arrangement of which allows for the rapid change of gear ratios.

quick return [MECH ENG] A device used in a reciprocating machine to make the return stroke faster than the power stroke.

quill [DES ENG] A hollow shaft into which another shaft is inserted in mechanical devices.

quill drive [MECH ENG] A drive in which the motor is mounted on a nonrotating hollow shaft surrounding the driving-wheel axle; pins on the armature mesh with spokes on the driving wheels, thereby transmitting motion to the wheels; used on electric locomotives.

quill gear [MECH ENG] A gear mounted on a hollow shaft.

quintal *See* hundredweight; metric centner.

Q unit [THERMO] A unit of energy, used in measuring the heat energy of fuel reserves, equal to 10^{18} British thermal units, or approximately 1.055×10^{21} joules.

R

R₂′ *See* Rayleigh number 2.

Ra₃ *See* Rayleigh number 3.

rabbet plane [DES ENG] A plane with the blade extending to the outer edge of one side that is open.

race [DES ENG] Either of the concentric pair of steel rings of a ball bearing or roller bearing.

rack [AERO ENG] A suspension device permanently fixed to an aircraft; it is designed for attaching, arming, and releasing one or more bombs; it may also be utilized to accommodate other items such as mines, rockets, torpedoes, fuel tanks, rescue equipment, sonobuoys, and flares. [DES ENG] *See* relay rack. [MECH ENG] A bar containing teeth on one face for meshing with a gear.

radial band pressure [MECH] The pressure which is exerted on the rotating band by the walls of the gun tube, and hence against the projectile wall at the band seat, as a result of the engraving of the band by the gun rifling.

radial bearing [MECH ENG] A bearing with rolling contact in which the direction of action of the load transmitted is radial to the axis of the shaft.

radial draw forming [MECH ENG] A metal-forming method in which tangential stretch and radial compression are applied gradually and simultaneously.

radial drill [MECH ENG] A drilling machine in which the drill spindle can be moved along a horizontal arm which itself can be rotated about a vertical pillar.

radial engine [MECH ENG] An engine characterized by radially arranged cylinders at equiangular intervals around the crankshaft.

radial-flow turbine [MECH ENG] A turbine in which the gases flow primarily in a radial direction.

radial force [MECH ENG] In machining, the force acting on the cutting tool in a direction opposite to depth of cut.

radial load [MECH ENG] The load perpendicular to the bearing axis.

radial locating [MECH ENG] One of the three locating problems in tooling to maintain the desired relationship between the workpiece, the cutter, and the body of the machine tool; the other two locating problems are concentric and plane locating.

radial motion [MECH] Motion in which a body moves along a line connecting it with an observer or reference point; for example, the motion of stars which move toward or away from the earth without a change in apparent position.

radial-ply [DES ENG] Pertaining to the construction of a tire in which the cords run straight across the tire, and an additional layered belt of fabric is placed around the circumference between the plies and the tread.

radial rake [MECH ENG] The angle between the cutter tooth face and a radial line passing through the cutting edge in a plane perpendicular to the cutter axis.

radial saw [MECH ENG] A power saw that has a circular blade suspended from a transverse head mounted on a rotatable overarm.

radial stress [MECH] Tangential stress at the periphery of an opening.

radial velocity [MECH] The component of the velocity of a body that is parallel to a line from an observer or reference point to the body; the radial velocities of stars are valuable in determining the structure and dynamics of the Galaxy. Also known as line-of-sight velocity.

radial wave equation [MECH] Solutions to wave equations with spherical symmetry can be found by separation of variables; the ordinary differential equation for the radial part of the wave function is called the radial wave equation.

radiant superheater [MECH ENG] A superheater designed to transfer heat from the products of combustion to the steam primarily by radiation.

radiant-type boiler [MECH ENG] A water-tube boiler in which boiler tubes form the boundary of the furnace.

radiating power *See* emittance.

radiation correction *See* cooling correction.

radiation loss [MECH ENG] Boiler heat loss to the atmosphere by conduction, radiation, and convection.

radiative transfer [THERMO] The transmission of heat by electromagnetic radiation.

radiator temperature drop [MECH ENG] In internal combustion engines, the difference in temperature of the coolant liquid entering and leaving the radiator.

radioactive heat [THERMO] Heat produced within a medium as a result of absorption of radiation from decay of radioisotopes in the medium, such as thorium-232, potassium-40, uranium-238, and uranium-235.

radio relay satellite *See* communications satellite.

radiosonde balloon [AERO ENG] A balloon used to carry a radiosonde aloft; it is considerably larger than a pilot balloon or a ceiling balloon.

radius cutter [MECH ENG] A formed milling cutter with teeth ground to produce a radius on the workpiece.

radius of gyration [MECH] The square root of the ratio of the moment of inertia of a body about a given axis to its mass.

rag bolt *See* barb bolt.

rail [MECH ENG] A high-pressure manifold in some fuel injection systems.

rail crane *See* locomotive crane.

railroad jack [MECH ENG] **1.** A hoist used for lifting locomotives. **2.** A portable jack for lifting heavy objects. **3.** A hydraulic jack, either powered or lever-operated.

Rajakaruna engine [MECH ENG] A rotary engine that uses a combustion chamber whose sides are pin-jointed together at their ends.

rake [DES ENG] A hand tool consisting of a long handle with a row of projecting prongs at one end; for example, the tool used for gathering leaves or grass on the ground. [MECH ENG] The angle between the tooth face or a tangent to the tooth face of a cutting tool at a given point and a reference plane or line.

ram [AERO ENG] The forward motion of an air scoop or air inlet through the air. [MECH ENG] A plunger, weight, or other guided structure for exerting pressure or drawing something by impact.

ram effect [MECH ENG] The increased air pressure in a jet engine or in the manifold of a piston engine, due to ram.

ramjet engine [AERO ENG] A type of jet engine with no mechanical compressor, consisting of a specially shaped tube or duct open at both ends, the air necessary for combustion being shoved into the duct and compressed by the forward motion of the engine; the air passes through a diffuser and is mixed with fuel and burned, the exhaust gases issuing in a jet from the rear opening.

ramjet exhaust nozzle [AERO ENG] The discharge nozzle in a ramjet engine; hot gas is ejected rearward through this nozzle.

ramp weight [AERO ENG] The static weight of a mission aircraft determined by adding operating weight, payload, flight plan fuel load, and fuel required for ground turbine power unit, taxi, runup, and takeoff.

ram recovery *See* recovery.

ram rocket [AERO ENG] **1.** A rocket motor mounted coaxially in the open front end of a ramjet, used to provide thrust at low speeds and to ignite the ramjet fuel. **2.** The entire unit or power plant consisting of the ramjet and such a rocket.

Ramsay-Shields-Eötvös equation [THERMO] An elaboration of the Eötvös rule which states that at temperatures not too near the critical temperature the molar surface energy of a liquid is proportional to $t_c - t - 6$ K, where t is the temperature and t_c is the critical temperature.

Ramsay-Young rule [THERMO] An empirical relationship which states that the ratio of the absolute temperatures at which two chemically similar liquids have the same vapor pressure is independent of this vapor pressure.

ram-type turret lathe [MECH ENG] A horizontal turret lathe in which the turret is mounted on a ram or slide which rides on a saddle.

random vibration [MECH] A varying force acting on a mechanical system which may be considered to be the sum of a large number of irregularly timed small shocks; induced typically by aerodynamic turbulence, airborne noise from rocket jets, and transportation over road surfaces.

range [MECH] The horizontal component of a projectile displacement at the instant it strikes the ground.

range control [AERO ENG] The operation of an aircraft to obtain the optimum flying time.

range control chart [AERO ENG] A graph kept in flight on which actual fuel consumption is plotted against distance flown for comparison with planned fuel consumption.

range deviation [MECH] Distance by which a projectile strikes beyond, or short of, the target; the distance as measured along the gun-target line or along a line parallel to the gun-target line.

Ranger program [AERO ENG] A series of nine spacecraft, launched in 1961–1965, designed to transmit photographs back to earth while on a collision course with the moon; the first six Rangers failed, but *Rangers 7, 8,* and *9* successfully transmitted high-resolution television pictures of the lunar surface up to the instant of impact.

Rankine body [FL MECH] A fluid flow pattern formed by combining a uniform stream with a source and a sink of equal strengths, with the line joining the source and sink along the stream direction.

Rankine cycle [THERMO] An ideal thermodynamic cycle consisting of heat addition at constant pressure, isentropic expansion, heat rejection at constant pressure, and isentropic compression; used as an ideal standard for the performance of heat-engine and heat-pump installations operating with a condensable vapor as the working fluid, such as a steam power plant. Also known as steam cycle.

Rankine efficiency [MECH ENG] The efficiency of an ideal engine operating on the Rankine cycle under specified conditions of steam temperature and pressure.

Rankine-Hugoniot equations [THERMO] Equations, derived from the laws of conservation of mass, momentum, and energy, which relate the velocity of a shock wave and the pressure, density, and enthalpy of the transmitting fluid before and after the shock wave passes.

Rankine temperature scale [THERMO] A scale of absolute temperature; the temperature in degrees Rankine (°R) is equal to $\frac{9}{5}$ of the temperature in kelvins and to the temperature in degrees Fahrenheit plus 459.67.

Rankine vortex [FL MECH] A vortex with a vertical axis and circular motion, in which the motion is that of a rotating solid cylinder inside some fixed radius, and the circulation is constant outside this radius.

rapid traverse [MECH ENG] A machine tool mechanism which rapidly repositions the workpiece while no cutting takes place.

rarefaction wave [FL MECH] A pressure wave or rush of air or water induced by rarefaction; it travels in the opposite direction to that of a shock wave directly following an explosion. Also known as suction wave.

rarefied gas [FL MECH] A gas whose pressure is much less than atmospheric pressure.

rasp [DES ENG] A metallic tool with a rough surface of small points used for shaping and finishing metal, plaster, stone, and wood; designed in a number of useful curved shapes.

ratchet [DES ENG] A wheel, usually toothed, operating with a catch or a pawl so as to rotate in only a single direction.

ratchet coupling [MECH ENG] A coupling between two shafts that uses a ratchet to allow the driven shaft to be turned in one direction only, and also to permit the driven shaft to overrun the driving shaft.

ratchet jack [DES ENG] A jack operated by a ratchet mechanism.

ratchet tool [DES ENG] A tool in which torque or force is applied in one direction only by means of a ratchet.

rate action *See* derivative action.

Rateau formula [FL MECH] A formula, $m = A_2 p_1 (16.367 - 0.96 \log p_1)/1000$, for determining the discharge m of saturated steam in pounds per second through a well-rounded convergent orifice; A_2 is the area of the orifice in square inches, and p_1 the reservoir pressure in pounds per square inch.

rate climb [AERO ENG] The climb of an aircraft to higher altitudes at a constant rate.

rate control [CONT SYS] A form of control in which the position of a controller determines the rate or velocity of motion of a controlled object. Also known as velocity control.

rated capacity [MECH ENG] The maximum capacity for which a boiler is designed, measured in pounds of steam per hour delivered at specified conditions of pressure and temperature.

rated engine speed [MECH ENG] The rotative speed of an engine specified as the allowable maximum for continuous reliable performance.

rate descent [AERO ENG] An aircraft descent from higher altitudes at a constant rate.

rated horsepower [MECH ENG] The normal maximum, allowable, continuous power output of an engine, turbine motor, or other prime mover.

rated load [MECH ENG] The maximum load a machine is designed to carry.

rated relieving capacity [DES ENG] The measured relieving capacity for which the pressure relief device is rated in accordance with the applicable code or standard.

rate gyroscope [MECH ENG] A gyroscope that is suspended in just one gimbal whose bearings form its output axis and which is restrained by a spring; rotation of the gyroscope frame about an axis perpendicular to both spin and output axes produces precession of the gimbal within the bearings proportional to the rate of rotation.

rate integrating gyroscope [MECH ENG] A single-degree-of-freedom gyro having primarily viscous restraint of its spin axis about the output axis; an output signal is produced by gimbal angular displacement, relative to the base, which is proportional to the integral of the angular rate of the base about the input axis.

rate of approach [AERO ENG] The relative speed of two aircraft when the distance between them is decreasing.

rate of change of acceleration [MECH] Time rate of change of acceleration; this rate is a factor in the design of some items of ammunition that undergo large accelerations.

rate of climb [AERO ENG] Ascent of aircraft per unit time, usually expressed as feet per minute.

rate-of-climb indicator [AERO ENG] A device used to indicate changes in the vertical position of an aircraft by comparing the actual outside air pressure to a reference volume that lags the outside pressure because a calibrated restrictor imposes a lag-time constant to the reference pressure volume. Also known as rate-of-descent indicator; vertical speed indicator.

rate of departure [AERO ENG] The relative speed of two aircraft when the distance between them is increasing.

rate-of-descent indicator *See* rate-of-climb indicator.

rate of flow *See* flow rate.

rate of return [AERO ENG] Aircraft relative to its base, either fixed or moving.

rate servomechanism *See* velocity servomechanism.

ratio control system [CONT SYS] Control system in which two process variables are kept at a fixed ratio, regardless of the variation of either of the variables, as when flow rates in two separate fluid conduits are held at a fixed ratio.

ratio of expansion [MECH ENG] The ratio of the volume of steam in the cylinder of an engine when the piston is at the end of a stroke to that when the piston is in the cutoff position.

rato [AERO ENG] A rocket system providing additional thrust for takeoff of an aircraft. Derived from rocket-assisted takeoff.

rattail file [DES ENG] A round tapering file used for smoothing or enlarging holes.

Rayleigh flow [FL MECH] An idealized type of gas flow in which heat transfer may occur, satisfying the assumptions that the flow takes place in constant-area cross section and is frictionless and steady, that the gas is perfect and has constant specific heat, that the composition of the gas does not change, and that there are no devices in the system which deliver or receive mechanical work.

Rayleigh number 1 [FL MECH] A dimensionless number used in studying the breakup of liquid jets, equal to Weber number 2. Symbolized N_{Ra1}.

Rayleigh number 2 [THERMO] A dimensionless number used in studying free convection, equal to the product of the Grashof number and the Prandtl number. Symbolized R_2'.

Rayleigh number 3 [THERMO] A dimensionless number used in the study of combined free and forced convection in vertical tubes, equal to Rayleigh number 2 times the

Nusselt number times the tube diameter divided by its entry length. Symbolized Ra_3.

Rayleigh's dissipation function   [MECH] A function which enters into the equations of motion of a system undergoing small oscillations and represents frictional forces which are proportional to velocities; given by a positive definite quadratic form in the time derivatives of the coordinates. Also known as dissipation function.

Rayleigh-Taylor instability   [FL MECH] The instability of the interface separating two fluids having different densities when the lighter fluid is accelerated toward the heavier fluid.

Rayleigh wave   [MECH] A wave which propagates on the surface of a solid; particle trajectories are ellipses in planes normal to the surface and parallel to the direction of propagation. Also known as surface wave.

Re$_r$ *See* rotating Reynolds number.

reach rod   [MECH ENG] A rod motion in a link used to transmit motion from the reversing rod to the lifting shaft.

reactant ratio   [AERO ENG] The ratio of the weight flow of oxidizer to fuel in a rocket engine.

reaction   [CONT SYS] *See* positive feedback.   [MECH] The equal and opposite force which results when a force is exerted on a body, according to Newton's third law of motion.

reaction engine   [AERO ENG] An engine that develops thrust by its reaction to a substance ejected from it; specifically, such an engine that ejects a jet or stream of gases created by the addition of energy to the gases in the engine. Also known as reaction motor.

reaction motor   *See* reaction engine.

reaction propulsion   [AERO ENG] Propulsion by means of reaction to a jet of gas or fluid projected rearward, as by a jet engine, rocket engine, or rocket motor.

reaction turbine   [MECH ENG] A power-generation prime mover utilizing the steady-flow principle of fluid acceleration, where nozzles are mounted on the moving element.

reaction wheel   [MECH ENG] A device capable of storing angular momentum which may be used in a space ship to provide torque to effect or maintain a given orientation.

real fluid flow   [FL MECH] The flow in which effects of tangential or shearing forces are taken into account; these forces give rise to fluid friction, because they oppose the sliding of one particle past another.

real gas   [THERMO] A gas, as considered from the viewpoint in which deviations from the ideal gas law, resulting from interactions of gas molecules, are taken into account. Also known as imperfect gas.

realizability   [CONT SYS] Property of a transfer function that can be realized by a network that has only resistances, capacitances, inductances, and ideal transformers.

reamer   [DES ENG] A tool used to enlarge, shape, smooth, or otherwise finish a hole.

reaming bit   [DES ENG] A bit used to enlarge a borehole. Also known as broaching bit; pilot reaming bit.

Réaumur temperature scale   [THERMO] Temperature scale where water freezes at 0°R and boils at 80°R.

rebound clip   [DES ENG] A clip surrounding the back and one or two other leaves of a leaf spring, to distribute the load during rebounds.

rebound leaf [DES ENG] In a leaf spring, a leaf placed over the master leaf to limit the rebound and help carry the load imposed by it.

receiver [MECH ENG] An apparatus placed near the compressor to equalize the pulsations of the air as it comes from the compressor to cause a more uniform flow of air through the pipeline and to collect moisture and oil carried in the air.

receiving station [MECH ENG] The location or device on conveyor systems where bulk material is loaded or otherwise received onto the conveyor.

recessed tube wall [MECH ENG] A boiler furnace wall which has openings to partially expose waterwall tubes to the radiant combustion gases.

reciprocal strain ellipsoid [MECH] In elastic theory, an ellipsoid of certain shape and orientation which under homogeneous strain is transformed into a set of orthogonal diameters of the sphere.

reciprocating compressor [MECH ENG] A positive-displacement compressor having one or more cylinders, each fitted with a piston driven by a crankshaft through a connecting rod.

reciprocating drill *See* piston drill.

reciprocating engine *See* piston engine.

reciprocating flight conveyor [MECH ENG] A reciprocating beam or beams with hinged flights that advance materials along a conveyor trough.

reciprocating-plate feeder [MECH ENG] A back-and-forth shaking tray used to feed abrasive materials, such as pulverized coal, into process units.

reciprocating pump *See* piston pump.

reciprocating screen [MECH ENG] Horizontal solids-separation screen (sieve) oscillated back and forth by an eccentric gear; used for solids classification.

recoil *See* gun reaction.

reconnaissance drone [AERO ENG] An unmanned aircraft guided by remote control, with photographic or electronic equipment for providing information about an enemy or potential enemy.

reconnaissance spacecraft [AERO ENG] A satellite put into orbit about the earth and containing electronic equipment designed to pick up and transmit back to earth information pertaining to activities such as military.

recoverable shear [FL MECH] Measure of the elastic content of a fluid, related to elastic recovery (mechanicallike property of elastic recoil); found in unvulcanized, unfilled natural rubber, and certain polymer solutions, soap gels, and biological fluids.

recovery [AERO ENG] 1. The procedure or action that obtains when the whole of a satellite, or a section, instrumentation package, or other part of a rocket vehicle, is retrieved after a launch. 2. The conversion of kinetic energy to potential energy, such as in the deceleration of air in the duct of a ramjet engine. Also known as ram recovery. 3. In flying, the action of a lifting vehicle returning to an equilibrium attitude after a nonequilibrium maneuver.

recovery area [AERO ENG] An area in which a satellite, satellite package, or spacecraft is recovered after reentry.

recovery capsule [AERO ENG] A space capsule designed to be recovered after reentry.

recovery package [AERO ENG] A package attached to a reentry or other body designed for recovery, containing devices intended to locate the body after impact.

recovery temperature *See* adiabatic recovery temperature.

recovery vehicle [MECH ENG] A special-purpose vehicle equipped with winch, hoist, or boom for recovery of vehicles.

rectilinear motion [MECH] A continuous change of position of a body so that every particle of the body follows a straight-line path. Also known as linear motion.

Redler conveyor [MECH ENG] A conveyor in which material is dragged through a duct by skeletonized or U-shaped impellers which move the material in which they are submerged because the resistance to slip through the element is greater than the drag against the walls of the duct.

reduced frequency *See* Strouhal number.

reduced mass [MECH] For a system of two particles with masses m_1 and m_2 exerting equal and opposite forces on each other and subject to no external forces, the reduced mass is the mass m such that the motion of either particle, with respect to the other as origin, is the same as the motion with respect to a fixed origin of a single particle with mass m acted on by the same force; it is given by $m = m_1 m_2/ (m_1 + m_2)$.

reduced-order controller [CONT SYS] A control algorithm in which certain modes of the structure to be controlled are ignored, to enable control commands to be computed with sufficient rapidity.

reduced pressure [THERMO] The ratio of the pressure of a substance to its critical pressure.

reduced property *See* reduced value.

reduced temperature [THERMO] The ratio of the temperature of a substance to its critical temperature.

reduced value [THERMO] The actual value of a quantity divided by the value of that quantity at the critical point. Also known as reduced property.

reduced volume [THERMO] The ratio of the specific volume of a substance to its critical volume.

reduction gear [MECH ENG] A gear train which lowers the output speed.

redundancy [MECH] A statically indeterminate structure.

reel [DES ENG] A revolving spool-shaped device used for storage of hose, rope, cable, wire, magnetic tape, and so on.

reentry [AERO ENG] The event when a spacecraft or other object comes back into the sensible atmosphere after being in space.

reentry angle [AERO ENG] That angle of the reentry body trajectory and the sensible atmosphere at which the body reenters the atmosphere.

reentry body [AERO ENG] That part of a space vehicle that reenters the atmosphere after flight above the sensible atmosphere.

reentry nose cone [AERO ENG] A nose cone designed especially for reentry, consists of one or more chambers protected by a heat sink.

reentry trajectory [AERO ENG] That part of a rocket's trajectory that begins at reentry and ends at the target or at the surface.

reentry vehicle [AERO ENG] Any payload-carrying vehicle designed to leave the sensible atmosphere and then return through it to earth.

reentry window [AERO ENG] The area, at the limits of the earth's atmosphere, through which a spacecraft in a given trajectory can pass to accomplish a successful reentry for a landing in a desired region.

reference dimension [DES ENG] In dimensioning, a dimension without tolerance used for informational purposes only, and does not govern machining operations in any way; it is indicated on a drawing by writing the abbreviation REF directly following or under the dimension.

reference plane [MECH ENG] The plane containing the axis and the cutting point of a cutter.

reflector satellite [AERO ENG] Satellite so designed that radio or other waves bounce off its surface.

refractory-lined firebox boiler [MECH ENG] A horizontal fire-tube boiler with the front portion of the shell located over a refractory furnace; the rear of the shell contains the first-pass tubes, and the second-pass tubes are located in the upper part of the shell.

refrigerated truck [MECH ENG] An insulated truck equipped and used as a refrigerator to transport fresh perishable or frozen products.

refrigeration [MECH ENG] The cooling of a space or substance below the environmental temperature.

refrigeration condenser [MECH ENG] A vapor condenser in a refrigeration system, where the refrigerant is liquefied and discharges its heat to the environment.

refrigeration cycle [THERMO] A sequence of thermodynamic processes whereby heat is withdrawn from a cold body and expelled to a hot body.

refrigeration system [MECH ENG] A closed-flow system in which a refrigerant is compressed, condensed, and expanded to produce cooling at a lower temperature level and rejection of heat at a higher temperature level for the purpose of extracting heat from a controlled space.

refrigerator [MECH ENG] An insulated, cooled compartment.

refrigerator car [MECH ENG] An insulated freight car constructed and used as a refrigerator.

regelation [THERMO] Phenomenon in which ice (or any substance which expands upon freezing) melts under intense pressure and freezes again when this pressure is removed; accounts for phenomena such as the slippery nature of ice and the motion of glaciers.

regeneration *See* positive feedback.

regenerative air heater [MECH ENG] An air heater in which the heat-transferring members are alternately exposed to heat-surrendering gases and to air.

regenerative cycle [MECH ENG] *See* bleeding cycle. [THERMO] An engine cycle in which low-grade heat that would ordinarily be lost is used to improve the cyclic efficiency.

regenerative engine [AERO ENG] 1. A jet or rocket engine that utilizes the heat of combustion to preheat air or fuel entering the combustion chamber. 2. Specifically, to a type of rocket engine in which one of the propellants is used to cool the engine by passing through a jacket prior to combustion.

regenerative feedback *See* positive feedback.

regenerative pump [MECH ENG] Rotating-vane device that uses a combination of mechanical impulse and centrifugal force to produce high liquid heads at low volumes. Also known as turbine pump.

regenerator [MECH ENG] A device used with hot-air engines and gas-burning furnaces which transfers heat from effluent gases to incoming air or gas.

register [MECH ENG] The portion of a burner which directs the flow of air used in the combustion process.

register control [CONT SYS] Automatic control of the position of a printed design with respect to reference marks or some other part of the design, as in photoelectric register control.

regular lay [DES ENG] The lay of a wire rope in which the wires in the strand are twisted in directions opposite to the direction of the strands.

regular-lay left twist *See* left-laid.

regulating system *See* automatic control system.

regulation [CONT SYS] The process of holding constant a quantity such as speed, temperature, voltage, or position by means of an electronic or other system that automatically corrects errors by feeding back into the system the condition being regulated; regulation thus is based on feedback, whereas control is not.

regulator [CONT SYS] A device that maintains a desired quantity at a predetermined value or varies it according to a predetermined plan.

regulatory control function [CONT SYS] That level in the functional decomposition of a large-scale control system which interfaces with the plant to implement the decisions of the optimizing controller inputted in the form of set points, desired trajectories, or targets. Also known as direct control function.

Rehbock weir formula [FL MECH] Probably the most accurate formula for the rate of flow of water over a rectangular suppressed weir; it includes a correction for the velocity of approach for normal, or fairly uniform, velocity distribution in the upstream channel; the formula is $Q = [3.234 + 5.347/(320h - 3) + 0.428h/d_0]lh^{3/2}$, where Q is the flow rate in cubic feet per second, l is the width of the weir in feet, h is the head of water above the crest of the weir in feet, and d_0 is the height of weir or depth of water at zero head in feet.

reheating [THERMO] A process in which the gas or steam is reheated after a partial isentropic expansion to reduce moisture content. Also known as resuperheating.

relative density *See* specific gravity.

relative gain array [CONT SYS] An analytical device used in process control multivariable applications, based on the comparison of single-loop control to multivariable control; expressed as an array (for all possible input-output pairs) of the ratios of a measure of the single-loop behavior between an input-output variable pair, to a related measure of the behavior of the same input-output pair under some idealization of multivariable control.

relative Mach number *See* Mach number.

relative momentum [MECH] The momentum of a body in a reference frame in which another specified body is fixed.

relative motion [MECH] The continuous change of position of a body with respect to a second body or to a reference point that is fixed. Also known as apparent motion.

relative roughness factor [FL MECH] Roughness of pipe-wall interior (distance from peaks to valleys) divided by pipe internal diameter; used to modify Reynolds number calculations for fluid flow through pipes.

relative velocity [MECH] The velocity of a body with respect to a second body; that is, its velocity in a reference frame where the second body is fixed.

relaxation [MECH] 1. Relief of stress in a strained material due to creep. 2. The lessening of elastic resistance in an elastic medium under an applied stress resulting in permanent deformation.

relay control system [CONT SYS] A control system in which the error signal must reach a certain value before the controller reacts to it, so that the control action is discontinuous in amplitude.

relay rack [DES ENG] A standardized steel rack designed to hold 19-inch (48.26-centimeter) panels of various heights, on which are mounted radio receivers, amplifiers, and other units of electronic equipment. Also known as rack.

relay satellite *See* communications satellite.

release [MECH ENG] A mechanical arrangment of parts for holding or freeing a device or mechanism as required.

release adiabat [MECH] A curve or locus of points which defines the succession of states through which a mass that has been shocked to a high-pressure state passes while monotonically returning to zero pressure.

release altitude [AERO ENG] Altitude of an aircraft above the ground at the time of release of bombs, rockets, missiles, tow targets, and so forth.

relief [MECH ENG] 1. A passage made by cutting away one side of a tailstock center so that the facing or parting tool may be advanced to or almost to the center of the work. 2. Clearance provided around the cutting edge by removal of tool material.

relief angle [MECH ENG] The angle between a relieved surface and a tangential plane at a cutting edge.

relief frame [MECH ENG] A frame placed between the slide valve of a steam engine and the steam chest cover; reduces pressure on the valve and thereby reduces friction.

relieving [MECH ENG] Treating an embossed metal surface with an abrasive to reveal the base-metal color on the elevations or highlights of the surface.

remaining velocity [MECH] Speed of a projectile at any point along its path of fire.

remote control [CONT SYS] Control of a quantity which is separated by an appreciable distance from the controlling quantity; examples include master-slave manipulators, telemetering, telephone, and television.

remotely piloted vehicle [AERO ENG] A robot aircraft, controlled over a two-wave radio link from a ground station or mother aircraft that can be hundreds of miles away; electronic guidance is generally supplemented by remote control television cameras feeding monitor receivers at the control station. Abbreviated RPV.

rendezvous [AERO ENG] 1. The event of two or more objects meeting with zero relative velocity at a preconceived time and place. 2. The point in space at which such an event takes place, or is to take place.

repeated load [MECH] A force applied repeatedly, causing variation in the magnitude and sometimes in the sense, of the internal forces.

replacement bit *See* reset bit.

repulsion [MECH] A force which tends to increase the distance between two bodies having like electric charges, or the force between atoms or molecules at very short distances which keeps them apart. Also known as repulsive force.

repulsive force *See* repulsion.

required thickness [DES ENG] The thickness calculated by recognized formulas for boiler or pressure vessel construction before corrosion allowance is added.

resealing pressure [MECH ENG] The inlet pressure at which leakage stops after a pressure relief valve is closed.

research rocket [AERO ENG] A rocket-propelled vehicle used to collect scientific data.

reserve aircraft [AERO ENG] Those aircraft which have been accumulated in excess of immediate needs for active aircraft and are retained in the inventory against possible future needs.

reset action [CONT SYS] Floating action in which the final control element is moved at a speed proportional to the extent of proportional-position action.

reset bit [DES ENG] A diamond bit made by reusing diamonds salvaged from a used bit and setting them in the crown attached to a new bit blank. Also known as replacement bit.

residual mode [CONT SYS] A characteristic motion of a structure which is deliberately ignored in the control algorithm of an active control system for the structure in the process of model reduction.

residual stress *See* internal stress.

resilience [MECH] **1.** Ability of a strained body, by virtue of high yield strength and low elastic modulus, to recover its size and form following deformation. **2.** The work done in deforming a body to some predetermined limit, such as its elastic limit or breaking point, divided by the body's volume.

resinoid wheel [DES ENG] A grinding wheel bonded with a synthetic resin.

resistance [FL MECH] *See* fluid resistance. [MECH] In damped harmonic motion, the ratio of the frictional resistive force to the speed. Also known as damping coefficient; damping constant; mechanical resistance.

resistance coefficient 1 [FL MECH] A dimensionless number used in the study of flow resistance, equal to the resistance force in flow divided by ½ the product of fluid density, the square of fluid velocity, and the square of a characteristic length. Symbolized c_f.

resistance coefficient 2 *See* Darcy number 1.

resisting moment [MECH] A moment produced by internal tensile and compressive forces that balances the external bending moment on a beam.

resonance vibration [MECH] Forced vibration in which the frequency of the disturbing force is very close to the natural frequency of the system, so that the amplitude of vibration is very large.

resonant jet [AERO ENG] A pulsejet engine, exhibiting intensification of power under the rhythm of explosions and compression waves within the engine.

response [CONT SYS] A quantitative expression of the output of a device or system as a function of the input. Also known as system response.

response characteristic [CONT SYS] The response as a function of an independent variable, such as direction or frequency, often presented in graphical form.

response time [CONT SYS] The time required for the output of a control system or element to reach a specified fraction of its new value after application of a step input or disturbance.

restart [AERO ENG] The act of firing a stage of a rocket after a previous powered flight.

resultant of forces [MECH] A system of at most a single force and a single couple whose external effects on a rigid body are identical with the effects of the several actual forces that act on that body.

resultant rake [MECH ENG] The angle between the face of a cutting tooth and an axial plane through the tooth point measured in a plane at right angles to the cutting edge.

resuperheating *See* reheating.

retaining ring [DES ENG] **1.** A shoulder inside a reaming shell that prevents the core lifter from entering the core barrel. **2.** A steel ring between the races of a ball bearing to maintain the correct distribution of the balls in the races.

retarder [MECH ENG] 1. A braking device used to control the speed of railroad cars moving along the classification tracks in a hump yard. 2. A strip inserted in a tube of a fire-tube boiler to increase agitation of the hot gases flowing therein.

retarding conveyor [MECH ENG] Any type of conveyor used to restrain the movement of bulk materials, packages, or objects where the incline is such that the conveyed material tends to propel the conveying medium.

retroaction *See* positive feedback.

retrofire time [AERO ENG] The computed starting time and duration of firing of retro-rockets to decrease the speed of a recovery capsule and make it reenter the earth's atmosphere at the correct point for a planned landing.

retrorocket [AERO ENG] A rocket fitted on or in a spacecraft, satellite, or the like to produce thrust opposed to forward motion. Also known as braking rocket.

return bend [DES ENG] A pipe fitting, equal to two ells, used to connect parallel pipes so that fluid flowing into one will return in the opposite direction through the other.

return connecting rod [MECH ENG] A connecting rod whose crankpin end is located on the same side of the crosshead as the cylinder.

return difference [CONT SYS] The difference between 1 and the loop transmittance.

return-flow burner [MECH ENG] A mechanical oil atomizer in a boiler furnace which regulates the amount of oil to be burned by the portion of oil recirculated to the point of storage.

return idler [MECH ENG] The idler or roller beneath the cover plates on which the conveyor belt rides after the load which it was carrying has been dumped.

reversal speed [AERO ENG] The speed of an aircraft above which the aeroelastic loads will exceed the control surface loading of a given flight control system; the resultant load will act in the reverse direction from the control surface loading, causing the control system to act in a direction opposite to that desired.

reverse Brayton cycle [THERMO] A refrigeration cycle using air as the refrigerant but with all system pressures above the ambient. Also known as dense-air refrigeration cycle.

reverse Carnot cycle [THERMO] An ideal thermodynamic cycle consisting of the processes of the Carnot cycle reversed and in reverse order, namely, isentropic expansion, isothermal expansion, isentropic compression, and isothermal compression.

reverse feedback *See* negative feedback.

reverse lay [DES ENG] The lay of a wire rope with strands alternating in a right and left lay.

reverse pitch [MECH ENG] A pitch on a propeller blade producing thrust in the direction opposite to the normal one.

reversible engine [THERMO] An ideal engine which carries out a cycle of reversible processes.

reversible path [THERMO] A path followed by a thermodynamic system such that its direction of motion can be reversed at any point by an infinitesimal change in external conditions; thus the system can be considered to be at equilibrium at all points along the path.

reversible-pitch propeller [MECH ENG] A type of controllable-pitch propeller; of either controllable or constant speed, it has provisions for reducing the pitch to and beyond the zero value, to the negative pitch range.

reversible process [THERMO] An ideal thermodynamic process which can be exactly reversed by making an indefinitely small change in the external conditions. Also known as quasistatic process.

reversible steering gear [MECH ENG] A steering gear for a vehicle which permits road shock and wheel deflections to come through the system and be felt in the steering control.

reversible tramway *See* jig back.

revolution [MECH] The motion of a body around a closed orbit.

revolution per minute [MECH] A unit of angular velocity equal to the uniform angular velocity of a body which rotates through an angle of 360° (2π radians), so that every point in the body returns to its original position, in 1 minute. Abbreviated rpm.

revolution per second [MECH] A unit of angular velocity equal to the uniform angular velocity of a body which rotates through an angle of 360° (2π radians), so that every point in the body returns to its original position, in 1 second. Abbreviated rps.

revolving-block engine [MECH ENG] Any of various engines which combine recipro-cating piston motion with rotational motion of the entire engine block.

revolving shovel [MECH ENG] A digging machine, mounted on crawlers or on rubber tires, that has the machinery deck and attachment on a vertical pivot so that it can swing freely.

reyn [FL MECH] A unit of dynamic viscosity equal to the dynamic viscosity of a fluid in which there is a tangential force of 1 poundal per square foot resisting the flow of two parallel fluid layers past each other when their differential velocity is 1 foot per second per foot of separation; equal to approximately 14.8816 poise.

Reynolds criterion [FL MECH] The principle that the type of fluid motion, that is, laminar flow or turbulent flow, in geometrically similar flow systems depends only on the Reynolds number; for example, in a pipe, laminar flow exists at Reynolds numbers less than 2000, turbulent flow at numbers above about 3000.

Reynolds equation [FL MECH] A form of the Navier-Stokes equation which is $\rho\partial u/\partial t = (\partial/\partial x)(p_{xx} - \rho u^2) + (\partial/\partial y)(p_{xy} - \rho uv) + (\partial/\partial z)(p_{xz} - \rho uw)$, where ρ is the fluid density, u, v, and w are the components of the fluid velocity, and p_{xx}, p_{xy}, and p_{xz} are normal and shearing stresses.

Reynolds number [FL MECH] A dimensionless number which is significant in the design of a model of any system in which the effect of viscosity is important in controlling the velocities or the flow pattern of a fluid; equal to the density of a fluid, times its velocity, times a characteristic length, divided by the fluid viscosity. Symbolized N_{Re}. Also known as Damköhler number V (DaV).

Reynolds stress [FL MECH] The net transfer of momentum across a surface in a tur-bulent fluid because of fluctuations in fluid velocity. Also known as eddy stress.

Reynolds stress tensor [FL MECH] A tensor whose components are the components of the Reynolds stress across three mutually perpendicular surfaces.

rhe [FL MECH] **1.** A unit of dynamic fluidity, equal to the dynamic fluidity of a fluid whose dynamic viscosity is 1 centipoise. **2.** A unit of kinematic fluidity, equal to the kinematic fluidity of a fluid whose kinematic viscosity is 1 centistoke.

rheogoniometry [MECH] Rheological tests to determine the various stress and shear actions on Newtonian and non-Newtonian fluids.

rheology [MECH] The study of the deformation and flow of matter, especially non-Newtonian flow of liquids and plastic flow of solids.

rheopectic fluid [FL MECH] A fluid for which the structure builds up on shearing; this phenomenon is regarded as the reverse of thixotropy.

rib [AERO ENG] A transverse structural member that gives cross-sectional shape and strength to a portion of an airfoil.

ribbed-clamp coupling [DES ENG] A rigid coupling which is split longitudinally and bored to shaft diameter, with a shim separating the two halves.

ribbon conveyor [MECH ENG] A type of screw conveyor which has an open space between the shaft and a ribbon-shaped flight, used for wet or sticky materials which would otherwise build up on the spindle.

ribbon mixer [MECH ENG] Device for the mixing of particles, slurries, or pastes of solids by the revolution of an elongated helicoid (spiral) ribbon of metal.

ribbon parachute [AERO ENG] A type of parachute having a canopy consisting of an arrangement of closely spaced tapes; this parachute has high porosity with attendant stability and slight opening shock.

Richardson number [FL MECH] A dimensionless number used in studying the strati-fied flow of multilayer systems; equal to the acceleration of gravity times the density gradient of a fluid, divided by the product of the fluid's density and the square of its velocity gradient at a wall. Symbolized N_{Ri}.

riddle [DES ENG] A sieve used for sizing or for removing foreign material from foundry sand or other granular materials.

riffler [DES ENG] A small, curved rasp or file for filing interior surfaces or enlarging holes.

rifle [DES ENG] A drill core that has spiral grooves on its outside surface.

rifling [MECH ENG] The technique of cutting helical grooves inside a rifle barrel to impart a spinning motion to a projectile around its long axis.

rift saw [DES ENG] **1.** A saw for cutting wood radially from the log. **2.** A circular saw divided into toothed arms for sawing flooring strips from cants.

rig [MECH ENG] A tripod, derrick, or drill machine complete with auxiliary and acces-sory equipment needed to drill.

rigging [AERO ENG] The shroud lines attached to a parachute.

right-cut tool [DES ENG] A single-point lathe tool which has the cutting edge on the right side when viewed face up from the point end.

right-hand cutting tool [DES ENG] A cutter whose flutes twist in a clockwise direction.

right-handed [DES ENG] **1.** Pertaining to screw threads that allow coupling only by turning in a clockwise direction. **2.** *See* right-laid.

right-hand screw [DES ENG] A screw that advances when turned clockwise.

righting lever [FL MECH] The horizontal distance from the center of mass of a floating body, slightly displaced from the equilibrium position, to a vertical line passing through the center of buoyancy.

right-laid [DES ENG] Rope or cable construction in which strands are twisted counter-clockwise. Also known as right-handed.

right lang lay [DES ENG] Rope or cable in which the individual wires or fibers and the strands are twisted to the right.

rigid body [MECH] An idealized extended solid whose size and shape are definitely fixed and remain unaltered when forces are applied.

rigid-body dynamics [MECH] The study of the motions of a rigid body under the influence of forces and torques.

rigid coupling [MECH ENG] A mechanical fastening of shafts connected with the axes directly in line.

rigidity [MECH] The quality or state of resisting change in form.

rim [DES ENG] **1.** The outer part of a wheel, usually connected to the hub by spokes. **2.** An outer edge or border, sometimes raised or projecting.

rim clutch [MECH ENG] A frictional contact clutch having surface elements that apply pressure to the rim either externally or internally.

ring [DES ENG] A tie member or chain link; tension or compression applied through the center of the ring produces bending moment, shear, and normal force on radial sections.

ring and circle shear [DES ENG] A rotary shear designed for cutting circles and rings where the edge of the metal sheet cannot be used as a start.

ringbolt [DES ENG] An eyebolt with a ring passing through the eye.

ring crusher [MECH ENG] Solids-reduction device with a rotor having loose crushing rings held outwardly by centrifugal force, which crush the feed by impact with the surrounding shell.

ring gage [DES ENG] A cylindrical ring of steel whose inside diameter is finished to gage tolerance and is used for checking the external diameter of a cylindrical object.

ring gear [MECH ENG] The ring-shaped gear in an automobile differential that is driven by the propeller shaft pinion and transmits power through the differential to the line axle.

ringing [CONT SYS] An oscillatory transient occurring in the output of a system as a result of a sudden change in input.

ring jewel [DES ENG] A type of jewel used as a pivot bearing in a time-keeping device, gyro, or instrument.

ring job [MECH ENG] Installation of new piston rings on a piston.

ring lifter *See* split-ring core lifter.

ringlock nail [DES ENG] A nail ringed with grooves to provide greater holding power.

ring-oil [MECH ENG] To oil (a bearing) by conveying the oil to the point to be lubricated by means of a ring, which rests upon and turns with the journal, and dips into a reservoir containing the lubricant.

ring-roller mill [MECH ENG] A grinding mill in which material is fed past spring-loaded rollers that apply force against the sides of a revolving bowl. Also known as roller mill.

rip panel [AERO ENG] A part of a manned free balloon; it is the panel to which the ripcord is attached and extends about ¼ to ⅓ of the circumference of the balloon along one of its meridians; it is torn open when the ripcord is pulled so that all the gas in the balloon escapes.

ripping bar [DES ENG] A steel bar with a chisel at one end and a curved claw for pulling nails at the other.

ripping punch [DES ENG] A tool with a rectangular cutting edge, used in a punch press to crosscut metal plates.

ripple *See* capillary wave.

ripsaw [MECH ENG] A heavy-tooth power saw used for cutting wood with the grain.

rise time [CONT SYS] The time it takes for the output of a system to change from a specified small percentage (usually 5 or 10%) of its steady-state increment to a specified large percentage (usually 90 or 95%).

Rittinger's law [MECH ENG] The law that energy needed to reduce the size of a solid particle is directly proportional to the resultant increase in surface area.

rivet [DES ENG] A short rod with a head formed on one end; it is inserted through aligned holes in parts to be joined, and the protruding end is pressed or hammered to form a second head.

riveting hammer [MECH ENG] A hammer used for driving rivets.

Roberts' linkage [MECH ENG] A type of approximate straight-line mechanism which provided, early in the 19th century, a practical means of making straight metal guides for the slides in a metal planner.

Robins-Messiter system [MECH ENG] A stacking conveyor system in which material arrives on a conveyor belt and is fed to one or two wing conveyors.

robot [CONT SYS] A completely self-controlled electronic, electric, or mechanical device.

Roche lobes [MECH] 1. Regions of space surrounding two massive bodies revolving around each other under their mutual gravitational attraction, such that the gravitational attraction of each body dominates the lobe surrounding it. 2. In particular, the effective potential energy (referred to a system of coordinates rotating with the bodies) is equal to a constant V_0 over the surface of the lobes, and if a particle is inside one of the lobes and if the sum of its effective potential energy and its kinetic energy is less than V_0, it will remain inside the lobe.

rockair [AERO ENG] A high-altitude sounding system consisting of a small solid-propellant research rocket carried aloft by an aircraft; the rocket is fired while the aircraft is in vertical ascent.

rock channeler [MECH ENG] A machine used in quarrying for cutting an artificial seam in a mass of stone.

rock drill [MECH ENG] A machine for boring relatively short holes in rock for blasting purposes; motive power may be compressed air, steam, or electricity.

rocker arm [MECH ENG] In an internal combustion engine, a lever that is pivoted near its center and operated by a pushrod at one end to raise and depress the valve stem at the other end.

rocker cam [MECH ENG] A cam that moves with a rocking motion.

rocket [AERO ENG] 1. Any kind of jet propulsion capable of operating independently of the atmosphere. 2. A complete vehicle driven by such a propulsive system.

rocket airplane [AERO ENG] An airplane using a rocket or rockets for its chief or only propulsion.

rocket assist [AERO ENG] An assist in thrust given an airplane or missile by use of a rocket motor or rocket engine during flight or during takeoff.

rocket chamber [AERO ENG] A chamber for the combustion of fuel in a rocket; in particular, that section of the rocket engine in which combustion of propellants takes place.

rocket engine [AERO ENG] A reaction engine that contains within itself, or carries along with itself, all the substances necessary for its operation or for the consumption or combustion of its fuel, not requiring intake of any outside substance and hence capable of operation in outer space. Also known as rocket motor.

rocket igniter [AERO ENG] An igniter for the propellant in a rocket.

rocket launcher [AERO ENG] A device for launching a rocket, wheel-mounted, motorized, or fixed for use on the ground, rocket launchers are mounted on aircraft, as under the wings, or are installed below or on the decks of ships.

rocket motor *See* rocket engine.

rocket nose section [AERO ENG] The extreme forward portion of a rocket, designed to contain instrumentation, spotting charges, fusing or arming devices, and the like, but does not contain the payload.

rocket propulsion [AERO ENG] Reaction propulsion by a rocket engine.

rocket ramjet [AERO ENG] A ramjet engine having a rocket mounted within the ramjet duct, the rocket being used to bring the ramjet up to the necessary operating speed. Also known as ducted rocket.

rocketry [AERO ENG] 1. The science or study of rockets, embracing theory, research, development, and experimentation. 2. The art and science of using rockets, especially rocket ammunition.

rocket sled [AERO ENG] A sled that runs on a rail or rails and is accelerated to high velocities by a rocket engine; the sled is used in determining g tolerances and for developing crash survival techniques.

rocket-sled testing [AERO ENG] A method of subjecting structures and devices to high accelerations or decelerations and aerodynamic flow phenomena under controlled conditions; the test object is mounted on a sled chassis running on precision steel rails and accelerated by rockets or decelerated by water scoops.

rocket staging [AERO ENG] The use of successive rocket sections or stages, each having its own engine or engines; each stage is a complete rocket vehicle in itself.

rocket thrust [AERO ENG] The thrust of a rocket engine.

rocket tube [AERO ENG] 1. A launching tube for rockets. 2. A tube or nozzle through which rocket gases are ejected.

rocking furnace [MECH ENG] A horizonal, cylindrical melting furnace that is rolled back and forth on a geared cradle.

rocking valve [MECH ENG] An engine valve in which a disk or cylinder turns in its seat to permit fluid flow.

rockoon [AERO ENG] A high-altitude sounding system consisting of a small solid-propellant research rocket carried aloft by a large plastic balloon.

rod [DES ENG] 1. A bar whose end is slotted, tapered, or screwed for the attachment of a drill bit. 2. A thin, round bar of metal or wood. [MECH] *See* perch.

rod bit [DES ENG] A bit designed to fit a reaming shell that is threaded to couple directly to a drill rod.

rod coupling [DES ENG] A double-pin-thread coupling used to connect two drill rods together.

rod mill [MECH ENG] A pulverizer operated by the impact of heavy metal rods.

rod string [MECH ENG] Drill rods coupled to form the connecting link between the core barrel and bit in the borehole and the drill machine at the collar of the borehole.

Rogallo wing [AERO ENG] A glider folded inside a spacecraft; to be deployed during the spacecraft's reentry like a parachute, gliding the spacecraft to a landing.

rolamite mechanism [MECH ENG] An elemental mechanism consisting of two rollers contained by two parallel planes and bounded by a fixed S-shaped band under tension.

roll [MECH] Rotational or oscillatory movement of an aircraft or similar body about a longitudinal axis through the body; it is called roll for any degree of such rotation. [MECH ENG] A cylinder mounted in bearings; used for such functions as shaping, crushing, moving, or printing work passing by it.

roll acceleration [MECH] The angular acceleration of an aircraft or missile about its longitudinal or X axis.

roll axis [MECH] A longitudinal axis through an aircraft, rocket, or similar body, about which the body rolls.

roll crusher [MECH ENG] A crusher having one or two toothed rollers to reduce the material.

roller [DES ENG] A cylindrical device for transmitting motion and force by rotation.

roller bearing [MECH ENG] A shaft bearing characterized by parallel or tapered steel rollers confined between outer and inner rings.

roller bit *See* cone rock bit.

roller cam follower [MECH ENG] A follower consisting of a rotatable wheel at the end of the shaft.

roller chain [MECH ENG] A chain drive assembled from roller links and pin links.

roller conveyor [MECH ENG] A gravity conveyor having a track of parallel tubular rollers set at a definite grade, usually on antifriction bearings, at fixed locations, over which package goods which are sufficiently rigid to prevent sagging between rollers are moved by gravity or propulsion.

roller leveling [MECH ENG] Leveling flat stock by passing it through a machine having a series of rolls whose axes are staggered about a mean parallel path by a decreasing amount.

roller mill *See* ring-roller mill.

roller pulverizer [MECH ENG] A pulverizer operated by the crushing action of rotating rollers.

roller stamping die [MECH ENG] An engraved roller used for stamping designs and other markings on sheet metal.

rolling [MECH] Motion of a body across a surface combined with rotational motion of the body so that the point on the body in contact with the surface is instantaneously at rest.

rolling contact [MECH] Contact between bodies such that the relative velocity of the two contacting surfaces at the point of contact is zero.

rolling-contact bearing [MECH ENG] A bearing composed of rolling elements interposed between an outer and inner ring.

rolling friction [MECH] A force which opposes the motion of any body which is rolling over the surface of another.

roll mill [MECH ENG] A series of rolls operating at different speeds for grinding and crushing.

roll threading [MECH ENG] Threading a metal workpiece by rolling it either between grooved circular rolls or between grooved straight lines.

rood [MECH] A unit of area, equal to $\frac{1}{4}$ acre, or 10,890 square feet, or 1011.7141056 square meters.

roofing nail [DES ENG] A nail used for attaching paper or shingle to roof boards; usually short with a barbed shank and a large flat head.

root [DES ENG] The bottom of a screw thread.

root circle [DES ENG] A hypothetical circle defined at the bottom of the tooth spaces of a gear.

root fillet [DES ENG] The rounded corner at the angle of a gear tooth flank and the bottom land.

root locus plot [CONT SYS] A plot in the complex plane of values at which the loop transfer function of a feedback control system is a negative number.

Roots blower [MECH ENG] A compressor in which a pair of hourglass-shaped members rotate within a casing to deliver large volumes of gas at relatively low pressure increments.

rope-and-button conveyor [MECH ENG] A conveyor consisting of an endless wire rope or cable with disks or buttons attached at intervals.

rope drive [MECH ENG] A system of ropes running in grooved pulleys or sheaves to transmit power over distances too great for belt drives.

rope sheave [DES ENG] A grooved wheel, usually made of cast steel or heat-treated alloy steel, used for rope drives.

rope socket [DES ENG] A drop-forged steel device, with a tapered hole, which can be fastened to the end of a wire cable or rope and to which a load may be attached.

rose bit [DES ENG] A hardened steel or alloy noncore bit with a serrated face to cut or mill out bits, casing, or other metal objects lost in the hole.

rose chucking reamer [DES ENG] A machine reamer with a straight or tapered shank and a straight or spiral flute; cutting is done at the ends of the teeth only; produces a rough hole since there are few teeth.

rose reamer [DES ENG] A reamer designed to cut on the beveled leading ends of the teeth rather than on the sides.

Rossby diagram [THERMO] A thermodynamic diagram, named after its designer, with mixing ratio as abscissa and potential temperature as ordinate; lines of constant equivalent potential temperature are added.

Rossby number [FL MECH] The nondimensional ratio of the inertial force to the Coriolis force for a given flow of a rotating fluid, given as $R_0 = U/fL$, where U is a characteristic velocity, f the Coriolis parameter (or, if the system is cylindrical rather than spherical, twice the system's rotation rate), and L a characteristic length.

Rossby parameter [FL MECH] The northward variation of the Coriolis parameter, arising from the sphericity of the earth. Also known as Rossby term.

Rossby regime [FL MECH] A type of flow pattern in a rotating fluid with differential radial heating in which the major radial transport of shear and momentum is effected by horizontal eddies of low wave-number; this regime occurs for low values of the Rossby number (of the order of 0.1).

Rossby term *See* Rossby parameter.

Ross feeder [MECH ENG] A chute for conveying bulk materials by means of a screen of heavy endless chains hung on a sprocket shaft; rotation of the shaft causes materials to slide.

rotary [MECH ENG] **1.** A rotary machine, such as a rotary printing press or a rotary well-drilling machine. **2.** The turntable and its supporting and rotating assembly in a well-drilling machine.

rotary actuator [MECH ENG] A device that converts electric energy into controlled rotary force; usually consists of an electric motor, gear box, and limit switches.

rotary air heater [MECH ENG] A regenerative air heater in which heat-transferring members are moved alternately through the gas and air streams.

rotary annular extractor [MECH ENG] Vertical, cylindrical shell with an inner, rotating cylinder; liquids to be contacted flow countercurrently through the annular space between the rotor and shell; used for liquid-liquid extraction processes.

rotary atomizer [MECH ENG] A hydraulic atomizer having the pump and nozzle combined.

rotary belt cleaner [MECH ENG] A series of blades symmetrically spaced about the axis of rotation and caused to scrape or beat against the conveyor belt for the purpose of cleaning.

rotary blower [MECH ENG] Positive-displacement, rotating-impeller, air-movement device; can be straight-lobe, screw, sliding-vane, or liquid-piston type.

rotary boring   [MECH ENG] A system of boring in which rock penetration is achieved by the rotation of the hollow cutting tool.

rotary bucket   [MECH ENG] A 12- to 96-inch-diameter (30- to 244-centimeter-diameter) posthole augerlike device, the bottom end of which is equipped with cutting teeth used to rotary-drill large-diameter shallow holes to obtain samples of soil lying above the groundwater level.

rotary compressor   [MECH ENG] A positive-displacement machine in which compression of the fluid is effected directly by a rotor and without the usual piston, connecting rod, and crank mechanism of the reciprocating compressor.

rotary crane   [MECH ENG] A crane consisting of a boom pivoted to a fixed or movable structure.

rotary crusher   [MECH ENG] Solids-reduction device in which a high-speed rotating cone on a vertical shaft forces solids against a surrounding shell.

rotary cutter   [MECH ENG] Device used to cut tough or fibrous materials by the shear action between two sets of blades, one set on a rotating holder, the other stationary on the surrounding casing.

rotary drill   [MECH ENG] Any of various drill machines that rotate a rigid, tubular string of rods to which is attached a rock cutting bit, such as an oil well drilling apparatus.

rotary drilling   [MECH ENG] The act or process of drilling a borehole by means of a rotary-drill machine, such as in drilling an oil well.

rotary dryer   [MECH ENG] A cylindrical furnace slightly inclined to the horizontal and rotated on suitable bearings; moisture is removed by rising hot gases.

rotary engine   [MECH ENG] A positive displacement engine (such as a steam or internal combustion type) in which the thermodynamic cycle is carried out in a mechanism that is entirely rotary and without the more customary structural elements of a reciprocating piston, connecting rods, and crankshaft.

rotary excavator   *See* bucket-wheel excavator.

rotary feeder   [MECH ENG] Device in which a rotating element or vane discharges powder or granules at a predetermined rate.

rotary filter   *See* drum filter.

rotary furnace   [MECH ENG] A heat-treating furnace of circular construction which rotates the workpiece around the axis of the furnace during heat treatment; workpieces are transported through the furnace along a circular path.

rotary-percussive drill   [MECH ENG] Drilling machine which operates as a rotary machine by the action of repeated blows to the bit.

rotary pump   [MECH ENG] A displacement pump that delivers a steady flow by the action of two members in rotational contact.

rotary roughening   [MECH ENG] A metal preparation technique in which the workpiece surface is roughened by a cutting tool.

rotary shear   [MECH ENG] A sheet-metal cutting machine having two rotary-disk cutters mounted on parallel shafts and driven in unison.

rotary shot drill   [MECH ENG] A rotary drill used to drill blastholes.

rotary swager   [MECH ENG] A machine for reducing diameter or wall thickness of a bar or tube by delivering hammerlike blows to the surface of the work supported on a mandrel.

rotary table   [MECH ENG] A milling machine attachment consisting of a round table with T-shaped slots and rotated by means of a handwheel actuating a worm and worm gear.

rotary vacuum filter   *See* drum filter.

rotary valve   [MECH ENG] A valve for the admission or release of working fluid to or from an engine cylinder where the valve member is a ported piston that turns on its axis.

rotating coordinate system   [MECH] A coordinate system whose axes as seen in an inertial coordinate system are rotating.

rotating-cylinder method   [FL MECH] A method of measuring the viscosity of a fluid in which the fluid fills the space between two concentric cylinders, and the torque on the stationary inner cylinder is measured when the outer cylinder is rotated at constant speed.

rotating Reynolds number   [FL MECH] A nondimensional number arising in problems of a rotating viscous fluid and, in particular, in problems involving the agitation of such a fluid by an impeller, equal to the product of the square of the impeller's diameter and its angular velocity divided by the kinematic viscosity of the fluid. Symbolized Re_r.

rotation   [MECH] Also known as rotational motion. 1. Motion of a rigid body in which either one point is fixed, or all the points on a straight line are fixed. 2. Angular displacement of a rigid body. 3. The motion of a particle about a fixed point.

rotational energy   [MECH] The kinetic energy of a rigid body due to rotation.

rotational flow   [FL MECH] Flow of a fluid in which the curl of the fluid velocity is not zero, so that each minute particle of fluid rotates about its own axis. Also known as rotational motion.

rotational impedance   [MECH] A complex quantity, equal to the phasor representing the alternating torque acting on a system divided by the phasor representing the resulting angular velocity in the direction of the torque at its point of application. Also known as mechanical rotational impedance.

rotational inertia   *See* moment of inertia.

rotational motion   *See* rotation; rotational flow.

rotational reactance   [MECH] The imaginary part of the rotational impedance. Also known as mechanical rotational reactance.

rotational resistance   [MECH] The real part of rotational impedance; it is responsible for dissipation of energy. Also known as mechanical rotational resistance.

rotational stability   [MECH] Property of a body for which a small angular displacement sets up a restoring torque that tends to return the body to its original position.

rotational strain   [MECH] Strain in which the orientation of the axes of strain is changed.

rotational wave   *See* shear wave.

rotation coefficients   [MECH] Factors employed in computing the effects on range and deflection which are caused by the rotation of the earth; they are published only in firing tables involving comparatively long ranges.

rotation moment   *See* torque.

rotator   [MECH] A rotating rigid body.

rotor   [AERO ENG] An assembly of blades designed as airfoils that are attached to a helicopter or similar aircraft and rapidly rotated to provide both lift and thrust. [MECH ENG] *See* impeller.

rotor balancing   *See* shaft balancing.

rough air   [AERO ENG] An aviation term for turbulence encountered in flight.

rough burning [AERO ENG] Pressure fluctuations frequently observed at the onset of burning and at the combustion limits of a ramjet or rocket.

rough grinding [MECH ENG] Preliminary grinding without regard to finish.

rough machining [MECH ENG] Preliminary machining without regard to finish.

roughness [FL MECH] Distance from peaks to valleys in pipe-wall irregularities; used to modify Reynolds number calculations for fluid flow through pipes.

roughness factor [FL MECH] A correction factor used in fluid-flow calculations to allow for flow resistance caused by the roughness of the surface over which the fluid must flow.

roughness-width cutoff [MECH ENG] The maximum width of surface irregularities included in roughness height measurements.

rough turning [MECH ENG] The removal of excess stock from a workpiece as rapidly and efficiently as possible.

round-face bit [DES ENG] Any bit with a rounded cutting face.

round file [DES ENG] A file having a circular cross section.

round-head bolt [DES ENG] A bolt having a rounded head at one end.

roundnose chisel [DES ENG] A chisel having a rounded cutting edge.

roundnose tool [DES ENG] A large-radius-nose cutting tool generally used in finishing operations.

round strand rope [DES ENG] A rope composed generally of six strands twisted together or laid to form the rope around a core of hemp, sisal, or manila, or, in a wire-cored rope, around a central strand composed of individual wires.

router plane [DES ENG] A plane for cutting grooves and smoothing the bottom of grooves.

Routh's rule of inertia [MECH] The moment of inertia of a body about an axis of symmetry equals $M(a^2 + b^2)/n$, where M is the body's mass, a and b are the lengths of the body's two other perpendicular semiaxes, and n equals 3, 4, or 5 depending on whether the body is a rectangular parallelepiped, elliptic cylinder, or ellipsoid, respectively.

rpm *See* revolution per minute.

rps *See* revolution per second.

rubber belt [DES ENG] A conveyor belt that consists essentially of a rubber-covered fabric; fabric is cotton, or nylon or other synthetic fiber, with steel-wire reinforcement.

rubber-covered steel conveyor [DES ENG] A steel conveyor band with a cover of rubber bonded to the steel.

rubber wheel [DES ENG] A grinding wheel made with rubber as the bonding agent.

Runge vector [MECH] A vector which describes certain unchanging features of a non-relativistic two-body interaction obeying an inverse-square law, either in classical or quantum mechanics; its constancy is a reflection of the symmetry inherent in the inverse-square interaction.

running block *See* traveling block.

running fit [DES ENG] The intentional difference in dimensions of mating mechanical parts that permits them to move relative to each other.

running gear [MECH ENG] The means employed to support a truck and its load and to provide rolling-friction contact with the running surface.

rupture disk *See* burst disk.

rupture disk device [MECH ENG] A nonreclosing pressure relief device which relieves the inlet static pressure in a system through the bursting of a disk.

Rzeppa joint [MECH ENG] A special application of the Bendix-Weiss universal joint in which four large balls are transmitting elements, while a center ball acts as a spacer; it transmits constant angular velocity through a single universal joint.

S

s *See* second.

S *See* stoke.

S-2 *See* Tracker.

Sabathé's cycle [MECH ENG] An internal combustion engine cycle in which part of the combustion is explosive and part at constant pressure.

saber saw [MECH ENG] A portable saw consisting of an electric motor, a straight saw blade with reciprocating mechanism, a handle, baseplate, and other essential parts.

Sabrejet [AERO ENG] An airplane widely used by the U.S. Air National Guard and free world countries; it has a range of beyond 1000 miles (1600 kilometers) and a speed of over 650 miles (1050 kilometers) per hour, carries two 1000-pound (450-kilogram) bombs or sixteen 5-inch (12.7-centimeter) rockets, or a combination plus two additional 1000-pound bombs in lieu of fuel tanks; has a crew of one. Designated F-86.

saddle [DES ENG] A support shaped to fit the object being held.

saddle-type turret lathe [MECH ENG] A turret lathe designed without a ram and with the turret mounted directly on a support (saddle) which slides on the bedways of the lathe.

safe load [MECH] The stress, usually expressed in tons per square foot, which a soil or foundation can safely support.

safety chuck [DES ENG] Any drill chuck on which the heads of the set screws do not protrude beyond the outer periphery of the chuck.

safety factor *See* factor of safety.

safety flange [DES ENG] A type of flange with tapered sides designed to keep a wheel intact in the event of accidental breakage.

safety hoist [MECH ENG] A hoisting gear that does not continue running when tension is released.

safety hook [DES ENG] A hoisting hook with a spring-loaded latch that prevents the load from accidentally slipping off the hook.

safety stop [MECH ENG] **1.** On a hoisting apparatus, a device by which the load may be prevented from falling. **2.** An automatic device on a hoisting engine designed to prevent overwinding.

SAMOS program [AERO ENG] A series of military reconnaissance satellites carrying cameras and other surveillance equipment; each satellite has a jettisonable camera package that is ejected from orbit and recovered in midair.

sampled-data control system [CONT SYS] A form of control system in which the signal appears at one or more points in the system as a sequence of pulses or numbers usually equally spaced in time.

sampler [CONT SYS] A device, used in sampled-data control systems, whose output is a series of impulses at regular intervals in time; the height of each impulse equals the value of the continuous input signal at the instant of the impulse.

sampling interval [CONT SYS] The time between successive sampling pulses in a sampled-data control system.

sander [MECH ENG] **1.** An electric machine used to sand the surface of wood, metal, or other material. **2.** A device attached to a locomotive or electric rail car which sands the rails to increase friction on the driving wheels.

sand mill [MECH ENG] Variation of a ball-type size-reduction mill in which grains of sand serve as grinding balls.

sand pump [MECH ENG] A pump, usually a centrifugal type, capable of handling sand- and gravel-laden liquids without clogging or wearing unduly; used to extract mud and cuttings from a borehole. Also known as sludge pump.

sand reel [MECH ENG] A drum, operated by a band wheel, for raising or lowering the sand pump or bailer during drilling operations.

sand slinger [MECH ENG] A machine which delivers sand to and fills molds at high speed by centrifugal force.

sand wheel [MECH ENG] A wheel fitted with steel buckets around the circumference for lifting sand or sludge out of a sump to stack it at a higher level.

sandwich construction [DES ENG] Composite construction of alloys, plastics, wood, or other materials consisting of a foam or honeycomb layer laminated and glued between two hard outer sheets. Also known as sandwich laminate.

Sargent cycle [THERMO] An ideal thermodynamic cycle consisting of four reversible processes: adiabatic compression, heating at constant volume, adiabatic expansion, and isobaric cooling.

SAS *See* stability augmentation system.

satellite *See* artificial satellite.

satellite tracking [AERO ENG] Determination of the positions and velocities of satellites through radio and optical means.

satelloid [AERO ENG] A vehicle that revolves about the earth or other celestial body, but at such altitudes as to require sustaining thrust to balance drag.

saturated vapor [THERMO] A vapor whose temperature equals the temperature of boiling at the pressure existing on it.

saturation specific humidity [THERMO] A thermodynamic function of state; the value of the specific humidity of saturated air at the given temperature and pressure.

saturation vapor pressure [THERMO] The vapor pressure of a thermodynamic system, at a given temperature, wherein the vapor of a substance is in equilibrium with a plane surface of that substance's pure liquid or solid phase.

Saturn [AERO ENG] One of the very large launch vehicles built primarily for the Apollo program; begun by Army Ordnance but turned over to the National Aeronautics and Space Administration for the manned space flight program to the moon.

Saunders air-lift pump [MECH ENG] A device for raising water from a well by the introduction of compressed air below the water level in the well.

Savonius rotor [MECH ENG] A rotor composed of two offset semicylindrical elements rotating about a vertical axis.

Savonius windmill [MECH ENG] A windmill composed of two semicylindrical offset cups rotating about a vertical axis.

saw [DES ENG] Any of various tools consisting of a thin, usually steel, blade with continuous cutting teeth on the edge.

saw bit [DES ENG] A bit having a cutting edge formed by teeth shaped like those in a handsaw.

saw gumming [MECH ENG] Grinding away the punch marks in the spaces between the teeth in saw manufacture.

sawmill [MECH ENG] A machine for cutting logs with a saw or a series of saws.

sawtooth barrel *See* basket.

sawtooth crusher [MECH ENG] Solids crusher in which feed is broken down between two sawtoothed shafts rotating at different speeds.

sax [DES ENG] A tool for chopping away the edges of roof slates; it has a pick at one end for making nail holes.

Saybolt Furol viscosity [FL MECH] The time in seconds for 60 milliliters of fluid to flow through a capillary tube in a Saybolt Furol viscosimeter at specified temperatures between 70 and 210°F (21 and 99°C); used for high-viscosity petroleum oils, such as transmission and gear oils, and heavy fuel oils.

Saybolt Seconds Universal [FL MECH] A unit of measurement for Saybolt Universal viscosity. Abbreviated SSU.

Saybolt Universal viscosity [FL MECH] The time in seconds for 60 milliliters of fluid to flow through a capillary tube in a Saybolt Universal viscosimeter at a given temperature.

Sc, *See* turbulent Schmidt number.

scale effect [AERO ENG] The necessary corrections applied to measurements of a model in a wind tunnel to ascertain corresponding values for a full-sized object. [FL MECH] An effect in fluid flow that results from changing the scale, but not the shape, of a body around which the flow passes; this effect is relevant to wind tunnel experiments.

scale-up [DES ENG] Design process in which the data of an experimental-scale operation (model or pilot plant) is used for the design of a large (scaled-up) unit, usually of commercial size.

scaling [MECH] Expressing the terms in an equation of motion in powers of nondimensional quantities (such as a Reynolds number), so that terms of significant magnitude under conditions specified in the problem can be identified, and terms of insignificant magnitude can be dropped.

scalpel [DES ENG] A small, straight, very sharp knife (or detachable blade for a knife), used for dissecting.

scarf joint [DES ENG] A joint made by the cutting of overlapping mating parts so that the joint is not enlarged and the patterns are complementary, and securing them by glue, fasteners, welding, or other joining method.

scavenging [MECH ENG] Removal of spent gases from an internal combustion engine cylinder and replacement by a fresh charge or air.

scfh [FL MECH] Cubic feet per hour of gas flow at specified standard conditions of temperature and pressure.

scfm [FL MECH] Cubic feet per minute of gas flow at specified standard conditions of temperature and pressure.

Scheffel engine [MECH ENG] A type of multirotor engine that uses nine approximately equal rotors turning in the same clockwise sense.

Schmidt number 1 *See* Prandtl number.

Schneider recoil system [MECH ENG] A recoil system for artillery, employing the hydropneumatic principle without a floating piston.

Schuler pendulum [MECH] Any apparatus which swings, because of gravity, with a natural period of 84.4 minutes, that is, with the same period as a hypothetical simple pendulum whose length is the earth's radius; the pendulum arm remains vertical despite any motion of its pivot, and the apparatus is therefore useful in navigation.

scissor engine *See* cat-and-mouse engine.

scissor jack [MECH ENG] A lifting jack driven by a horizontal screw; the linkages of the jack are parallelograms whose horizontal diagonals are lengthened or shortened by the screw.

scoop [DES ENG] 1. Any of various ladle-, shovel-, or bucketlike utensils or containers for moving liquid or loose materials. 2. A funnel-shaped opening for channeling a fluid into a desired path. [MECH ENG] A large shovel with a scoop-shaped blade.

Scorpion [AERO ENG] An all-weather interceptor aircraft with twin turbojet engines; its armament consists of air-to-air rockets with nuclear or nonnuclear warheads. Designated F-89.

scotch boiler [MECH ENG] A fire-tube boiler with one or more cylindrical internal furnaces enveloped by a boiler shell equipped with five tubes in its upper part; heat is transferred to water partly in the furnace area and partly in passage of hot gases through the tubes. Also known as dry-back boiler; scotch marine boiler (marine usage).

Scotch derrick *See* stiffleg derrick.

Scotch yoke [MECH ENG] A type of four-bar linkage; it is employed to convert a steady rotation into a simple harmonic motion.

scouring [MECH ENG] Mechanical finishing or cleaning of a hard surface by using an abrasive and low pressure.

Scout [AERO ENG] A four-stage all-solid-propellant rocket, used as a space probe and orbital test vehicle; first launched July 1, 1960, with a 150-pound (68-kilogram) payload.

scramble [AERO ENG] To take off as quickly as possible (usually followed by course and altitude instructions).

scramjet [AERO ENG] Essentially a ramjet engine, intended for flight at hypersonic speeds. Derived from supersonic combustion ramjet.

scraper conveyor [MECH ENG] A type of flight conveyor in which the element (chain and flight) for moving materials rests on a trough.

scraper hoist [MECH ENG] A drum hoist that operates the scraper of a scraper loader.

scraper loader [MECH ENG] A machine used for loading coal or rock by pulling a scoop through the material to an apron or ramp, where the load is discharged onto a car or conveyor.

scraper ring [MECH ENG] A piston ring that scrapes oil from a cylinder wall to prevent it from being burnt.

screaming [AERO ENG] A form of combustion instability, especially in a liquid-propellant rocket engine, of relatively high frequency, characterized by a high-pitched noise.

screeching [AERO ENG] A form of combustion instability, especially in an afterburner, of relatively high frequency, characterized by a harsh, shrill noise.

screen deck [DES ENG] A surface provided with apertures of specified size, used for screening purposes.

screw [DES ENG] 1. A cylindrical body with a helical groove cut into its surface. 2. A fastener with continuous ribs on a cylindrical or conical shank and a slotted, recessed, flat, or rounded head. Also known as screw fastener.

screw compressor [MECH ENG] A rotary-element gas compressor in which compression is accomplished between two intermeshing, counterrotating screws.

screw conveyor [MECH ENG] A conveyor consisting of a helical screw that rotates upon a single shaft within a stationary trough or casing, and which can move bulk material along a horizontal, inclined, or vertical plane. Also known as auger conveyor; spiral conveyor; worm conveyor.

screw displacement [MECH] A rotation of a rigid body about an axis accompanied by a translation of the body along the same axis.

screwdriver [DES ENG] A tool for turning and driving screws in place; a thin, wedge-shaped or fluted end enters the slot or recess in the head of the screw.

screw elevator [MECH ENG] A type of screw conveyor for vertical delivery of pulverized materials.

screw fastener *See* screw.

screwfeed [MECH ENG] A system or combination of gears, ratchets, and friction devices in the swivel head of a diamond drill, which controls the rate at which a bit penetrates a rock formation.

screw feeder [MECH ENG] A mechanism for handling bulk (pulverized or granulated solids) materials, in which a rotating helicoid screw moves the material forward, toward and into a process unit.

screw machine [MECH ENG] A lathe for making relatively small, turned metal parts in large quantities.

screw press [MECH ENG] A press having the slide operated by a screw mechanism.

screw propeller [MECH ENG] A marine and airplane propeller consisting of a streamlined hub attached outboard to a rotating engine shaft on which are mounted two to six blades; the blades form helicoidal surfaces in such a way as to advance along the axis about which they revolve.

screw pump [MECH ENG] A pump that raises water by means of helical impellers in the pump casing.

screw rivet [DES ENG] A short rod threaded along the length of the shaft that is set without access to the point.

screw spike [DES ENG] A large nail with a helical thread on the upper portion of the shank; used to fasten railroad rails to the ties.

screwstock [MECH ENG] Free-machining bar, rod, or wire.

screw thread [DES ENG] A helical ridge formed on a cylindrical core, as on fasteners and pipes.

screw-thread gage [DES ENG] Any of several devices for determining the pitch, major, and minor diameters, and the lead, straightness, and thread angles of a screw thread.

screw-thread micrometer [DES ENG] A micrometer used to measure pitch diameter of a screw thread.

scriber [DES ENG] A sharp-pointed tool used for drawing lines on metal workpieces.

scroll gear [DES ENG] A variable gear resembling a scroll with teeth on one face.

scrub [AERO ENG] To cancel a scheduled firing, either before or during countdown.

scuffle hoe [DES ENG] A hoe having two sharp edges so that it can be pushed and pulled.

scythe [DES ENG] A tool with a long curved blade attached at a more or less right angle to a long handle with grips for both hands; used for cutting grass as well as grain and other crops.

Sea Bat [AERO ENG] An antisubmarine warfare helicopter equipped with active-passive sonar, acoustic homing torpedoes, and instrument–night flight capability. Designated SH-34G.

sealed cabin [AERO ENG] The occupied space of an aircraft or spacecraft characterized by walls which do not allow gaseous exchange between the inner atmosphere and its surrounding atmosphere, and containing its own mechanisms for maintenance of the inside atmosphere.

Seale rope [DES ENG] A wire rope with six or eight strands, each having a large wire core covered by nine small wires, which, in turn, are covered by nine large wires.

seaplane [AERO ENG] An airplane that takes off from and alights on the water; it is supported on the water by pontoons, or floats, or by a hull which is a specially designed fuselage. Also known as airboat.

seat [MECH ENG] The fixed, pressure-containing portion of a valve which comes into contact with the moving portions of that valve.

seating-lock locking fastener [DES ENG] A locking fastener that locks only when firmly seated and is therefore free-running on the bolt.

sec *See* second.

second [MECH] The fundamental unit of time equal to 9,192,631,770 periods of the radiation corresponding to the transition between the two hyperfine levels of the ground state of an atom of cesium-133. Abbreviated s: sec.

secondary air [MECH ENG] Combustion air introduced over the burner flame to enhance completeness of combustion.

secondary creep [MECH] The change in shape of a substance under a minimum and almost constant differential stress, with the strain-time relationship a constant. Also known as steady-state creep.

secondary crusher [MECH ENG] Any of a group of crushing and pulverizing machines used after the primary treatment to further reduce the particle size of shale or other rock.

secondary flow [FL MECH] A field of fluid motion which can be considered as superposed on a primary field of motion through the action of friction usually in the vicinity of solid boundaries. Also known as frictional secondary flow.

secondary grinding [MECH ENG] A further grinding of material previously reduced to sand size.

secondary stress [MECH] A self-limiting normal or shear stress which is caused by the constraint of a structure and which is expected to cause minor distortions that would not result in a failure of the structure.

second law of motion *See* Newton's second law.

second law of thermodynamics [THERMO] A general statement of the idea that there is a preferred direction for any process; there are many equivalent statements of the law, the best known being those of Clausius and of Kelvin.

second-level controller [CONT SYS] A controller which influences the actions of first-level controllers, in a large-scale control system partitioned by plant decomposition, to compensate for subsystem interactions so that overall objectives and constraints of the system are satisfied. Also known as coordinator.

second-order transition [THERMO] A change of state through which the free energy of a substance and its first derivatives are continuous functions of temperature and pressure, or other corresponding variables.

second pilot [AERO ENG] A pilot, not necessarily qualified on type, who is responsible for assisting the first pilot to fly the aircraft and is authorized as second pilot.

sectional conveyor [MECH ENG] A belt conveyor that can be lengthened or shortened by the addition or the removal of interchangeable sections.

sectional core barrel [DES ENG] A core barrel whose length can be increased by coupling unit sections together.

sectional header boiler [MECH ENG] A horizontal boiler in which tubes are assembled in sections into front and rear headers; the latter, in turn, are connected to the boiler drum by vertical tubes.

section modulus [MECH] The ratio of the moment of inertia of the cross section of a beam undergoing flexure to the greatest distance of an element of the beam from the neutral axis.

sector gear [DES ENG] 1. A toothed device resembling a portion of a gear wheel containing the center bearing and a part of the rim with its teeth. 2. A gear having such a device as its chief essential feature. [MECH ENG] A gear system employing such a gear as a principal part.

seepage [FL MECH] The slow movement of water or other fluid through a porous medium.

segment saw [MECH ENG] A saw consisting of steel segments attached around the edge of a flange and used for cutting veneer.

seiche [FL MECH] An oscillation of a fluid body in response to the disturbing force having the same frequency as the natural frequency of the fluid system.

selective transmission [MECH ENG] A gear transmission with a single lever for changing from one gear ratio to another; used in automotive vehicles.

selector [MECH ENG] 1. The part of the gearshift in an automotive transmission that selects the required gearshift bar. 2. The lever with which a driver operates an automatic gearshift.

self-centering chuck [MECH ENG] A drill chuck that, when closed, automatically positions the drill rod in the center of the drive rod of a diamond-drill swivel head.

self-energizing brake [MECH ENG] A brake designed to reinforce the power applied to it, such as a hand brake.

self-excited vibration *See* self-induced vibration.

self-induced vibration [MECH] The vibration of a mechanical system resulting from conversion, within the system, of nonoscillatory excitation to oscillatory excitation. Also known as self-excited vibration.

self-locking nut [DES ENG] A nut having an inherent locking action, so that it cannot readily be loosened by vibration.

self-organizing function [CONT SYS] That level in the functional decomposition of a large-scale control system which modifies the modes of control action or the structure of the control system in response to changes in system objectives, contingency events, and so forth.

self-propelled [MECH ENG] Pertaining to a vehicle given motion by means of a self-contained motor.

self-starter [MECH ENG] An attachment for automatically starting an internal combustion engine.

self-tapping screw [DES ENG] A screw with a specially hardened thread that makes it possible for the screw to form its own internal thread in sheet metal and soft materials when driven into a hole. Also known as sheet-metal screw; tapping screw.

self-tuning regulator [CONT SYS] A type of adaptive control system composed of two loops, an inner loop which consists of the process and an ordinary linear feedback regulator, and an outer loop which is composed of a recursive parameter estimator and a design calculation, and which adjusts the parameters of the regulator. Abbreviated STR.

sellers hob [MECH ENG] A hob that turns on the centers of a lathe, the work being fed to it by the lathe carriage.

Selwood engine [MECH ENG] A revolving-block engine in which two curved pistons opposed 180° run in toroidal tracks, forcing the entire engine block to rotate.

semiautomatic transmission [MECH ENG] An automobile transmission that assists the driver to shift from one gear to another.

semiclosed-cycle gas turbine [MECH ENG] A heat engine in which a portion of the expanded gas is recirculated.

semidiesel engine [MECH ENG] **1.** An internal combustion engine of a type resembling the diesel engine in using heavy oil as fuel but employing a lower compression pressure and spraying it under pressure, against a hot (uncooled) surface or spot, or igniting it by the precombustion or supercompression of a portion of the charge in a separate member or uncooled portion of the combustion chamber. **2.** A true diesel engine that uses a means other than compressed air for fuel injection.

semifloating axle [MECH ENG] A supporting member in motor vehicles which carries torque and wheel loads at its outer end.

semimonocoque [AERO ENG] A fuselage structure in which longitudinal members (stringers) as well as rings or frames which run circumferentially around the fuselage reinforce the skin and help carry the stress. Also known as stiffened-shell fuselage.

sems [DES ENG] A preassembled screw and washer combination.

sensible heat *See* enthalpy.

sensible-heat factor [THERMO] The ratio of space sensible heat to space total heat; used for air-conditioning calculations. Abbreviated SHF.

sensible-heat flow [THERMO] The heat given up or absorbed by a body upon being cooled or heated, as the result of the body's ability to hold heat; excludes latent heats of fusion and vaporization.

sensitivity function [CONT SYS] The ratio of the fractional change in the system response of a feedback-compensated feedback control system to the fractional change in an open-loop parameter, for some specified parameter variation.

separation [AERO ENG] The action of a fallaway section or companion body as it casts off from the remaining body of a vehicle, or the action of the remaining body as it leaves a fallaway section behind it.

separation theorem [CONT SYS] A theorem in optimal control theory which states that the solution to the linear quadratic Gaussian problem separates into the optimal deterministic controller (that is, the optimal controller for the corresponding problem without noise) in which the state used is obtained as the output of an optimal state estimator.

separator *See* cage.

sepatrix [CONT SYS] A curve in the phase plane of a control system representing the solution to the equations of motion of the system which would cause the system to move to an unstable point.

series compensation *See* cascade compensation.

service brake [MECH ENG] The brake used for ordinary driving in an automotive vehicle; usually foot-operated.

service ceiling [AERO ENG] The height at which, under standard atmospheric conditions, an aircraft is unable to climb faster than a specified rate (100 feet or 30 meters per minute in the United States, Great Britain, and Canada).

servo *See* servomotor.

servo brake [MECH ENG] 1. A brake in which the motion of the vehicle is used to increase the pressure on one of the shoes. 2. A brake in which the force applied by the operator is augmented by a power-driven mechanism.

servolink [CONT SYS] A power amplifier, usually mechanical, by which signals at a low power level are made to operate control surfaces requiring relatively large power inputs, for example, a relay and motor-driven actuator.

servomechanism [CONT SYS] An automatic feedback control system for mechanical motion; it applies only to those systems in which the controlled quantity or output is mechanical position or one of its derivatives (velocity, acceleration, and so on). Also known as servo system.

servomotor [CONT SYS] The electric, hydraulic, or other type of motor that serves as the final control element in a servomechanism; it receives power from the amplifier element and drives the load with a linear or rotary motion. Also known as servo.

servo system *See* servomechanism.

sessile drop method [FL MECH] A method of measuring surface tension in which the depth and mass of a drop resting on a surface that it does not wet are measured; from this, the shape of the drop and, in turn, the surface tension are determined.

set *See* permanent set.

setback [MECH] The relative rearward movement of component parts in a projectile, missile, or fuse undergoing forward acceleration during its launching; these movements, and the setback force which causes them, are used to promote events which participate in the arming and eventual functioning of the fuse.

setback force [MECH] The rearward force of inertia which is created by the forward acceleration of a projectile or missile during its launching phase; the forces are directly proportional to the acceleration and mass of the parts being accelerated.

set bit [DES ENG] A bit insert with diamonds or other cutting media.

set forward [MECH] Relative forward movement of component parts which occurs in a projectile, missile, or bomb in flight when impact occurs; the effect is due to inertia and is opposite in direction to setback.

set forward force [MECH] The forward force of inertia which is created by the deceleration of a projectile, missile, or bomb when impact occurs; the forces are directly proportional to the deceleration and mass of the parts being decelerated. Also known as impact force.

set forward point [MECH] A point on the expected course of the target at which it is predicted the target will arrive at the end of the time of flight.

set hammer [DES ENG] 1. A hammer used as a shaping tool by blacksmiths. 2. A hollow-face tool used in setting rivets.

set point [CONT SYS] The value selected to be maintained by an automatic controller.

set pressure [MECH ENG] The inlet pressure at which a relief valve begins to open as required by the code or standard applicable to the pressure vessel to be protected.

set screw [DES ENG] A small headless machine screw, usually having a point at one end and a recessed hexagonal socket or a slot at the other end, used for such purposes as holding a knob or gear on a shaft.

setting angle [MECH ENG] The angle, usually 90°, between the straight portion of the tool shank of the machined portion of the work.

settling time *See* correction time.

settling velocity [FL MECH] The rate at which suspended solids subside and are deposited. Also known as fall velocity. [MECH] The velocity reached by a particle as it falls through a fluid, dependent on its size and shape, and the difference between its specific gravity and that of the settling medium; used to sort particles by grain size.

sewing machine [MECH ENG] A mechanism that stitches cloth, leather, book pages, or other material by means of a double-pointed or eye-pointed needle.

SFC *See* specific fuel consumption.

SH-34G *See* Sea Bat.

shackle [DES ENG] An open or closed link of various shapes with extended legs; each leg has a transverse hole to accommodate a pin, bolt, or the like, which may or may not be furnished.

shackle bolt [DES ENG] A cylindrically shaped metal bar for connecting the ends of a shackle.

shaft [MECH ENG] A cylindrical piece of metal used to carry rotating machine parts, such as pulleys and gears, to transmit power or motion.

shaft balancing [DES ENG] The process of redistributing the mass attached to a rotating body in order to reduce vibrations arising from centrifugal force. Also known as rotor balancing.

shaft coupling *See* coupling.

shaft hopper [MECH ENG] A hopper that feeds shafts or tubes to grinders, threaders, screw machines, and tube benders.

shaft horsepower [MECH ENG] The output power of an engine, motor, or other prime mover; or the input power to a compressor or pump.

shafting [MECH ENG] The cylindrical machine element used to transmit rotary motion and power from a driver to a driven element; for example, a steam turbine driving a ship's propeller.

shake table *See* vibration machine.

shaking screen [MECH ENG] A screen used in separating material into desired sizes; has an eccentric drive or an unbalanced rotating weight to produce shaking.

shank [DES ENG] 1. The end of a tool which fits into a drawing holder, as on a drill. 2. *See* bit blank.

shank-type cutter [DES ENG] A cutter having a shank to fit into the machine tool spindle or adapter.

shape coding [DES ENG] The use of special shapes for control knobs, to permit recognition and sometimes also position monitoring by sense of touch.

shape factor [FL MECH] The quotient of the area of a sphere equivalent to the volume of a solid particle divided by the actual surface of the particle; used in calculations of gas flow through beds of granular solids.

shaper [MECH ENG] A machine tool for cutting flat-on-flat, contoured surfaces by reciprocating a single-point tool across the workpiece.

shaping [MECH ENG] A machining process in which a reciprocating single-point tool moves over the work in straight, parallel lines to produce a flat surface.

shaping dies [MECH ENG] A set of dies for bending, pressing, or otherwise shaping a material to a desired form.

sharp V thread [DES ENG] A screw thread having a sharp crest and root; the included angle is usually 60°.

shattering [FL MECH] One theory to explain homogenization or globule fractionation in milk; it is the effect that occurs when whole milk under high velocity strikes a flat surface, such as an impact ring. [MECH] The breaking up into highly irregular, angular blocks of a very hard material that has been subjected to severe stresses.

shave hook [DES ENG] A plumber's or metalworker's tool composed of a sharp-edged steel plate on a shank; used for scraping metal.

shaving [MECH ENG] 1. Cutting off a thin layer from the surface of a workpiece. 2. Trimming uneven edges from stampings, forgings, and tubing.

shear [DES ENG] A cutting tool having two opposing blades between which a material is cut. [MECH] See shear strain.

shear angle [MECH ENG] The angle made by the shear plane with the work surface.

shear center See center of twist.

shear diagram [MECH] A diagram in which the shear at every point along a beam is plotted as an ordinate.

shear drag See shear resistance.

shear fracture [MECH] A fracture resulting from shear stress.

shearing [MECH ENG] Separation of material by the cutting action of shears.

shearing die [MECH ENG] A die with a punch for shearing the work from the stock.

shearing instability See Helmholtz instability.

shearing machine [MECH ENG] A machine for cutting cloth or bars, sheets, or plates of metal or other material.

shearing punch [MECH ENG] A punch that cuts material by shearing it, with minimal crushing effect.

shearing stress [MECH] A stress in which the material on one side of a surface pushes on the material on the other side of the surface with a force which is parallel to the surface. Also known as shear stress; tangential stress.

shearing tool [DES ENG] A cutting tool (for a lathe, for example) with a considerable angle between its face and a line perpendicular to the surface being cut.

shear modulus See modulus of elasticity in shear.

shear pin [DES ENG] 1. A pin or wire provided in a fuse design to hold parts in a fixed relationship until forces are exerted on one or more of the parts which cause shearing of the pin or wire; the shearing is usually accomplished by setback or set forward (impact) forces; the shear member may be augmented during transportation by an additional safety device. 2. In a propellant-actuated device, a locking member which is released by shearing. 3. In a power train, such as a winch, any pin, as through a gear and shaft, which is designed to fail at a predetermined force in order to protect a mechanism.

shear plane [MECH] A confined zone along which fracture occurs in metal cutting.

shear rate [FL MECH] The relative velocities in laminar flow of parallel adjacent layers of a fluid body under shear force.

shear resistance [FL MECH] A tangential stress caused by fluid viscosity and taking place along a boundary of a flow in the tangential direction of local motion. Also known as shear drag.

shear spinning [MECH ENG] A sheet-metal-forming process which forms parts with rotational symmetry over a mandrel with the use of a tool or roller in which deformation is carried out with a roller in such a manner that the diameter of the original blank does not change but the thickness of the part decreases by an amount dependent on the mandrel angle.

shear strain [MECH] Also known as shear. 1. A deformation of a solid body in which a plane in the body is displaced parallel to itself relative to parallel planes in the body; quantitatively, it is the displacement of any plane relative to a second plane, divided by the perpendicular distance between planes. 2. The force causing such deformation.

shear strength [MECH] 1. The maximum shear stress which a material can withstand without rupture. 2. The ability of a material to withstand shear stress.

shear stress *See* shearing stress.

shear thickening [FL MECH] Viscosity increase of non-Newtonian fluids (for example, complex polymers, proteins, protoplasm) that undergo viscosity increases under conditions of shear stress (that is, viscometric flow).

shear thinning [FL MECH] Viscosity reduction of non-Newtonian fluids (for example, polymers and their solutions, most slurries and suspensions, lube oils with viscosity-index improvers) that undergo viscosity reductions under conditions of shear stress (that is, viscometric flow).

shear-viscosity function [FL MECH] The expression of the viscometric flow of a purely viscous, non-Newtonian fluid in terms of velocity gradient and shear stress of the flowing fluid.

shear wave [MECH] A wave that causes an element of an elastic medium to change its shape without changing its volume. Also known as rotational wave.

sheave [DES ENG] A grooved wheel or pulley.

SHED *See* solar heat exchanger drive.

sheepsfoot roller [DES ENG] A cylindrical steel drum to which knob-headed spikes are fastened; used for compacting earth.

sheepskin wheel [DES ENG] A polishing wheel made of sheepskin disks or wedges either quilted or glued together.

sheet cavitation [FL MECH] A type of cavitation in which cavities form on a solid boundary and remain attached as long as the conditions that led to their formation remain unaltered. Also known as steady-state cavitation.

shell [DES ENG] 1. The case of a pulley block. 2. A thin hollow cylinder. 3. A hollow hemispherical structure. 4. The outer wall of a vessel or tank.

shellac wheel [DES ENG] A grinding wheel having the abrasive bonded with shellac.

shell clearance [DES ENG] The difference between the outside diameter of a bit or core barrel and the outside set or gage diameter of a reaming shell.

shell pump [MECH ENG] A simple pump for removing wet sand or mud; consists of a hollow cylinder with a ball or clack valve at the bottom.

shell reamer [DES ENG] A machine reamer consisting of two parts, the arbor and the replaceable reamer, with straight or spiral flutes; designed as a sizing or finishing reamer.

SHF *See* sensible-heat factor.

shift [MECH ENG] To change the ratio of the driving to the driven gears to obtain the desired rotational speed or to avoid overloading and stalling an engine or a motor.

shimmy [MECH] Excessive vibration of the front wheels of a wheeled vehicle causing a jerking motion of the steering wheel.

shingle lap [DES ENG] A lap joint of tapered sections, the bottom of each section overlapping the top of the section below it.

shingle nail [DES ENG] A nail about a half to a full gage thicker than a common nail of the same length.

ship auger [DES ENG] An auger consisting of a spiral body having a single cutting edge, with or without a screw; there is no spur at the outer end of the cutting edge.

shock [MECH] A pulse or transient motion or force lasting thousandths to tenths of a second which is capable of exciting mechanical resonances; for example, a blast produced by explosives.

shock absorber [MECH ENG] A spring, a dashpot, or a combination of the two, arranged to minimize the acceleration of the mass of a mechanism or portion thereof with respect to its frame or support.

shock isolation [MECH ENG] The application of isolators to alleviate the effects of shock on a mechanical device or system.

shock mount [MECH ENG] A mount used with sensitive equipment to reduce or prevent transmission of shock motion to the equipment.

shock strut [AERO ENG] The primary working part of any landing gear, which supplies the force as the airplane sinks toward the ground, turning the flight path from one intersecting the ground to one parallel to the ground.

shock tube [FL MECH] A long tube divided into two parts by a diaphragm; the volume on one side of the diaphragm constitutes the compression chamber, the other side is the expansion chamber; a high pressure is developed by suitable means in the compression chamber, and the diaphragm ruptured; the shock wave produced in the expansion chamber can be used for the calibration of air blast gages, or the chamber can be instrumented for the study of the characteristics of the shock wave.

shoe [MECH ENG] 1. A metal block used as a form or support in various bending operations. 2. A replaceable piece used to break rock in certain crushing machines. 3. *See* brake shoe.

shoe brake [MECH ENG] A type of brake in which friction is applied by a long shoe, extending over a large portion of the rotating drum; the shoe may be external or internal to the drum.

short-baseline system [AERO ENG] A trajectory measuring system using a baseline the length of which is very small compared with the distance of the object being tracked.

shortcoming [DES ENG] An imperfection or malfunction occurring during the life cycle of equipment, which should be reported and which must be corrected to increase efficiency and to render the equipment completely serviceable.

short takeoff and landing [AERO ENG] The ability of an aircraft to clear a 50-foot (15-meter) obstacle within 1500 feet (450 meters) of commencing takeoff, or in landing, to stop within 1500 feet after passing over a 50-foot obstacle. Abbreviated STOL.

short ton *See* ton.

shot [AERO ENG] An act or instance of firing a rocket, especially from the earth's surface.

shot bit [DES ENG] A short length of heavy-wall steel tubing with diagonal slots cut in the flat-faced bottom edge.

shot feed [MECH ENG] A device to introduce chilled-steel shot, at a uniform rate and in the proper quantities, into the circulating fluid flowing downward through the rods or pipe connected to the core barrel and bit of a shot drill.

shothole drill [MECH ENG] A rotary or churn drill for drilling shotholes.

shoulder [DES ENG] The portion of a shaft, a stepped object, or a flanged object that shows an increase of diameter.

shoulder screw [DES ENG] A screw with an unthreaded cylindrical section, or shoulder, between threads and screwhead; the shoulder is larger in diameter than the threaded section and provides an axis around which close-fitting moving parts operate.

shovel [DES ENG] A hand tool having a flattened scoop at the end of a long handle for moving soil, aggregate, cement, or other similar material. [MECH ENG] A mechanical excavator.

shovel dozer *See* tractor loader.

shovel loader [MECH ENG] A loading machine mounted on wheels, with a bucket hinged to the chassis which scoops up loose material, elevates it, and discharges it behind the machine.

shrink fit [DES ENG] A tight interference fit between mating parts made by shrinking-on, that is, by heating the outer member to expand the bore for easy assembly and then cooling so that the outer member contracts.

shrink forming [DES ENG] Forming metal wherein the piece undergoes shrinkage during cooling following the application of heat, cold upset, or pressure.

shrink ring [DES ENG] A heated ring placed on an assembly of parts, which on subsequent cooling fixes them in position by contraction.

shut height [MECH ENG] The distance in a press between the bottom of the slide and the top of the bed, indicating the maximum die height that can be accommodated.

shutoff [AERO ENG] In rocket propulsion, the intentional termination of burning by command from the ground or from a self-contained guidance system.

shutoff head [MECH ENG] The pressure developed in a centrifugal or axial flow pump when there is zero flow through the system.

shuttle [MECH ENG] A back-and-forth motion of a machine which continues to face in one direction.

shuttle conveyor [MECH ENG] Any conveyor in a self-contained structure movable in a defined path parallel to the flow of the material.

Siacci method [MECH] An accurate and useful method for calculation of trajectories of high-velocity missiles with low quadrant angles of departure; basic assumptions are that the atmospheric density anywhere on the trajectory is approximately constant, and the angle of departure is less than about 15°.

sickle [DES ENG] A hand tool consisting of a hooked metal blade with a short handle, used for cutting grain or other agricultural products.

side direction [MECH] In stress analysis, the direction perpendicular to the plane of symmetry of an object.

sidehill bit [DES ENG] A drill bit which is set off-center so that it cuts a hole of larger diameter than that of the bit.

side milling [MECH ENG] Milling with a side-milling cutter to machine one vertical surface.

side-milling cutter [DES ENG] A milling cutter with teeth on one or both sides as well as around the periphery.

side rake [MECH ENG] The angle between the tool face and a reference plane for a single-point turning tool.

side relief angle [DES ENG] The angle that the portion of the flanks of a cutting tool below the cutting edge makes with a plane normal to the base.

side rod [MECH ENG] 1. A rod linking the crankpins of two adjoining driving wheels on the same side of a locomotive; distributes power from the main rod to the driving wheels. 2. One of the rods linking the piston-rod crossheads and the side levers of a side-lever engine.

sieve mesh [DES ENG] The standard opening in sieve or screen, defined by four boundary wires (warp and woof); the laboratory mesh is square and is defined by the shortest distance between two parallel wires as regards aperture (quoted in micrometers or millimeters), and by the number of parallel wires per linear inch as regards mesh; 60-mesh equals 60 wires per inch.

sigma function [THERMO] A property of a mixture of air and water vapor, equal to the difference between the enthalpy and the product of the specific humidity and the enthalpy of water (liquid) at the thermodynamic wet-bulb temperature; it is constant for constant barometric pressure and thermodynamic wet-bulb temperature.

silent stock support [MECH ENG] A flexible metal guide tube in which the stock tube of an automatic screw machine rotates; it is covered with a casing which deadens sound and prevents transfer of noise and vibration.

silicate grinding wheel [DES ENG] A mild-acting grinding wheel where the abrasive grain is bonded with sodium silicate and fillers.

silo [AERO ENG] A missile shelter that consists of a hardened vertical hole in the ground with facilities either for lifting the missile to a launch position, or for direct launch from the shelter.

simple engine [MECH ENG] An engine (such as a steam engine) in which expansion occurs in a single phase, after which the working fluid is exhausted.

simple harmonic motion *See* harmonic motion.

simple machine [MECH ENG] Any of several elementary machines, one or more being incorporated in every mechanical machine; usually, only the lever, wheel and axle, pulley (or block and tackle), inclined plane, and screw are included, although the gear drive and hydraulic press may also be considered simple machines.

simple pendulum [MECH] A device consisting of a small, massive body suspended by an inextensible object of negligible mass from a fixed horizontal axis about which the body and suspension are free to rotate.

simplex pump [MECH ENG] A pump with only one steam cylinder and one water cylinder.

simultaneity [MECH] Two events have simultaneity, relative to an observer, if they take place at the same time according to a clock which is fixed relative to the observer.

sine bar [DES ENG] A device consisting of a steel straight edge with two cylinders of equal diameter attached near the ends with their centers equidistant from the straightedge; used to measure angles accurately and to lay out work at a desired angle in relationship to a surface.

singing [CONT SYS] An undesired, self-sustained oscillation in a system or component, at a frequency in or above the passband of the system or component; generally due to excessive positive feedback.

singing margin [CONT SYS] The difference in level, usually expressed in decibels, between the singing point and the operating gain of a system or component.

singing point [CONT SYS] The minimum value of gain of a system or component that will result in singing.

single acting [MECH ENG] Acting in one direction only, as a single-acting plunger, or a single-acting engine (admitting the working fluid on one side of the piston only).

single-action press [MECH ENG] A press having a single slide.

single-block brake [MECH ENG] A friction brake consisting of a short block fitted to the contour of a wheel or drum and pressed up against the surface by means of a lever on a fulcrum; used on railroad cars.

single-cut file [DES ENG] A file with one set of parallel teeth, extending diagonally across the face of the file.

single-degree-of-freedom gyro [MECH] A gyro the spin reference axis of which is free to rotate about only one of the orthogonal axes, such as the input or output axis.

single-layer bit *See* surface-set bit.

single-loop feedback [CONT SYS] A system in which feedback may occur through only one electrical path.

single-loop servomechanism [CONT SYS] A servomechanism which has only one feedback loop. Also known as servo loop.

single-piece milling [MECH ENG] A milling method whereby one part is held and milled in one machine cycle.

single-stage compressor [MECH ENG] A machine that effects overall compression of a gas or vapor from suction to discharge conditions without any sequential multiplicity of elements, such as cylinders or rotors.

single-stage pump [MECH ENG] A pump in which the head is developed by a single impeller.

single-stage rocket [AERO ENG] A rocket or rocket missile to which the total thrust is imparted in a single phase, by either a single or multiple thrust unit.

single thread [DES ENG] A screw thread having a single helix in which the lead and pitch are equal.

singular arc [CONT SYS] In an optimal control problem, that portion of the optimal trajectory in which the Hamiltonian is not an explicit function of the control inputs, requiring higher-order necessary conditions to be applied in the process of solution.

sink flow [FL MECH] **1.** In three-dimensional flow, a point into which fluid flows uniformly from all directions. **2.** In two-dimensional flow, a straight line into which fluid flows uniformly from all directions at right angles to the line.

sinter setting *See* mechanical setting.

sister hook [DES ENG] **1.** Either of a pair of hooks which can be fitted together to form a closed ring. **2.** A pair of such hooks.

sixth-power law [FL MECH] A law stating that the size of particles that can be carried by a stream is proportional to the sixth power of its velocity.

sixty degrees Fahrenheit British thermal unit *See* British thermal unit.

size block *See* gage block.

size dimension [DES ENG] In dimensioning, a specified value of a diameter, width, length, or other geometrical characteristic directly related to the size of an object.

size reduction [MECH ENG] The breaking of large pieces of coal, ore, or stone by a primary breaker, or of small pieces by grinding equipment.

sizing [MECH ENG] A finishing operation to correct surfaces and shapes to meet specified dimensions and tolerances.

sizing screen [DES ENG] A mesh sheet with standard-size apertures used to separate granular material into classes according to size; the Tyler standard screen is an example.

skew   [MECH ENG]   Gearing whose shafts are neither interesecting nor parallel.

skew level gear   [DES ENG]   A level gear whose axes are not in the same place.

skid   [AERO ENG]   The metal bar or runner used as part of the landing gear of helicopters and planes.   [MECH ENG]   A brake for a power machine.

skin   [AERO ENG]   The covering of a body, such as the covering of a fuselage, a wing, a hull, or an entire aircraft.

skin friction   [FL MECH]   A type of friction force which exists at the surface of a solid body immersed in a much larger volume of fluid which is in motion relative to the body.

skip   *See* skip hoist.

skip hoist   [MECH ENG]   A basket, bucket, or open car mounted vertically or on an incline on wheels, rails, or shafts and hoisted by a cable; used to raise materials. Also known as skip.

skip trajectory   [MECH]   A trajectory made up of ballistic phases alternating with skipping phases; one of the basic trajectories for the unpowered portion of the flight of a reentry vehicle or spacecraft reentering earth's atmosphere.

skip vehicle   [AERO ENG]   A reentry body which climbs after striking the sensible atmosphere in order to cool the body and to increase its range.

skiving   [MECH ENG]   1. Removal of material in thin layers or chips with a high degree of shear or slippage of the cutting tool. 2. A machining operation in which the cut is made with a form tool with its face at an angle allowing the cutting edge to progress from one end of the work to the other as the tool feeds tangentially past ten rotating workpieces.

Skyhawk   [AERO ENG]   A United States single-engine, turbojet attack aircraft designed to operate from aircraft carriers, and capable of delivering nuclear or nonnuclear weapons, providing troop support, or conducting reconnaissance missions; it can act as a tanker, and can itself be air refueled; it possesses a limited all-weather attack capability, and can operate from short, unprepared fields.

skyhook balloon   [AERO ENG]   A large plastic constant-level balloon for duration flying at very high altitudes, used for determining wind fields and measuring upper-atmospheric parameters.

Skylab   [AERO ENG]   The first United States space station, attended by three separate flight crews of three astronauts each in 1973 and 1974.

Skyraider   [AERO ENG]   A United States single reciprocating-engine, general-purpose attack aircraft designed to operate from aircraft carriers; it is capable of relatively long-range, low-level nuclear and nonnuclear weapons delivery, minelaying, reconnaissance, torpedo delivery, and troop support. Designated A-1.

Skyray   [AERO ENG]   A United States single-engine, single-pilot, supersonic, limited all-weather jet fighter designed for operating from aircraft carriers for interception and destruction of enemy aircraft; armament includes the Sidewinder. Designated F-6.

Skywarrior   [AERO ENG]   A United States twin-engine, turbojet, tactical, all-weather attack aircraft designed to operate from aircraft carriers, and capable of delivering nuclear or nonnuclear weapons, conducting reconnaissance, or minelaying missions; its range can be extended by in-flight refueling, and it has a crew of four. Designated A-3.

slabbing cutter   [MECH ENG]   A face-milling cutter used to make wide, rough cuts.

slab cutter   *See* plain milling cutter.

slackline cableway   [MECH ENG]   A machine, widely used in sand-and-gravel plants, employing an open-ended dragline bucket suspended from a carrier that runs upon a track cable, which can dig, elevate, and convey materials in one continuous operation.

slant depth [DES ENG] The distance between the crest and root of a screw thread measured along the angle forming the flank of the thread.

slat [AERO ENG] A movable auxiliary airfoil running along the leading edge of a wing, remaining against the leading edge in normal flight conditions, but lifting away from the wing to form a slot at certain angles of attack.

slat conveyor [MECH ENG] A conveyor consisting of horizontal slats on an endless chain.

sledgehammer [DES ENG] A large heavy hammer that is usually wielded with two hands; used for driving stakes or breaking stone.

sleeve bearing [MECH ENG] A machine bearing in which the shaft turns and is lubricated by a sleeve.

sleeve coupling [DES ENG] A hollow cylinder which fits over the ends of two shafts or pipes, thereby joining them.

sleeve valve [MECH ENG] An admission and exhaust valve on an internal-combustion engine consisting of one or two hollow sleeves that fit around the inside of the cylinder and move with the piston so that their openings align with the inlet and exhaust ports in the cylinder at proper stages in the cycle.

slenderness ratio [AERO ENG] A configuration factor expressing the ratio of a missile's length to its diameter.

slide [MECH ENG] The main reciprocating member of a mechanical press, guided in a press frame, to which the punch or upper die is fastened.

slider coupling [MECH ENG] A device for connecting shafts that are laterally misaligned. Also known as double-slider coupling; Oldham coupling.

slide rest [MECH ENG] An adjustable slide for holding a cutting tool, as on an engine lathe.

slide valve [MECH ENG] A sliding mechanism to cover and uncover ports for the admission of fluid, as in some steam engines.

sliding-block linkage [MECH ENG] A mechanism in which a crank and sliding block serve to convert rotary motion into translation, or vice versa.

sliding-chain conveyor [MECH ENG] A conveying machine to handle cases, cans, pipes, or similar products on the plain or modified links of a set of parallel chains.

sliding fit [DES ENG] A fit between two parts that slide together.

sliding friction [MECH] Rubbing of bodies in sliding contact.

sliding gear [DES ENG] A change gear in which speed changes are made by sliding gears along their axes, so as to place them in or out of mesh.

sliding-gear transmission [MECH ENG] A transmission system utilizing a pair of sliding gears.

sliding pair [MECH ENG] Two adjacent links, one of which is constrained to move in a particular path with respect to the other; the lower, or closed, pair is completely constrained by the design of the links of the pair.

sliding vector [MECH] A vector whose direction and line of application are prescribed, but whose point of application is not prescribed.

slip [FL MECH] The difference between the velocity of a solid surface and the mean velocity of a fluid at a point just outside the surface.

slip flow [FL MECH] A situation in which the mean free path of a gas is between 1 and 65% of the channel diameter; the gas layer next to the channel wall assumes a velocity of slip past the liquid, known as slip flow.

slip friction clutch [MECH ENG] A friction clutch designed to slip when too much power is applied to it.

slipper brake [MECH ENG] 1. A plate placed against a moving part to slow or stop it. 2. A plate applied to the wheel of a vehicle or to the track roadway to slow or stop the vehicle.

slip ratio [MECH ENG] For a screw propeller, relates the actual advance to the theoretic advance determined by pitch and spin.

slip velocity [FL MECH] The difference in velocities between liquids and solids (or gases and liquids) in the vertical flow of two-phase mixtures through a pipe because of the slip between the two phases.

slit [DES ENG] A long, narrow opening through which radiation or streams of particles enter or leave certain instruments.

slitter [MECH ENG] A synchronized feeder-knife variation of a rotary cutter; used for precision cutting of sheet material, such as metal, rubber, plastics, or paper, into strips.

slitting [MECH ENG] The passing of sheet or strip material (metal, plastic, paper, or cloth) through rotary knives.

slope conveyor [MECH ENG] A troughed belt conveyor used for transporting material on steep grades.

slope deviation [AERO ENG] The difference between planned and actual slopes of aircraft travel, expressed in either angular or linear measurement.

slope of fall [MECH] Ratio between the drop of a projectile and its horizontal movement; tangent of the angle of fall.

slot [AERO ENG] 1. An air gap between a wing and the length of a slat or certain other auxiliary airfoils, the gap providing space for airflow or room for the auxiliary airfoil to be depressed in such a manner as to make for smooth air passage on the surface. 2. Any of certain narrow apertures made through a wing to improve aerodynamic characteristics. [DES ENG] A narrow, vertical opening.

slotted-head screw [DES ENG] A screw fastener with a single groove across the diameter of the head.

slotted nut [DES ENG] A regular hexagon nut with slots cut across the flats of the hexagon so that a cotter pin or safety wire can hold it in place.

slotter [MECH ENG] A machine tool used for making a mortise or shaping the sides of an aperture.

slotting [MECH ENG] Cutting a mortise or a similar narrow aperture in a material using a machine with a vertically reciprocating tool.

slotting machine [MECH ENG] A vertically reciprocating planing machine, used for making mortises and for shaping the sides of openings.

slot washer [DES ENG] 1. A lock washer with an indentation on its edge through which a nail or screw can be driven to hold it in place. 2. A washer with a slot extending from its edge to the center hole to allow the washer to be removed without first removing the bolt.

sludge pump *See* sand pump.

slug [MECH] A unit of mass in the British gravitational system of units, equal to the mass which experiences an acceleration of 1 foot per second per second when a force of 1 pound acts on it; equal to approximately 32.1740 pound mass or 14.5939 kilograms. Also known as geepound.

slug bit *See* insert bit.

slug flow *See* piston flow.

small calorie *See* calorie.

Smith-McIntyre sampler [MECH ENG] A device for taking samples of sediment from the ocean bottom; the digging and hoisting mechanisms are independent: the digging bucket is forced into the sediment before the hoisting action occurs.

smokebox [MECH ENG] A chamber external to a boiler for trapping the unburned products of combustion.

smoke technique [FL MECH] A technique used to measure only very-low-speed air velocity; smoke enables the fluid motion to be observed with the eye, and the smoke is timed over a measured distance along an airway of constant cross section to determine the velocity of flow.

smoothing mill [MECH ENG] A revolving stone wheel used to cut and bevel glass or stone.

smoothing plane [DES ENG] A finely set hand tool, usually 5.5–10 inches (14.0–25.4 centimeters) long, for finishing small areas on wood.

snagging [MECH ENG] Removing surplus metal or large surface defects by using a grinding wheel.

snap fastener [DES ENG] A fastener consisting of a ball on one edge of an article that fits in a socket on an opposed edge, and used to hold edges together, such as those of a garment.

snap gage [DES ENG] A device with two flat, parallel surfaces spaced to control one limit of tolerance of an outside diameter or a length.

snap hook *See* spring hook.

snap ring [DES ENG] A form of spring used as a fastener; the ring is elastically deformed, put in place, and allowed to snap back toward its unstressed position into a groove or recess.

snatch block [DES ENG] A pulley frame or sheave with an eye through which lashing can be passed to fasten it to a scaffold or pole.

snow blower [MECH ENG] A machine that removes snow from a road surface or pavement using a screw-type blade to push the snow into the machine and from which it is ejected at some distance.

snowplow [MECH ENG] A device for clearing away snow, as from a road or railway track.

snubber [MECH ENG] A mechanical device consisting essentially of a drum, spring, and friction band, connected between axle and frame, in order to slow the recoil of the spring and reduce jolting.

soar [AERO ENG] To fly without loss of altitude, using no motive power other than updrafts in the atmosphere.

socket-head screw [DES ENG] A screw fastener with a geometric recess in the head into which an appropriate wrench is inserted for driving and turning, with consequent improved nontamperability.

socket wrench [DES ENG] A wrench with a socket to fit the head of a bolt or a nut.

soft landing [AERO ENG] The act of landing on the surface of a planet or moon without damage to any portion of the vehicle or payload, except possibly the landing gear.

solar engine [MECH ENG] An engine which converts thermal energy from the sun into electrical, mechanical, or refrigeration energy; may be used as a method of spacecraft propulsion, either directly by photon pressure on huge solar sails, or indirectly from solar cells or from a reflector-boiler combination used to heat a fluid.

solar heat exchanger drive [AERO ENG] A proposed method of spacecraft propulsion in which solar radiation is focused on an area occupied by a boiler to heat a working fluid that is expelled to produce thrust directly. Abbreviated SHED.

solar heating [MECH ENG] The conversion of solar radiation into heat for technological, comfort-heating, and cooking purposes.

solar power [MECH ENG] The conversion of the energy of the sun's radiation to useful work.

solar probe [AERO ENG] A space probe whose trajectory passes near the sun so that instruments on board may detect and transmit back to earth data about the sun.

solar propulsion [AERO ENG] Spacecraft propulsion with a system composed of a type of solar engine.

solar rocket [AERO ENG] A rocket designed to carry instruments to measure and transmit parameters of the sun.

solar sail [AERO ENG] A surface of a highly polished material upon which solar light radiation exerts a pressure. Also known as photon sail.

solar satellite [AERO ENG] A space vehicle designed to enter into orbit about the sun. Also known as sun satellite.

solar turboelectric drive [AERO ENG] A proposed method of spacecraft propulsion in which solar radiation is focused on an area occupied by a boiler to heat a working fluid that drives a turbine generator system, producing electrical energy. Abbreviated STED.

solenoid brake [MECH ENG] A device that retards or arrests rotational motion by means of the magnetic resistance of a solenoid.

solenoid valve [MECH ENG] A valve actuated by a solenoid, for controlling the flow of gases or liquids in pipes.

solid box [MECH ENG] A solid, unadjustable ring bearing lined with babbitt metal, used on light machinery.

solid coupling [MECH ENG] A flanged-face or a compression-type coupling used to connect two shafts to make a permanent joint and usually designed to be capable of transmitting the full load capacity of the shaft; a solid coupling has no flexibility.

solid cutter [DES ENG] A cutter made of a single piece of material.

solid die [DES ENG] A one-piece screw-cutting tool with internal threads.

solid injection system [MECH ENG] A fuel injection system for a diesel engine in which a pump forces fuel through a fuel line and an atomizing nozzle into the combustion chamber.

solid-propellant rocket engine [AERO ENG] A rocket engine fueled with a solid propellant; such motors consist essentially of a combustion chamber containing the propellant, and a nozzle for the exhaust jet.

solid rocket [AERO ENG] A rocket that is propelled by a solid-propellant rocket engine.

solid shafting [MECH ENG] A solid round bar that supports a roller and wheel of a machine.

sonic barrier [AERO ENG] A popular term for the large increase in drag that acts upon an aircraft approaching acoustic velocity; the point at which the speed of sound is attained and existing subsonic and supersonic flow theories are rather indefinite. Also known as sound barrier.

sonic drilling [MECH ENG] The process of cutting or shaping materials with an abrasive slurry driven by a reciprocating tool attached to an audio-frequency electromechanical transducer and vibrating at sonic frequency.

sonic sifter [MECH ENG] A high-speed vibrating apparatus used in particle size analysis.

sortie [AERO ENG] An operational flight by one aircraft.

sound barrier *See* sonic barrier.

sounding rocket [AERO ENG] A rocket that carries aloft equipment for making observations of or from the upper atmosphere.

source [THERMO] A device that supplies heat.

source flow [FL MECH] 1. In three-dimensional flow, a point from which fluid issues at a uniform rate in all directions. 2. In two-dimensional flow, a line normal to the planes of flow, from which fluid flows uniformly in all directions at right angles to the line.

southbound node *See* descending node.

Soyuz Program [AERO ENG] A crewed space-flight program begun in 1967 by the Soviet Union.

space capsule [AERO ENG] A container, crewed or uncrewed, used for carrying out an experiment or operation in space.

space cone [MECH] The cone in space that is swept out by the instantaneous axis of a rigid body during Poinsot motion. Also known as herpolhode cone.

spacecraft [AERO ENG] Devices, crewed or uncrewed, which are designed to be placed into an orbit about the earth or into a trajectory to another celestial body. Also known as space ship; space vehicle.

spacecraft launching [AERO ENG] The setting into motion of a space vehicle with sufficient force to cause it to leave the earth's atmosphere.

spacecraft propulsion [AERO ENG] The use of rocket engines to accelerate space vehicles.

space flight [AERO ENG] Travel beyond the earth's sensible atmosphere; space flight may be an orbital flight about the earth or it may be a more extended flight beyond the earth into space.

space-flight trajectory [AERO ENG] The track or path taken by a spacecraft.

space mission [AERO ENG] A journey by a vehicle, manned or unmanned, beyond the earth's atmosphere, usually for the purpose of collecting scientific data.

spaceport [AERO ENG] An installation used to test and launch spacecraft.

space power system [AERO ENG] An on-board assemblage of equipment to generate and distribute electrical energy on satellites and spacecraft.

space probe [AERO ENG] An instrumented vehicle, the payload of a rocket-launching system designed specifically for flight missions to other planets or the moon and into deep space, as distinguished from earth-orbiting satellites.

space reconnaissance [AERO ENG] Reconnaissance of the surface of a planet from a space ship or satellite.

space research [AERO ENG] Research involving studies of all aspects of environmental conditions beyond the atmosphere of the earth.

space satellite [AERO ENG] A vehicle, crewed or uncrewed, for orbiting the earth.

space ship *See* spacecraft.

space shuttle [AERO ENG] A spacecraft designed to travel from the earth to a space station and to return to earth.

space simulator [AERO ENG] **1.** Any device which simulates one or more parameters of the space environment and which is used to test space systems or components. **2.** Specifically, a closed chamber capable of reproducing approximately the vacuum and normal environments of space.

space station [AERO ENG] An autonomous, permanent facility in space for the conduct of scientific and technological research, earth-oriented applications, and astronomical observations.

space technology [AERO ENG] The systematic application of engineering and scientific disciplines to the exploration and utilization of outer space.

space vehicle *See* spacecraft.

space walk [AERO ENG] The movement of an astronaut outside the protected environment of a spacecraft during a space flight; the astronaut wears a spacesuit.

spade [DES ENG] A shovellike implement with a flat oblong blade; used for turning soil by pushing against the blade with the foot.

spade bolt [DES ENG] A bolt having a spade-shaped flattened head with a transverse hole, used to fasten shielded coils, capacitors, and other components to a chassis.

spade drill [DES ENG] A drill consisting of three main parts: a cutting blade, a blade holder or shank, and a device, such as a screw, which fastens the blade to the holder; used for cutting holes over 1 inch (2.54 centimeters) in diameter.

spade lug [DES ENG] An open-ended flat termination for a wire lead, easily slipped under a terminal nut.

span [AERO ENG] **1.** The dimension of a craft measured between lateral extremities; the measure of this dimension. **2.** Specifically, the dimension of an airfoil from tip to tip measured in a straight line.

spanner [DES ENG] A wrench with a semicircular head having a projection or hole at one end.

spar [AERO ENG] A principal spanwise member of the structural framework of an airplane wing, aileron, stabilizer, and such; it may be of one-piece design or a fabricated section.

spark-ignition combustion cycle *See* Otto cycle.

spark-ignition engine [MECH ENG] An internal combustion engine in which an electrical discharge ignites the explosive mixture of fuel and air.

spark knock [MECH ENG] The knock produced in an internal combustion engine precedes the arrival of the piston at the top dead-center position.

spark lead [MECH ENG] The amount by which the spark precedes the arrival of the piston at its top (compression) dead-center position in the cylinder of an internal combustion engine.

spark-over-initiated discharge machining [MECH ENG] An electromachining process in which a potential is impressed between the tool (cathode) and workpiece (anode) which are separated by a dielectric material; a heavy discharge current flows through the ionized path when the applied potential is sufficient to cause rupture of the dielectric.

spear [DES ENG] A rodlike fishing tool having a barbed-hook end, used to recover rope, wire line, and other materials from a borehole.

special flight [AERO ENG] An air transport flight, other than a scheduled service, set up to move a specific load.

specific energy [THERMO] The internal energy of a substance per unit mass.

specific fuel consumption [MECH ENG] The weight flow rate of fuel required to produce a unit of power or thrust, for example, pounds per horsepower-hour. Abbreviated SFC. Also known as specific propellant consumption.

specific gravity [MECH] The ratio of the density of a material to the density of some standard material, such as water at a specified temperature, for example, 4°C or 60°F, or (for gases) air at standard conditions of pressure and temperature. Abbreviated sp gr. Also known as relative density.

specific heat [THERMO] 1. The ratio of the amount of heat required to raise a mass of material 1 degree in temperature to the amount of heat required to raise an equal mass of a reference substance, usually water, 1 degree in temperature; both measurements are made at a reference temperature, usually at constant pressure or constant volume. 2. The quantity of heat required to raise a unit mass of homogeneous material one degree in temperature in a specified way; it is assumed that during the process no phase or chemical change occurs.

specific impulse [AERO ENG] A performance parameter of a rocket propellant, expressed in seconds, equal to the thrust in pounds divided by the weight flow rate in pounds per second. Also known as specific thrust.

specific propellant consumption *See* specific fuel consumption.

specific speed [MECH ENG] A number, N_s, used to predict the performance of centrifugal and axial pumps or hydraulic turbines: for pumps, $N_s = N\sqrt{Q}/H^{3/4}$; for turbines, $N_s = N\sqrt{P/H^{5/4}}$, where N_s is specific speed, N is the rotational speed in revolutions per minute, Q is the rate of flow in gallons per minute, H is head in feet, and P is shaft horsepower.

specific thrust *See* specific impulse.

specific viscosity [FL MECH] The specific viscosity of a polymer is the relative viscosity of a polymer solution of known concentration minus 1; usually determined at low concentration of the polymer; for example, 0.5 gram per 100 milliliters of solution, or less.

specific volume [MECH] The volume of a substance per unit mass; it is the reciprocal of the density. Abbreviated sp vol.

specific weight [MECH] The weight per unit volume of a substance.

speed [MECH] The time rate of change of position of a body without regard to direction; in other words, the magnitude of the velocity vector.

speed cone [MECH ENG] A cone-shaped pulley, or a pulley composed of a series of pulleys of increasing diameter forming a stepped cone.

speed density metering [AERO ENG] A type of aircraft carburetion in which the fuel feed is regulated by the parameters of engine feed and intake manifold pressure.

speed lathe [MECH ENG] A light, pulley-driven lathe, usually without a carriage or back gears, used for work in which the tool is controlled by hand.

speed reducer [MECH ENG] A train of gears placed between a motor and the machinery which it will drive, to reduce the speed with which power is transmitted.

spherical pendulum [MECH] A simple pendulum mounted on a pivot so that its motion is not confined to a plane; the bob moves over a spherical surface.

spherical stress [MECH] The portion of the total stress that corresponds to an isotropic hydrostatic pressure; its stress tensor is the unit tensor multiplied by one-third the trace of the total stress tensor.

spike [DES ENG] A large nail, especially one longer than 3 inches (7.6 centimeters), and often of square section.

spin [MECH] Rotation of a body about its axis.

spin compensation [MECH] Overcoming or reducing the effect of projectile rotation in decreasing the penetrating capacity of the jet in shaped-charge ammunition.

spin-decelerating moment [MECH] A couple about the axis of the projectile, which diminishes spin.

spindle [DES ENG] A short, slender or tapered shaft.

Spinnbarkeit relaxation [FL MECH] A rheological effect illustrated by the pulling away of liquid threads when an object that has been immersed in a viscoelastic fluid is pulled out.

spinning [MECH ENG] Shaping and finishing sheet metal by rotating the workpiece over a mandrel and working it with a round-ended tool. Also known as metal spinning.

spinning machine [MECH ENG] 1. A machine that winds insulation on electric wire. 2. A machine that shapes metal hollow ware.

spin rocket [AERO ENG] A small rocket that imparts spin to a larger rocket vehicle or spacecraft.

spin stabilization [AERO ENG] Directional stability of a spacecraft obtained by the action of gyroscopic forces which result from spinning the body about its axis of symmetry.

spiral bevel gear [DES ENG] Bevel gear with curved, oblique teeth to provide gradual engagement and bring more teeth together at a given time than an equivalent straight bevel gear.

spiral chute [DES ENG] A gravity chute in the form of a continuous helical trough spiraled around a column for conveying materials to a lower level.

spiral conveyor See screw conveyor.

spiral gear [MECH ENG] A helical gear that transmits power from one shaft to another, nonparallel shaft.

spiral-jaw clutch [MECH ENG] A modification of the square-jaw clutch permitting gradual meshing of the mating faces, which have a helical section.

spiral pipe [DES ENG] Strong, lightweight steel pipe with a single continuous welded helical seam from end to end.

spiral spring [DES ENG] A spring bar or wire wound in an Archimedes spiral in a plane; each end is fastened to the force-applying link of the mechanism.

spiral welded pipe [DES ENG] A steel pipe made of long strips of steel plate fitted together to form helical seams, which are welded.

spirit level See level.

Spiroid gear [DES ENG] A trade name of the Illinois Tool Works, it resembles a hypoid-type bevel gear but performs like a worm mesh; used to connect skew shafts.

splashdown [AERO ENG] 1. The landing of a spacecraft or missile on water. 2. The moment of impact of a spacecraft on water.

spline [DES ENG] One of a number of equally spaced keys cut integral with a shaft, or similarly, keyways in a hubbed part; the mated pair permits the transmission of rotation or translatory motion along the axis of the shaft.

spline broach [MECH ENG] A broach for cutting straight-sided splines, or multiple keyways in holes.

splined shaft [DES ENG] A shaft with longitudinal gearlike ridges along its interior or exterior surface.

split-altitude profile [AERO ENG] Flight profile at two separate altitudes.

split barrel [DES ENG] A core barrel that is split lengthwise so that it can be taken apart and the sample removed.

split-barrel sampler [DES ENG] A drive-type soil sampler with a split barrel.

split bearing [DES ENG] A shaft bearing composed of two pieces bolted together.

split flap [AERO ENG] A hinged plate forming the rear upper or lower portion of an airfoil; the lower portion may be deflected downward to give increased lift and drag; the upper portion may be raised over a portion of the wing for the purpose of lateral control.

split link [DES ENG] A metal link in the shape of a two-turn helix pressed together.

split pin [DES ENG] A pin with a split at one end so that it can spread to hold it in place.

split-ring core lifter [DES ENG] A hardened steel ring having an open slit, an outside taper, and an inside or outside serrated surface; in its expanded state it allows the core to pass through it freely, but when the drill string is lifted, the outside taper surface slides downward into the bevel of the bit or reaming shell, causing the ring to contract and grip tightly the core which it surrounds. Also known as core catcher; core gripper; core lifter; ring lifter; split-ring lifter; spring lifter.

split-ring lifter *See* split-ring core lifter.

split-ring piston packing [MECH ENG] A metal ring mounted on a piston to prevent leakage along the cylinder wall.

split shovel [DES ENG] A shovel containing parallel troughs separated by slots; used for sampling ground ore.

spoiler [AERO ENG] A plate, series of plates, comb, tube, bar, or other device that projects into the airstream about a body to break up or spoil the smoothness of the flow, especially such a device that projects from the upper surface of an airfoil, giving an increased drag and a decreased lift.

spoke [DES ENG] A bar or rod radiating from the center of a wheel.

spontaneous process [THERMO] A thermodynamic process which takes place without the application of an external agency, because of the inherent properties of a system.

spool [MECH ENG] **1.** The drum of a hoist. **2.** The movable part of a slide-type hydraulic valve.

spool-type roller conveyor [MECH ENG] A type of roller conveyor in which the rolls are of conical or tapered shape with the diameter at the ends of the roll larger than that at the center.

spoon [DES ENG] A slender rod with a cup-shaped projection at right angles to the rod, used for scraping drillings out of a borehole.

spot drilling [MECH ENG] Drilling a small hole or indentation in the surface of a material to serve as a centering guide in later machining operations.

spot facing [MECH ENG] A finished circular surface around the top of a hole to seat a bolthead or washer, or to allow flush mounting of mating parts.

spot hover [AERO ENG] To remain stationary relative to a point on the ground while airborne.

spray angle [FL MECH] The angle formed by the cone of liquid leaving a nozzle orifice.

spray chamber [MECH ENG] A compartment in an air conditioner where humidification is conducted.

spray dryer [MECH ENG] A machine for drying an atomized mist by direct contact with hot gases.

spray gun [MECH ENG] An apparatus shaped like a gun which delivers an atomized mist of liquid.

spray nozzle [MECH ENG] A device in which a liquid is subdivided to form a stream (mist) of small drops.

spreader [MECH ENG] **1.** A tool used in sharpening machine drill bits. **2.** A machine which spreads dumped material with its blades.

spreader stoker [MECH ENG] A coal-burning system where mechanical feeders and distributing devices form a thin fuel bed on a traveling grate, intermittent-cleaning dump grate, or reciprocating continuous-cleaning grate.

spreading coefficient [THERMO] The work done in spreading one liquid over a unit area of another, equal to the surface tension of the stationary liquid, minus the surface tension of the spreading liquid, minus the interfacial tension between the liquids.

Sprengel pump [MECH ENG] An air pump that exhausts by trapping gases between drops of mercury in a tube.

spring [MECH ENG] An elastic, stressed, stored-energy machine element that, when released, will recover its basic form or position. Also known as mechanical spring.

spring bolt [DES ENG] A bolt which must be retracted by pressure and which is shot into place by a spring when the pressure is released.

spring clip [DES ENG] **1.** A U-shaped fastener used to attach a leaf spring to the axle of a vehicle. **2.** A clip that grips an inserted part under spring pressure; used for electrical connections.

spring collet [DES ENG] A bushing that surrounds and holds the end of the work in a machine tool; the bushing is slotted and tapered, and when the collet is slipped over it, the slot tends to close and the bushing thereby grips the work.

spring cotter [DES ENG] A cotter made of an elastic metal that has been bent double to form a split pin.

spring coupling [MECH ENG] A flexible coupling with resilient parts.

spring die [DES ENG] An adjustable die consisting of a hollow cylinder with internal cutting teeth, used for cutting screw threads.

spring hammer [MECH ENG] A machine-driven hammer actuated by a compressed spring or by compressed air.

spring hook [DES ENG] A hook closed at the end by a spring snap. Also known as snap hook.

spring-joint caliper [DES ENG] An outside or inside caliper having a heavy spring joining the legs together at the top; legs are opened and closed by a knurled nut.

spring lifter See split-ring core lifter.

spring-loaded regulator [MECH ENG] A pressure-regulator valve for pressure vessels or flow systems; the regulator is preloaded by a calibrated spring to open (or close) at the upper (or lower) limit of a preset pressure range.

spring modulus [MECH] The additional force necessary to deflect a spring an additional unit distance; if a certain spring has a modulus of 100 newtons per centimeter, a 100-newton weight will compress it 1 centimeter, a 200-newton weight 2 centimeters, and so on.

spring pin [MECH ENG] An iron rod which is mounted between spring and axle on a locomotive, and which maintains a regulated pressure on the axle.

spring stop-nut locking fastener [DES ENG] A locking fastener that functions by a spring action clamping down on the bolt.

sprocket [DES ENG] A tooth on the periphery of a wheel or cylinder to engage in the links of a chain, the perforations of a motion picture film, or other similar device.

sprocket chain [MECH ENG] A continuous chain which meshes with the teeth of a sprocket and thus can transmit mechanical power from one sprocket to another.

sprocket wheel [DES ENG] A wheel with teeth or cogs, used for a chain drive or to engage the blocks on a cable.

sprung axle [MECH ENG] A supporting member for carrying the rear wheels of an automobile.

sprung weight [MECH ENG] The weight of a vehicle which is carried by the springs, including the frame, radiator, engine, clutch, transmission, body, load, and so forth.

spud [DES ENG] **1.** A diamond-point drill bit. **2.** An offset type of fishing tool used to clear a space around tools stuck in a borehole. **3.** Any of various spade- or chisel-shaped tools or mechanical devices.

spur gear [DES ENG] A toothed wheel with radial teeth parallel to the axis.

Sputnik program [AERO ENG] A series of Soviet earth-orbiting satellites; Sputnik I, launched on October 4, 1957, was the first artificial satellite.

sp vol *See* specific volume.

sq *See* square.

square [MECH] Denotes a unit of area; if x is a unit of length, a square x is the area of a square whose sides have a length of $1x$: for example, a square meter, or a meter squared, is the area of a square whose sides have a length of 1 meter. Abbreviated sq.

square engine [MECH ENG] An engine in which the stroke is equal to the cylinder bore.

square-head bolt [DES ENG] A cylindrical threaded fastener with a square head.

square-jaw clutch [MECH ENG] A type of positive clutch consisting of two or more jaws of square section which mesh together when they are aligned.

square key [DES ENG] A machine key of square, usually uniform, but sometimes tapered, cross section.

square mesh [DES ENG] A wire-cloth textile mesh count that is the same in both directions.

square-nose bit *See* flat-face bit.

square thread [DES ENG] A screw thread having a square cross section; the width of the thread is equal to the pitch or distance between threads.

square wheel [DES ENG] A wheel with a flat spot on its rim.

squaring shear [MECH ENG] A machine tool consisting of one fixed cutting blade and another mounted on a reciprocating crosshead; used for cutting sheet metal or plate.

squeegee [DES ENG] A device consisting of a handle with a blade of rubber or leather set transversely at one end and used for spreading, pushing, or wiping liquids off or across a surface.

squeeze roll [MECH ENG] A roller designed to exert pressure on material passing between it and a similar roller.

SST *See* supersonic transport.

SSU *See* Saybolt Seconds Universal.

St *See* stoke; Stokes number.

stabilator [AERO ENG] A one-piece horizontal tail that is swept back and movable; movement is controlled by motion of the pilot's control stick; usually used in supersonic aircraft.

stability [CONT SYS] The property of a system for which any bounded input signal results in a bounded output signal. [FL MECH] The resistance to overturning or mixing in the water column, resulting from the presence of a positive (increasing downward) density gradient. [MECH] *See* dynamic stability.

stability augmentation system [AERO ENG] Automatic control devices which supplement a pilot's manipulation of the aircraft controls and are used to modify inherent aircraft handling qualities. Abbreviated SAS. Also known as stability augmentors.

stability augmentors *See* stability augmentation system.

stability criterion [CONT SYS] A condition which is necessary and sufficient for a system to be stable, such as the Nyquist criterion, or the condition that poles of the system's overall transmittance lie in the left half of the complex-frequency plane.

stability exchange principle [CONT SYS] In a linear system, which is either dynamically stable or unstable depending on the value of a parameter, the complex frequency varies with the parameter in such a way that its real and imaginary parts pass through zero simultaneously; the principle is often violated.

stability matrix *See* stiffness matrix.

stabilization *See* compensation.

stabilized feedback *See* negative feedback.

stabilized flight [AERO ENG] Maintenance of desired orientation in flight.

stabilized platform *See* stable platform.

stabilizer [AERO ENG] Any airfoil or any combination of airfoils considered as a single unit, the primary function of which is to give stability to an aircraft or missile.

stable platform [AERO ENG] A gimbal-mounted platform, usually containing gyros and accelerometers, the purpose of which is to maintain a desired orientation in inertial space independent of craft motion. Also known as stabilized platform.

stack effect [MECH ENG] The pressure difference between the confined hot gas in a chimney or stack and the cool outside air surrounding the outlet.

stacker [MECH ENG] A machine for lifting merchandise on a platform or fork and arranging it in tiers; operated by hand, or electric or hydraulic mechanisms.

stacker-reclaimer [MECH ENG] Equipment which transports and builds up material stockpiles, and recovers and transports material to processing plants.

stacking [AERO ENG] The holding pattern of aircraft awaiting their turn to approach and land at an airport.

stage [AERO ENG] A self-propelled separable element of a rocket vehicle or spacecraft.

staged crew [AERO ENG] An aircrew specifically positioned at an intermediate airfield to take over aircraft operating on an air route, thus relieving a complementary crew of flying fatigue and speeding up the traffic rate of the aircraft concerned.

stage loader *See* feeder conveyor.

stagger-tooth cutter [MECH ENG] Side-milling cutter with successive teeth having alternating helix angles.

staging [AERO ENG] The process or operation during the flight of a rocket vehicle whereby a full stage or half stage is disengaged from the remaining body and made free to decelerate or be propelled along its own flightpath.

stagnation point [FL MECH] A point in a field of flow about a body where the fluid particles have zero velocity with respect to the body.

stagnation pressure *See* dynamic pressure.

stagnation temperature *See* adiabatic recovery temperature.

stall [AERO ENG] **1.** The action or behavior of an airplane (or one of its airfoils) when by the separation of the airflow, as in the case of insufficient airspeed or of an excessive angle of attack, the airplane or airfoil tends to drop; the condition existing during this behavior. **2.** A flight performance in which an airplane is made to lose flying speed and to drop by pointing the nose steeply upward. **3.** An act or instance of stalling.

stall flutter [AERO ENG] A type of dynamic instability that takes place when the separation of flow around an airfoil occurs during the whole or part of each cycle of a flutter motion.

stalling angle of attack *See* critical angle of attack.

stalling Mach number [AERO ENG] The Mach number of an aircraft when the coefficient of lift of the aerodynamic surfaces is the maximum obtainable for the pressure altitude, true airspeed, and angle of attack under which the craft is operated.

stall warning indicator [AERO ENG] A device that determines the critical angle of attack for a given aircraft; usually operates from vane sensors, airflow sensors, tabs on leading edges of wings, and computing devices such as accelerometers or airspeed indicators.

stamping [MECH ENG] Almost any press operation including blanking, shearing, hot or cold forming, drawing, bending, and coining.

standard atmosphere *See* atmosphere.

standard ballistic conditions [MECH] A set of ballistic conditions arbitrarily assumed as standard for the computation of firing tables.

standard fit [DES ENG] A fit whose allowance and tolerance are standardized.

standard free energy increase [THERMO] The increase in Gibbs free energy in a chemical reaction, when both the reactants and the products of the reaction are in their standard states.

standard gage [DES ENG] A highly accurate gage used only as a standard for working gages.

standard gravity [MECH] A value of the acceleration of gravity equal to 9.80665 meters per second per second.

standard heat of formation [THERMO] The heat needed to produce one mole of a compound from its elements in their standard state.

standard hole [DES ENG] A hole with zero allowance plus a specified tolerance; fit allowance is provided for by the shaft in the hole.

standardization [DES ENG] The adoption of generally accepted uniform procedures, dimensions, materials, or parts that directly affect the design of a product or a facility.

standardized product [DES ENG] A product that conforms to specifications resulting from the same technical requirements.

standard load [DES ENG] A load which has been preplanned as to dimensions, weight, and balance, and designated by a number or some classification.

standard rate turn [AERO ENG] A turn in an aircraft in which heading changes at the rate of 3° per second.

standard shaft [DES ENG] A shaft with zero allowance minus a specified tolerance.

standard ton *See* ton.

standard trajectory [MECH] Path through the air that it is calculated a projectile will follow under given conditions of weather, position, and material, including the particular fuse, projectile, and propelling charge that are used; firing tables are based on standard trajectories.

standard wire rope [DES ENG] Wire rope made of six wire strands laid around a sisal core. Also known as hemp-core cable.

standing derrick *See* gin pole.

standing operating procedure [AERO ENG] A set of instructions covering those features of operations which lend themselves to a definite or standardized procedure without loss of effectiveness; the procedure is applicable unless prescribed otherwise in a particular case; thus, the flexibility necessary in special situations is retained.

Stanton diagram [FL MECH] The plot of the airflow friction coefficient against the Reynolds number.

Stanton number [THERMO] A dimensionless number used in the study of forced convection, equal to the heat-transfer coefficient of a fluid divided by the product of the specific heat at constant pressure, the fluid density, and the fluid velocity. Symbolized N_{St}. Also known as Margoulis number (M).

staple [DES ENG] A U-shaped loop of wire with points at both ends; used as a fastener.

star drill [DES ENG] A tool with a star-shaped point, used for drilling in stone or masonry.

Starfighter [AERO ENG] A United States supersonic, single-engine, turbojet-powered, tactical and air superiority fighter; the tactical version employs cannons or nuclear weapons for attack against surface targets, and is capable of providing close support for ground forces; the interceptor version employs Sidewinders or cannons.

Stark number *See* Stefan number.

Starlifter [AERO ENG] A United States large cargo transport powered by four turbofan engines, capable of intercontinental range with heavy payloads and airdrops.

starting friction *See* static friction.

starting resistance [MECH ENG] The force needed to produce an oil film on the journal bearings of a train when it is at a standstill.

starting taper [DES ENG] A slight end taper on a reamer to aid in starting.

start-to-leak pressure [MECH ENG] The amount of inlet pressure at which the first bubble occurs at the outlet of a safety relief valve with a resilient disk when the valve is subjected to an air test under a water seal.

state [CONT SYS] A minimum set of numbers which contain enough information about a system's history to enable its future behavior to be computed.

state equations [CONT SYS] Equations which express the state of a system and the output of a system at any time as a single valued function of the system's input at the same time and the state of the system at some fixed initial time.

state estimator *See* observer.

state feedback [CONT SYS] A class of feedback control laws in which the control inputs are explicit memoryless functions of the dynamical system state, that is, the control inputs at a given time t_a are determined by the values of the state variables at t_a and do not depend on the values of these variables at earlier times $t \geqslant t_a$.

state observer *See* observer.

state of strain [MECH] A complete description, including the six components of strain, of the deformation within a homogeneously deformed volume.

state of stress [MECH] A complete description, including the six components of stress, of a homogeneously stressed volume.

state parameter *See* thermodynamic function of state.

state space [CONT SYS] The set of all possible values of the state vector of a system.

state transition equation [CONT SYS] The equation satisfied by the $n \times n$ state transition matrix $\Phi(t, t_0)$: $\partial \Phi(t, t_0)/\partial t = A(t) \, \Phi(t, t_0)$, $\Phi(t_0, t_0) = I$; here I is the unit $n \times n$ matrix, and $A(t)$ is the $n \times n$ matrix which appears in the vector differential equation $dx(t)/dt = A(t)x(t)$ for the n-component state vector $x(t)$.

state transition matrix [CONT SYS] A matrix $\Phi(t, t_0)$ whose product with the state vector x at an initial time t_0 gives the state vector at a later time t; that is, $x(t) = \Phi(t, t_0)x(t_0)$.

state variable [CONT SYS] One of a minimum set of numbers which contain enough information about a system's history to enable computation of its future behavior. [THERMO] *See* thermodynamic function of state.

state vector [CONT SYS] A column vector whose components are the state variables of a system.

static equilibrium *See* equilibrium.

static firing [AERO ENG] The firing of a rocket engine in a hold-down position to measure thrust and to accomplish other tests.

static fluid column [FL MECH] An unchanging height of fluid in a vertical pipe, well bore, process vessel, or tank.

static friction [MECH] **1.** The force that resists the initiation of sliding motion of one body over the other with which it is in contact. **2.** The force required to move one of the bodies when they are at rest. Also known as starting friction.

static gearing ratio [AERO ENG] The ratio of the control-surface deflection in degrees to angular displacement of the missile which caused the deflection of the control surface.

static gel buildup [FL MECH] A method used to infer the degree of thixotropy of a fluid by viscometric measurement of its gel strength.

static head [FL MECH] Pressure of a fluid due to the head of fluid above some reference point.

static level [FL MECH] Elevation of the water level or a pressure surface at rest.

static line [AERO ENG] A line attached to a parachute pack and to a strop or anchor cable in an aircraft so that when the load is dropped the parachute is deployed automatically.

static load [MECH] A nonvarying load; the basal pressure exerted by the weight of a mass at rest, such as the load imposed on a drill bit by the weight of the drill-stem equipment or the pressure exerted on the rocks around an underground opening by the weight of the superimposed rocks. Also known as dead load.

static moment [MECH] **1.** A scalar quantity (such as area or mass) multiplied by the perpendicular distance from a point connected with the quantity (such as the centroid of the area or the center of mass) to a reference axis. **2.** The magnitude of some vector (such as force, momentum, or a directed line segment) multiplied by the length of a perpendicular dropped from the line of action of the vector to a reference point.

static pressure   [FL MECH] 1. The normal component of stress, the force per unit area, exerted across a surface moving with a fluid, especially across a surface which lies in the direction of fluid flow. 2. The average of the normal components of stress exerted across three mutually perpendicular surfaces moving with a fluid.

static reaction   [MECH] The force exerted on a body by other bodies which are keeping it in equilibrium.

statics   [MECH] The branch of mechanics which treats of force and force systems abstracted from matter, and of forces which act on bodies in equilibrium.

static test   [AERO ENG] In particular, a test of a rocket or other device in a stationary or hold-down position, either to verify structural design criteria, structural integrity, and the effects of limit loads, or to measure the thrust of a rocket engine.

stationary cone classifier   [MECH ENG] In a pulverizer directly feeding a coal furnace, a device which returns oversize coal to the pulverizing zone.

stationary engine   [MECH ENG] A permanently placed engine, as in a power house, factory, or mine.

stationary orbit   [AERO ENG] A circular, equatorial orbit in which the satellite revolves about the primary body at the angular rate at which the primary body rotates on its axis; from the primary body, the satellite thus appears to be stationary over a point on the primary body; a stationary orbit must be synchronous, but the reverse need not be true.

stationary satellite   [AERO ENG] A satellite in a stationary orbit.

station time   [AERO ENG] Time at which crews, passengers, and cargo are to be on board air transport and ready for the flight.

stator   [MECH ENG] A stationary machine part in or about which a rotor turns.

statute mile   *See* mile.

stave   [DES ENG] 1. A rung of a ladder. 2. Any of the narrow wooden strips or metal plates placed edge to edge to form the sides, top, or lining of a vessel or structure, such as a barrel.

staybolt   [DES ENG] A bolt with a thread along the entire length of the shaft; used to attach machine parts that are under pressure to separate.

stay time   [AERO ENG] In rocket engine usage, the average value of the time spent by each gas molecule or atom within the chamber volume.

steady flow   [FL MECH] Fluid flow in which all the conditions at any one point are constant with respect to time.

steady-state cavitation   *See* sheet cavitation.

steady-state conduction   [THERMO] Heat conduction in which the temperature and heat flow at each point does not change with time.

steady-state creep   *See* secondary creep.

steady-state error   [CONT SYS] The error that remains after transient conditions have disappeared in a control system.

steady-state vibration   [MECH] Vibration in which the velocity of each particle in the system is a continuous periodic quantity.

steam accumulator   [MECH ENG] A pressure vessel in which water is heated by steam during off-peak demand periods and regenerated as steam when needed.

steam attemperation [MECH ENG] The control of the maximum temperature of superheated steam by water injection or submerged cooling.

steam boiler [MECH ENG] A pressurized system in which water is vaporized to steam by heat transferred from a source of higher temperature, usually the products of combustion from burning fuels. Also known as steam generator.

steam condenser [MECH ENG] A device to maintain vacuum conditions on the exhaust of a steam prime mover by transfer of heat to circulating water or air at the lowest ambient temperature.

steam cycle *See* Rankine cycle.

steam drive [MECH ENG] Any device which uses power generated by the pressure of expanding steam to move a machine or a machine part.

steam dryer [MECH ENG] A device for separating liquid from vapor in a steam supply system.

steam engine [MECH ENG] A thermodynamic device for the conversion of heat in steam into work, generally in the form of a positive displacement, piston and cylinder mechanism.

steam-generating furnace *See* boiler furnace.

steam generator *See* steam boiler.

steam hammer [MECH ENG] A forging hammer in which the ram is raised, lowered, and operated by a steam cylinder.

steam-heated evaporator [MECH ENG] A structure using condensing steam as a heat source on one side of a heat-exchange surface to evaporate liquid from the other side.

steam heating [MECH ENG] A system that used steam as the medium for a comfort or process heating operation.

steam jacket [MECH ENG] A casing applied to the cylinders and heads of a steam engine, or other space, to keep the surfaces hot and dry.

steam-jet cycle [MECH ENG] A refrigeration cycle in which water is used as the refrigerant; high-velocity steam jets provide a high vacuum in the evaporator, causing the water to boil at low temperature and at the same time compressing the flashed vapor up to the condenser pressure level.

steam-jet ejector [MECH ENG] A fluid acceleration vacuum pump or compressor using the high velocity of a steam jet for entrainment.

steam locomotive [MECH ENG] A railway propulsion power plant using steam generally in a reciprocating, noncondensing engine.

steam nozzle [MECH ENG] A streamlined flow structure in which heat energy of steam is converted to the kinetic form.

steam point [THERMO] The boiling point of pure water whose isotopic composition is the same as that of sea water at standard atmospheric pressure; it is assigned a value of 100°C on the International Practical Temperature Scale of 1968.

steam pump [MECH ENG] A pump driven by steam acting on the coupled piston rod and plunger.

steam reheater [MECH ENG] A steam boiler component in which heat is added to intermediate-pressure steam, which has given up some of its energy in expansion through the high-pressure turbine.

steam roller [MECH ENG] A road roller driven by a steam engine.

steam separator [MECH ENG] A device for separating a mixture of the liquid and vapor phases of water. Also known as steam purifier.

steam shovel   [MECH ENG] A power shovel operated by steam.

steam superheater   [MECH ENG] A boiler component in which sensible heat is added to the steam after it has been evaporated from the liquid phase.

steam trap   [MECH ENG] A device which drains and removes condensate automatically from steam lines.

steam-tube dryer   [MECH ENG] Rotary dryer with steam-heated tubes running the full length of the cylinder and rotating with the dryer shell.

steam turbine   [MECH ENG] A prime mover for the conversion of heat energy of steam into work on a revolving shaft, utilizing fluid acceleration principles in jet and vane machinery.

STED   *See* solar turboelectric drive.

steel-cable conveyor belt   [DES ENG] A rubber conveyor belt in which the carcass is composed of a single plane of steel cables.

steel-clad rope   [DES ENG] A wire rope made from flat strips of steel wound helically around each of the six strands composing the rope.

Steelflex coupling   [MECH ENG] A flexible coupling made with two grooved steel hubs keyed to their respective shafts and connected by a specially tempered alloy-steel member called the grid.

steering arm   [MECH ENG] An arm that transmits turning motion from the steering wheel of an automotive vehicle to the drag link.

steering brake   [MECH ENG] Means of turning, stopping, or holding a tracked vehicle by braking the tracks individually.

steering gear   [MECH ENG] The mechanism, including gear train and linkage, for the directional control of a vehicle or ship.

steering wheel   [MECH ENG] A hand-operated wheel for controlling the direction of the wheels of an automotive vehicle or of the rudder of a ship.

Stefan number   [THERMO] A dimensionless number used in the study of radiant heat transfer, equal to the Stefan-Boltzmann constant times the cube of the temperature times the thickness of a layer divided by the layer's thermal conductivity. Symbolized St. Also known as Stark number (Sk).

Steiner's theorem   *See* parallel axis theorem.

stem correction   [THERMO] A correction which must be made in reading a thermometer in which part of the stem, and the thermometric fluid within it, is at a temperature which differs from the temperature being measured.

stem-winding   [MECH ENG] Pertaining to a timepiece that is wound by an internal mechanism turned by an external knob and stem (the winding button of a watch).

step bearing   [MECH ENG] A device which supports the bottom end of a vertical shaft. Also known as pivot bearing.

step-by-step system   [CONT SYS] A control system in which the drive motor moves in discrete steps when the input element is moved continuously.

step-climb profile   [AERO ENG] The aircraft climbs a specified number of feet whenever its weight reaches a predetermined amount, thus stepping to an optimum altitude as gross weight decreases.

step gage   [DES ENG] 1. A plug gage containing several cylindrical gages of increasing diameter mounted on the same axis. 2. A gage consisting of a body in which a blade slides perpendicularly; used to measure steps and shoulders.

stepped cone pulley   [DES ENG] A one-piece pulley with several diameters to engage transmission belts and thereby provide different speed ratios.

stepped gear wheel [DES ENG] A gear wheel containing two or more sets of teeth on the same rim, with adjacent sets slightly displaced to form a series of steps.

stepped screw [DES ENG] A screw from which sectors have been removed, the remaining screw surfaces forming steps.

step pulley [MECH ENG] A series of pulleys of various diameters combined in a single concentric unit and used to vary the velocity ratio of shafts. Also known as cone pulley.

step response [CONT SYS] The behavior of a system when its input signal is zero before a certain time and is equal to a constant nonzero value after this time.

step rocket *See* multistage rocket.

stère [MECH] A unit of volume equal to 1 cubic meter; it is used mainly in France, and in measuring timber volumes.

sterhydraulic [MECH ENG] Pertaining to a hydraulic press in which motion or pressure is produced by the introduction of a solid body into a cylinder filled with liquid.

stern attack [AERO ENG] In air intercept, an attack by an interceptor aircraft which terminates with a heading crossing angle of 45° or less.

sthène [MECH] The force which, when applied to a body whose mass is 1 metric ton, results in an acceleration of 1 meter per second per second; equal to 1000 newtons. Formerly known as funal.

stiction [MECH] Friction that tends to prevent relative motion between two movable parts at their null position.

stiffened-shell fuselage *See* semimonocoque.

stiffleg derrick [MECH ENG] A derrick consisting of a mast held in the vertical position by a fixed tripod of steel or timber legs. Also known as derrick crane; Scotch derrick.

stiffness [MECH] The ratio of a steady force acting on a deformable elastic medium to the resulting displacement.

stiffness coefficient [MECH] The ratio of the force acting on a linear mechanical system, such as a spring, to its displacement from equilibrium.

stiffness constant [MECH] Any one of the coefficients of the relations in the generalized Hooke's law used to express stress components as linear functions of the strain components. Also known as elastic constant.

stiffness matrix [MECH] A matrix $\mathbf{K}$ used to express the potential energy V of a mechanical system during small displacements from an equilibrium position, by means of the equation $V = \frac{1}{2}\mathbf{q}^T\mathbf{K}\mathbf{q}$, where $\mathbf{q}$ is the vector whose components are the generalized components of the system with respect to time and $\mathbf{q}^T$ is the transpose of $\mathbf{q}$. Also known as stability matrix.

stigma [MECH] A unit of length used mainly in nuclear measurements, equal to 10^{-12} meter. Also known as bicron.

Stillson wrench [DES ENG] A trademark for an adjustable pipe wrench consisting of an L-shaped jaw in a sleeve, the sleeve being pivoted to a handle; pressure on the handle increases the grip of the jaw.

stimulus [CONT SYS] A signal that affects the controlled variable in a control system.

Stirling cycle [THERMO] A regenerative thermodynamic power cycle using two isothermal and two constant volume phases.

Stirling engine [MECH ENG] An engine in which work is performed by the expansion of a gas at high temperature; heat for the expansion is supplied through the wall of the piston cylinder.

stockage objective [AERO ENG] The maximum quantities of material to be maintained on hand to sustain current operations; it will consist of the sum of stocks represented by the operating level and the safety level.

stocking cutter [MECH ENG] **1.** A gear cutter having side rake or curved edges to rough out the gear-tooth spaces before they are formed by the regular gear cutter. **2.** A concave gear cutter ganged beside a regular gear cutter and used to finish the periphery of a gear blank by milling ahead of the regular cutter.

stoke [FL MECH] A unit of kinematic viscosity, equal to the kinematic viscosity of a fluid with a dynamic viscosity of 1 poise and a density of 1 gram per cubic centimeter. Symbol St (formerly S). Also known as lentor (deprecated usage); stokes.

stoker [MECH ENG] A mechanical means, as used in a furnace, for feeding coal, removing refuse, controlling air supply, and mixing with combustibles for efficient burning.

stokes *See* stoke.

Stokes drift [FL MECH] The drift of particles in a gravity wave, which arises from the fact that particle velocities are periodic with a mean which is not zero.

Stokes flow [FL MECH] Fluid flow in which the Reynolds number is very small, so that the nonlinear terms in the Navier-Stokes equations can be neglected.

Stokes number 1 [FL MECH] A dimensionless number used in the study of the dynamics of a particle in a fluid, equal to the product of the dynamic viscosity of the fluid and the particle's vibration time, divided by the product of the fluid density and a characteristic length. Symbol St.

Stokes' law [FL MECH] At low velocities, the frictional force on a spherical body moving through a fluid at constant velocity is equal to 6π times the product of the velocity, the fluid viscosity, and the radius of the sphere.

Stokes stream function [FL MECH] A one-component vector potential function used in analyzing and describing a steady, axially symmetric fluid flow; at any point it is equal to $\frac{1}{2}\pi$ times the mass rate of flow inside the surface generated by rotating the streamline on which the point is located about the axis of symmetry.

STOL *See* short takeoff and landing.

STOL aircraft [AERO ENG] Heavier-than-air craft that cannot take off and land vertically, but can operate within areas substantially more confined than those normally required by aircraft of the same size. Derived from short takeoff and landing aircraft.

stone [MECH] A unit of mass in common use in the United Kingdom, equal to 14 pounds or 6.35029318 kilograms.

stop nut [DES ENG] **1.** An adjustable nut that restricts the travel of an adjusting screw. **2.** A nut with a compressible insert that binds it so that a lock washer is not needed.

stove bolt [DES ENG] A coarsely-threaded bolt with a slotted head, which with a square nut is used to join metal parts.

STR *See* self-tuning regulator.

straddle milling [MECH ENG] Face milling of two parallel vertical surfaces of a workpiece simultaneously by using two side-milling cutters.

straight bevel gear [DES ENG] A simple form of bevel gear having straight teeth which, if extended inward, would come together at the intersection of the shaft axes.

straightedge [DES ENG] A strip of wood, plastic, or metal with one or more long edges made straight with a desired degree of accuracy.

straight-line mechanism [MECH ENG] A linkage so proportioned and constrained that some point on it describes over part of its motion a straight or nearly straight line.

straight strap clamp [DES ENG] A clamp made of flat stock with an elongated slot for convenient positioning; held in place by a T bolt and nut.

straight-tube boiler [MECH ENG] A water-tube boiler in which all the tubes are devoid of curvature and therefore require suitable connecting devices to complete the circulatory system. Also known as header-type boiler.

straight turning [MECH ENG] Work turned in a lathe so that the diameter is constant over the length of the workpiece.

straightway pump [MECH ENG] A pump with suction and discharge valves arranged to give a direct flow of fluid.

straight wheel [DES ENG] A grinding wheel whose sides or face are straight and not in any way changed from a cylindrical form.

strain [MECH] Change in length of an object in some direction per unit undistorted length in some direction, not necessarily the same; the nine possible strains form a second-rank tensor.

strain axis *See* principal axis of strain.

strain ellipsoid [MECH] A mathematical representation of the strain of a homogeneous body by a strain that is the same at all points or of unequal stress at a particular point. Also known as deformation ellipsoid.

strain energy [MECH] The potential energy stored in a body by virtue of an elastic deformation, equal to the work that must be done to produce this deformation.

strain rate [MECH] The time rate for the usual tensile test.

strain rosette [MECH] A pattern of intersecting lines on a surface along which linear strains are measured to find stresses at a point.

stranger [AERO ENG] In air intercept, an unidentified aircraft, bearing, distance, and altitude as indicated relative to an aircraft.

strap bolt [DES ENG] **1.** A bolt with a hook or flat extension instead of a head. **2.** A bolt with a flat center portion and which can be bent into a U shape.

strap hammer [MECH ENG] A heavy hammer controlled and operated by a belt drive in which the head is slung from a strap, usually of leather.

strap hinge [DES ENG] A hinge fastened to a door and the adjacent wall by a long hinge.

strategic airlift [AERO ENG] The continuous, sustained air movement of units, personnel, and materiel in support of all U.S. Department of Defense agencies between area commands.

strategic transport aircraft [AERO ENG] Aircraft designed primarily for the carriage of personnel or cargo over long distances.

stratified flow [FL MECH] A two-phase flow in which a liquid flows along the bottom of a pipe and gas flows separately above it.

stratified fluid [FL MECH] A fluid having density variation along the axis of gravity, usually implying upward decrease of density, that is, a stratification characterized by static stability.

Stratofortress [AERO ENG] A United States all-weather, intercontinental, strategic heavy bomber powered by eight turbojet engines; it is capable of delivering nuclear and nonuclear bombs, air-to-surface missiles, and decoys; its range is extended by in-flight refueling. Designated B-52.

Stratofreighter [AERO ENG] A United States strategic, aerial tanker-freighter powered by four reciprocating engines; it is equipped for inflight refueling of bombers and fighters. Designated KC-97.

Stratojet [AERO ENG] A United States all-weather strategic medium bomber; it is powered by six turbojet engines, has intercontinental range through in-flight refueling, and is capable of delivering nuclear and nonnuclear bombs. Designated B-47.

Stratotanker [AERO ENG] A United States multipurpose aerial tanker-transport powered by four turbojet engines; and equipped for high speed, high-altitude refueling of bombers and fighters. Designated KC-135.

streak line [FL MECH] A line within a fluid which, at a given instant, is formed by those fluid particles which at some previous instant have passed through a specified fixed point in the fluid; an example is the line of color in a flow into which a dye is continuously introduced through a small tube, all dyed fluid particles having passed the tube's end.

stream function *See* Lagrange stream function.

streamline [FL MECH] A line which is every where parallel to the direction of fluid flow at a given instant.

streamline flow [FL MECH] Flow of a fluid in which there is no turbulence: particles of the fluid follow well-defined continuous paths, and the flow velocity at a fixed point either remains constant or varies in a regular fashion with time.

streamlining [DES ENG] The contouring of a body to reduce its resistance to motion through a fluid.

stream takeoff [AERO ENG] Aircraft taking off in tail/column formation.

stream tube [FL MECH] In fluid flow, an imaginary tube whose wall is generated by streamlines passing through a closed curve.

street elbow [DES ENG] A pipe elbow with an internal thread at one end and an external thread at the other.

strength [MECH] The stress at which material ruptures or fails.

stress [MECH] The force acting across a unit area in a solid material in resisting the separation, compacting, or sliding that tends to be induced by external forces.

stress amplitude [MECH ENG] One half the algebraic difference between the maximum and minimum stress in one fatigue test cycle.

stress axis *See* principal axis of stress.

stress concentration [MECH] A condition in which a stress distribution has high localized stresses; usually induced by an abrupt change in the shape of a member; in the vicinity of notches, holes, changes in diameter of a shaft, and so forth, maximum stress is several times greater than where there is no geometrical discontinuity.

stress concentration factor [MECH] A theoretical factor K_t expressing the ratio of the greatest stress in the region of stress concentration to the corresponding nominal stress.

stress crack [MECH] An external or internal crack in a solid body (metal or plastic) caused by tensile, compressive, or shear forces.

stress difference [MECH] The difference between the greatest and the least of the three principal stresses.

stress ellipsoid [MECH] A mathematical representation of the state of stress at a point that is defined by the minimum, intermediate, and maximum stresses and their intensities.

stress function [MECH] A single function, such as the Airy stress function, or one of two or more functions, such as Maxwell's or Morera's stress functions, that uniquely define the stresses in an elastic body as a function of position.

stress intensity [MECH] Stress at a point in a structure due to pressure resulting from combined tension (positive) stresses and compression (negative) stresses.

stress lines *See* isostatics.

stress range [MECH] The algebraic difference between the maximum and minimum stress in one fatigue test cycle.

stress ratio [MECH] The ratio of minimum to maximum stress in fatigue testing, considering tensile stresses as positive and compressive stresses as negative.

stress-strain curve *See* deformation curve.

stress tensor [MECH] A second-rank tensor whose components are stresses exerted across surfaces perpendicular to the coordinate directions.

stress trajectories *See* isostatics.

stretch former [MECH ENG] A machine used to stretch form materials, such as metals and plastics.

stretch forming [MECH ENG] Shaping metals and plastics by applying tension to stretch the heated sheet or part, wrapping it around a die, and then cooling it. Also known as wrap forming.

strich *See* millimeter.

strike plate [DES ENG] A metal plate or box which is set in a door jamb and is either pierced or recessed to receive the bolt or latch of a lock.

striking velocity *See* impact velocity.

string [MECH] A solid body whose length is many times as large as any of its cross-sectional dimensions, and which has no stiffness.

string milling [MECH ENG] A milling method in which parts are placed in a row and milled consecutively.

strip-borer drill [MECH ENG] An electric or diesel skid- or caterpillar-mounted drill used at quarry or opencast sites to drill 3- to 6-inch-diameter (8- to 15-centimeter-diameter), horizontal blast holes up to 100 feet (30 meters) in length, without the use of flush water.

stroke [MECH ENG] The linear movement, in either direction, of a reciprocating mechanical part.

stroke-bore ratio [MECH ENG] The ratio of the distance traveled by a piston in a cylinder to the diameter of the cylinder.

Strouhal number [MECH] A dimensionless number used in studying the vibrations of a body past which a fluid is flowing; it is equal to a characteristic dimension of the body times the frequency of vibrations divided by the fluid velocity relative to the body; for a taut wire perpendicular to the fluid flow, with the characteristic dimension taken as the diameter of the wire, it has a value between 0.185 and 0.2. Symbolized S_r. Also known as reduced frequency.

structural deflections [MECH] The deformations or movements of a structure and its flexural members from their original positions.

structural drill [MECH ENG] A highly mobile diamond- or rotary-drill rig complete with hydraulically controlled derrick mounted on a truck, designed primarily for rapidly drilling holes to determine the structure in subsurface strata or for use as a shallow, slim-hole producer or seismograph drill.

structural weight *See* construction weight.

structure [AERO ENG] The construction or makeup of an airplane, spacecraft, or missile, including that of the fuselage, wings, empennage, nacelles, and landing gear, but not that of the power plant, furnishings, or equipment.

structure number [DES ENG] A number, generally from 0 to 15, indicating the spacing of abrasive grains in a grinding wheel relative to their grit size.

strut [AERO ENG] A bar supporting the wing or landing gear of an airplane.

Stuart windmill *See* Fales-Stuart windmill.

stub axle [MECH ENG] An axle carrying only one wheel.

Stubs gage [DES ENG] A number system for denoting the thickness of steel wire and drills.

stub tube [MECH ENG] A short tube welded to a boiler or pressure vessel to provide for the attachment of additional parts.

stud [DES ENG] 1. A rivet, boss, or nail with a large, ornamental head. 2. A short rod or bolt threaded at both ends without a head.

subatmospheric heating system [MECH ENG] A system which regulates steam flow into the main throttle valve under automatic thermostatic control and maintains a fixed vacuum differential between supply and return by means of a differential controller and a vacuum pump.

subcomponent [DES ENG] A part of a component having characteristics of the component.

subcritical flow *See* subsonic flow.

sublimation [THERMO] The process by which solids are transformed directly to the vapor state or vice versa without passing through the liquid phase.

sublimation cooling [THERMO] Cooling caused by the extraction of energy to produce sublimation.

sublimation curve [THERMO] A graph of the vapor pressure of a solid as a function of temperature.

sublimation energy [THERMO] The increase in internal energy when a unit mass, or 1 mole, of a solid is converted into a gas, at constant pressure and temperature.

sublimation point [THERMO] The temperature at which the vapor pressure of the solid phase of a compound is equal to the total pressure of the gas phase in contact with it; analogous to the boiling point of a liquid.

sublimation pressure [THERMO] The vapor pressure of a solid.

sublime [THERMO] To change from the solid to the gaseous state without passing through the liquid phase.

submersible pump [MECH ENG] A pump and its electric motor together in a protective housing which permits the unit to operate under water.

subsonic flight [AERO ENG] Movement of a vehicle through the atmosphere at a speed appreciably below that of sound waves; extends from zero (hovering) to a speed about 85% of sonic speed corresponding to ambient temperature.

subsonic flow [FL MECH] Flow of a fluid at a speed less than that of the speed of sound in the fluid. Also known as subcritical flow.

subsonic speed [FL MECH] A speed relative to surrounding fluid less than that of the speed of sound in the same fluid.

suction boundary layer control [AERO ENG] A technique that is used in addition to purely geometric means to control boundary layer flow; it consists of sucking away the retarded flow in the lower regions of the boundary through slots or perforations in the surface.

suction-cutter dredger [MECH ENG] A dredger in which rotary blades dislodge the material to be excavated, which is then removed by suction as in a sand-pump dredger.

suction head   *See* suction lift.

suction lift   [MECH ENG] The head, in feet, that a pump must provide on the inlet side to raise the liquid from the supply well to the level of the pump. Also known as suction head.

suction pump   [MECH ENG] A pump that raises water by the force of atmospheric pressure pushing it into a partial vacuum under the valved piston, which retreats on the upstroke.

suction stroke   [MECH ENG] The piston stroke that draws a fresh charge into the cylinder of a pump, compressor, or internal combustion engine.

suction wave   *See* rarefaction wave.

Sullivan angle compressor   [MECH ENG] A two-stage compressor in which the low-pressure cylinder is horizontal and the high-pressure cylinder is vertical; a compact compressor driven by a belt, or directly connected to an electric motor or diesel engine.

Sulzer two-cycle engine   [MECH ENG] An internal combustion engine utilizing the Sulzer Co. system for the effective scavenging and charging of the two-cycle diesel engine.

sump pump   [MECH ENG] A small, single-stage vertical pump used to drain shallow pits or sumps.

sun-and-planet motion   [MECH ENG] A train of two wheels moving epicyclically with a small wheel rotating a wheel on the central axis.

sun gear   *See* central gear.

sun-synchronous orbit   [AERO ENG] An earth orbit of a spacecraft so that the craft is always in the same direction relative to that of the sun; as a result, the spacecraft passes over the equator at the same spots at the same times.

superaerodynamics   [FL MECH] That branch of gas dynamics dealing with the flow of gases at such low density that the molecular mean free path is not negligibly small; under these conditions the gas no longer behaves as a continuous fluid.

supercentrifuge   [MECH ENG] A centrifuge built to operate at faster speeds than an ordinary centrifuge.

supercharger   [MECH ENG] An air pump or blower in the intake system of an internal combustion engine used to increase the weight of air charge and consequent power output from a given engine size.

supercharging   [MECH ENG] A method of introducing air for combustion into the cylinder of an internal combustion engine at a pressure in excess of that which can be obtained by natural aspiration.

supercobalt drill   [DES ENG] A drill made of 8% cobalt highspeed steel; used for drilling work-hardened stainless steels, silicon chrome, and certain chrome-nickel alloy steels.

supercompressibility factor   *See* compressibility factor.

supercooling   [THERMO] Cooling of a substance below the temperature at which a change of state would ordinarily take place without such a change of state occurring, for example, the cooling of a liquid below its freezing point without freezing taking place; this results in a metastable state.

supercritical   [THERMO] Property of a gas which is above its critical pressure and temperature.

supercritical flow   *See* supersonic flow.

supercritical wing   [AERO ENG] A wing developed to permit subsonic aircraft to maintain an efficient cruise speed very close to the speed of sound; the middle portion of the wing is relatively flat with substantial downward curvature near the trailing edge; in addition, the forward region of the lower surface is more curved than that of the upper surface with a substantial cusp of the rearward portion of the lower surface.

superheat [THERMO] Sensible heat in a gas above the amount needed to maintain the gas phase.

superheated vapor [THERMO] A vapor that has been heated above its boiling point.

superheater [MECH ENG] A component of a steam-generating unit in which steam, after it has left the boiler drum, is heated above its saturation temperature.

superheating [THERMO] Heating of a substance above the temperature at which a change of state would ordinarily take place without such a change of state occurring, for example, the heating of a liquid above its boiling point without boiling taking place; this results in a metastable state.

superimposed back pressure [MECH ENG] The static pressure at the outlet of an operating pressure relief device, resulting from pressure in the discharge system.

superposition integral [CONT SYS] An integral which expresses the response of a linear system to some input in terms of the impulse response or step response of the system; it may be thought of as the summation of the responses to impulses or step functions occurring at various times.

Super Sabre [AERO ENG] A United States supersonic, single-engine, turbojet-powered, tactical and air superiority fighter capable of delivering either nuclear or non-nuclear bombs, rockets, and Bullpup missiles against surface targets, or cannons and Sidewinder missiles against airborne targets; it is capable of providing close support for ground forces, and it can be refueled in flight. Designated F-100.

supersonic aerodynamics [FL MECH] The study of aerodynamics of supersonic speeds.

supersonic aircraft [AERO ENG] Aircraft capable of supersonic speeds.

supersonic airfoil [AERO ENG] An airfoil designed to produce lift at supersonic speeds.

supersonic compression ramjet *See* scramjet.

supersonic compressor [MECH ENG] A compressor in which a supersonic velocity is imparted to the fluid relative to the rotor blades, the stator blades, or both, producing oblique shock waves over the blades to obtain a high-pressure rise.

supersonic diffuser [MECH ENG] A diffuser designed to reduce the velocity and to increase the pressure of fluid moving at supersonic velocities.

supersonic flow [FL MECH] Flow of a fluid over a body at speeds greater than the speed of sound in the fluid, and in which the shock waves start at the surface of the body. Also known as supercritical flow.

supersonic inlet [AERO ENG] An inlet of a jet engine at which single, double, or multiple shock waves form.

supersonic nozzle *See* convergent-divergent nozzle.

supersonic transport [AERO ENG] A transport plane capable of flying at speeds higher than the speed of sound. Abbreviated SST.

superturbulent flow [FL MECH] The flow of water in which the energy loss by friction is so great that Reynolds criterion for the transition of laminar to turbulent flow does not apply.

surface carburetor [MECH ENG] A carburetor in which air is passed over the surface of gasoline to charge it with fuel.

surface condenser [MECH ENG] A heat-transfer device used to condense a vapor, usually steam under vacuum, by absorbing its latent heat in cooling fluid, ordinarily water.

surface-effect ship [MECH ENG] A transportation device with fixed side walls, which is supported by low-pressure, low-velocity air and operates on water only.

surface energy [FL MECH] The energy per unit area of an exposed surface of a liquid; generally greater than the surface tension, which equals the free energy per unit surface.

surface force [MECH] An external force which acts only on the surface of a body; an example is the force exerted by another object with which the body is in contact.

surface gage [DES ENG] 1. A scribing tool in an adjustable stand, used to mark off castings and to test the flatness of surfaces. 2. A gage for determining the distances of points on a surface from a reference plane.

surface grinder [MECH ENG] A grinding machine that produces a plane surface.

surface plate [DES ENG] A plate having a very accurate plane surface used for testing other surfaces or to provide a true surface for accurately measuring and locating testing fixtures.

surface-set bit [DES ENG] A bit containing a single layer of diamonds set so that the diamonds protrude on the surface of the crown. Also known as single-layer bit.

surface tension [FL MECH] The force acting on the surface of a liquid, tending to minimize the area of the surface; quantitatively, the force that appears to act across a line of unit length on the surface. Also known as interfacial force; interfacial tension; surface tensity.

surface tension number [FL MECH] A dimensionless number used in studying mass transfer in packed columns equal to the square of the dynamic viscosity of a fluid times the length of the perimeter of a packing element, divided by the product of the surface area of the packing element, the surface tension, and the density of the liquid. Symbol T_s.

surface tensity *See* surface tension.

surface vibrator [MECH ENG] A vibrating device used on the surface of a pavement or flat slab to consolidate the concrete.

surface wave [FL MECH] A wave that distorts the free surface that separates two fluid phases, usually a liquid and a gas. [MECH] *See* Rayleigh wave.

surge [FL MECH] A wave at the free surface of a liquid generated by the motion of a vertical wall, having a change in the height of the surface across the wavefront and violent eddy motion at the wavefront.

surge header *See* accumulator.

surge stress [MECH] The physical stress on process equipment or systems resulting from a sudden surge in fluid (gas or liquid) flow rate or pressure.

surveillance satellite [AERO ENG] A satellite whose function is to make systematic observations of the earth, usually by photographic means, for military intelligence or for other purposes.

survey foot [MECH] A unit of length, used by the U.S. Coast and Geodetic Survey, equal to 12/39.37 meter, or approximately 1.000002 feet.

Surveyor program [AERO ENG] A program in which unmanned spacecraft made soft landings on the moon to take photographs, analyze samples of soil, and obtain other data that could be transmitted back to earth for guidance in planning manned landings.

suspended transformation [THERMO] The cessation of change before true equilibrium is reached, or the failure of a system to change immediately after a change in conditions, such as in supercooling and other forms of metastable equilibrium.

suspended tray conveyor [MECH ENG] A vertical conveyor having pendant trays or other carriers on one or more endless chains.

suspension system [MECH ENG] A system of springs, shock absorbers, and other devices supporting the upper part of a motor vehicle on its running gear.

sustained oscillation [CONT SYS] Continued oscillation due to insufficient attenuation in the feedback path.

sustainer rocket engine [AERO ENG] A rocket engine that maintains the velocity of a rocket vehicle once it has achieved its programmed velocity by use of a booster or other engine.

sverdrup [FL MECH] A unit of volume transport equal to 1,000,000 cubic meters per second.

swage bolt [DES ENG] A bolt having indentations with which it can be gripped in masonry.

swamp buggy [MECH ENG] A wheeled vehicle that runs on land, mud, or through shallow water; used especially in swamps.

swash-plate pump [MECH ENG] A rotary pump in which the angle between the drive shaft and the plunger-carrying body is varied.

sweat cooling [AERO ENG] A technique for cooling combustion chambers or aerodynamically heated surfaces by forcing a coolant through a porous wall, resulting in film cooling at the interface. Also known as transpiration cooling.

sweepback [AERO ENG] 1. The backward slant of a leading or trailing edge of an airfoil. 2. The amount of this slant, expressed as the angle between a line perpendicular to the plane of symmetry and a reference line in the airfoil.

swell diameter [AERO ENG] In a body of revolution having an ogival portion, such as a projectile, the swell diameter is in the diameter of the maximum transverse section of the geometrical ogive.

sweptback wing [AERO ENG] An airplane wing on which both the leading and trailing edges have sweepback, the trailing edge forming an acute angle with the longitudinal axis of the airplane aft of the root. Also known as swept wing.

swept wing *See* sweptback wing.

swing-around trajectory [AERO ENG] A planetary round-trip trajectory which requires minimal propulsion at the destination planet, but instead uses the planet's gravitational field to effect the bulk of the necessary orbit change to return to earth.

swing-frame grinder [MECH ENG] A grinding machine hanging by a chain so that it may swing in all directions for surface grinding heavy work.

swing joint [DES ENG] A pipe joint in which the parts may be rotated relative to each other.

Swiss pattern file [DES ENG] A type of fine file used for precision filing of jewelry, instrument parts, and dies.

switchblade knife [DES ENG] A knife in which the blade is restrained by a spring and swings open when released by a pushbutton.

switching surface [CONT SYS] In feedback control systems employing bang-bang control laws, the surface in state space which separates a region of maximum control effort from one of minimum control effort.

swivel [DES ENG] A part that oscillates freely on a headed bolt or pin.

swivel block [DES ENG] A block with a swivel attached to its hook or shackle permitting it to revolve.

swivel coupling [MECH ENG] A coupling that gives complete rotary freedom to a deflecting wedge-setting assembly.

swivel head [MECH ENG] The assembly of a spindle, chuck, feed nut, and feed gears on a diamond-drill machine that surrounds, rotates, and advances the drill rods and drilling stem; on a hydraulic-feed drill the feed gears are replaced by a hydraulically actuated piston assembly.

swivel hook [DES ENG] A hook with a swivel connection to its base or eye.

swivel joint [DES ENG] A joint with a packed swivel that allows one part to move relative to the other.

swivel neck *See* water swivel.

swivel pin *See* kingpin.

symmetry axis *See* axis of symmetry.

synchromesh [MECH ENG] An automobile transmission device that minimizes clashing; acts as a friction clutch, bringing gears approximately to correct speed just before meshing.

synchronized shifting [MECH ENG] Changing speed gears, with the gears being brought to the same speed before the change can be made.

synchronous orbit [AERO ENG] 1. An orbit in which a satellite makes a limited number of equatorial crossing points which are then repeated in synchronism with some defined reference (usually earth or sun). 2. Commonly, the equatorial, circular, 24-hour case in which the satellite appears to hover over a specific point of the earth.

synchronous satellite *See* geostationary satellite.

Syncom [AERO ENG] One of a series of communication satellites placed in synchronous equatorial orbit; used for relaying television and radio communications over long distances.

synergic curve [AERO ENG] A curve plotted for the ascent of a rocket, space-air vehicle, or space vehicle, calculated to give optimum fuel economy and optimum velocity.

synoptic correlation *See* Eulerian correlation.

synthesis *See* system design.

system analysis [CONT SYS] The use of mathematics to determine how a set of inter-connected components whose individual characteristics are known will behave in response to a given input or set of inputs.

system bandwidth [CONT SYS] The difference between the frequencies at which the gain of a system is $\sqrt{2}/2$ (that is, 0.707) times its peak value.

system design [CONT SYS] A technique of constructing a system that performs in a specified manner, making use of available components. Also known as synthesis.

system response *See* response.

T

t *See* troy system.

T$_s$ *See* surface tension number.

tab-card cutter [DES ENG] A device for die-cutting card stock to uniform tabulating-card size.

table [MECH ENG] That part of a grinding machine which directly or indirectly supports the work being ground.

tablespoonful [MECH] A unit of volume used particularly in cookery, equal to 4 fluid drams or ½ fluid ounce; in the United States this is equal to approximately 14.7868 cubic centimeters, in the United Kingdom to approximately 14.2065 cubic centimeters. Abbreviated tbsp.

tack [DES ENG] A small, sharp-pointed nail with a broad flat head.

tackle [MECH ENG] Any arrangement of ropes and pulleys to gain a mechanical advantage.

tactical air-direction center [AERO ENG] An air operations installation under the overall control of the tactical air-control center, from which aircraft and air warning service functions of tactical air operations in an area of responsibility are directed.

tactical air force [AERO ENG] An air force charged with carrying out tactical air operations in coordination with ground or naval forces.

tactical airlift [AERO ENG] That airlift which provides the immediate and responsive air movement and delivery of combat troops and supplies directly into objective areas through air landing, extraction, airdrop, or other delivery techniques; and the air logistic support of all theater forces, including those engaged in combat operations, to meet specific theater objectives and requirements.

tactical air observer [AERO ENG] An officer trained as an air observer whose function is to observe from airborne aircraft and report on movement and disposition of friendly and enemy forces, on terrain and weather and hydrography, and to execute other missions as directed.

tactical air operations [AERO ENG] An air operation involving the employment of air power in coordination with ground or naval forces to gain and maintain air superiority, to prevent movement of enemy forces into and within the objective area and to seek out and destroy these forces and their supporting installations, and to join with ground or naval forces in operations within the objective area in order to assist directly in attainment of their immediate objective.

tactical air reconnaissance [AERO ENG] The use of air vehicles to obtain information concerning terrain, weather, and the disposition, composition, movement, installations, lines of communications, and electronic and communication emissions of enemy forces.

tactical air support [AERO ENG] Air operations carried out in coordination with surface forces which directly assist the land or naval battle.

tactical air transport [AERO ENG] The use of air transport in direct support of airborne assaults, carriage of air-transported forces, tactical air supply, evacuation of casualties from forward airdromes, and clandestine operations.

tactical transport aircraft [AERO ENG] Aircraft designed primarily for carrying personnel or cargo over short or medium distances.

tail [AERO ENG] 1. The rear part of a body, as of an aircraft or a rocket. 2. The tail surfaces of an aircraft or a rocket.

tail assembly *See* empennage.

tail fin [AERO ENG] A fin at the rear of a rocket or other body.

tail pulley [MECH ENG] A pulley at the tail of the belt conveyor opposite the normal discharge end; may be a drive pulley or an idler pulley.

tailstock [MECH ENG] A part of a lathe that holds the end of the work not being shaped, allowing it to rotate freely.

tail surface [AERO ENG] A stabilizing or control surface in the tail of an aircraft or missile.

takeoff [AERO ENG] Ascent of an aircraft or rocket at any angle, as the action of a rocket vehicle departing from its launch pad or the action of an aircraft as it becomes airborne.

takeoff assist [AERO ENG] 1. The extra thrust given to an airplane or missile during takeoff through the use of a rocket motor or other device. 2. The device used in such a takeoff.

takeoff weight [AERO ENG] The weight of an aircraft or rocket vehicle ready for takeoff, including the weight of the vehicle, the fuel, and the payload.

takeup [MECH ENG] A tensioning device in a belt-conveyor system for taking up slack of loose parts.

takeup pulley [MECH ENG] An adjustable idler pulley to accommodate changes in the length of a conveyor belt to maintain proper belt tension.

tandem [AERO ENG] The fore and aft configuration used in boosted missiles, long-range ballistic missiles, and satellite vehicles; stages are stacked together in series and are discarded at burnout of the propellant for each stage.

tandem-drive conveyor [MECH ENG] A conveyor having the conveyor belt in contact with two drive pulleys, both driven with the same motor.

tandem roller [MECH ENG] A steam- or gasoline-driven road roller in which the weight is divided between heavy metal rolls, of dissimilar diameter, one behind the other.

tangential acceleration [MECH] The component of linear acceleration tangent to the path of a particle moving in a circular path.

tangential helical-flow turbine *See* helical-flow turbine.

tangential stress *See* shearing stress.

tangential velocity [MECH] 1. The instantaneous linear velocity of a body moving in a circular path; its direction is tangential to the circular path at the point in question. 2. The component of the velocity of a body that is perpendicular to a line from an observer or reference point to the body.

tap [DES ENG] 1. A plug of accurate thread, form, and dimensions on which cutting edges are formed; it is screwed into a hole to cut an internal thread. 2. A threaded cone-shaped fishing tool.

tap bolt [DES ENG] A bolt with a head that can be screwed into a hole and held in place without a nut. Also known as tap screw.

tap drill [MECH ENG] A drill used to make a hole of a precise size for tapping.

tape-controlled machine [MECH ENG] A machine tool whose movements are automatically controlled by means of a magnetic or punched tape.

taper [AERO ENG] An airfoil feature in which either the thickness or the chord length or both decrease from the root to the tip.

taper bit [DES ENG] A long, cone-shaped noncoring bit used in drilling blastholes and in wedging and reaming operations.

tapered core bit [DES ENG] A core bit having a conical diamond-inset crown surface tapering from a borehole size at the bit face to the next larger borehole size at its upper, shank, or reaming-shell end.

tapered joint [DES ENG] A firm, leakproof connection between two pieces of pipe having the thread formed with a slightly tapering diameter.

tapered thread [DES ENG] A screw thread cut on the surface of a tapered part; it may be either a pine or box thread, or a V-, Acme, or square-screw thread.

tapered wheel [DES ENG] A flat-face grinding wheel with greater thickness at the hub than at the face.

taper-in-thickness ratio [AERO ENG] A gradual change in the thickness ratio along the wing span with the chord remaining constant.

taper key [DES ENG] A rectangular machine key that is slightly tapered along its length.

taper pin [DES ENG] A small, tapered self-holding peg or nail used to connect parts together.

taper pipe thread *See* pipe thread.

taper plug gage [DES ENG] An internal gage in the shape of a frustrum of a cone used to measure internal tapers.

taper reamer [DES ENG] A reamer whose fluted portion tapers toward the front end.

taper ring gage [DES ENG] An external gage having a conical internal contour; used to measure external tapers.

taper-rolling bearing [MECH ENG] A roller bearing capable of sustaining end thrust by means of tapered rollers and coned races.

taper shank [DES ENG] A cone-shaped part on a tool that fits into a tapered sleeve on a driving member.

taper tap [DES ENG] A threaded cone-shaped tool for cutting internal screw threads.

tappet [MECH ENG] A lever or oscillating member moved by a cam and intended to tap or touch another part, such as a push rod or valve system.

tappet rod [MECH ENG] A rod carrying a tappet or tappets, as one for opening or closing the valves in a steam or an internal combustion engine.

tapping [MECH ENG] Forming an internal screw thread in a hole or other part by means of a tap.

tare [MECH] The weight of an empty vehicle or container; subtracted from gross weight to ascertain net weight.

tare effect [FL MECH] In wind tunnel testing, the forces and moments due to support assembly and mutual interference between support assembly and model.

target acquisition [AERO ENG] The process of optically, manually, mechanically, or electronically orienting a tracking system in direction and range to lock on a target.

target approach point [AERO ENG] In air transport operations, a navigational checkpoint over which the final turn-in to the drop zone–landing zone is made.

target drone [AERO ENG] A pilotless aircraft controlled by radio from the ground or from a mother ship and used exclusively as a target for antiaircraft weapons.

target pattern [AERO ENG] The flight path of aircraft during the attack phase.

taut-line cableway [MECH ENG] A cableway whose operation is limited to the distance between two towers, usually 3000 feet (914 meters) apart, has only one carrier, and the traction cable is reeved at the carrier so that loads can be raised and lowered; the towers can be mounted on trucks or crawlers, and the machine shifted across a wide area.

Taylor effect [FL MECH] A phenomenon in which the relative motion of a homogeneous rotating liquid tends to be the same in all planes perpendicular to the axis of rotation.

Taylor number [FL MECH] A nondimensional number arising in problems of a rotating viscous fluid, written as $T = (f^2h^4)/v^2$, where f is the Coriolis parameter (or, for a cylindrical system, twice the rate of rotation of the system), h the depth of the fluid, and v the kinematic viscosity; the square root of the Taylor number is a rotating Reynolds number, and the fourth root is proportional to the ratio of the depth h to the depth of the Ekman layer.

T bolt [DES ENG] A bolt with a T-shaped head, made to fit into a T-shaped slot in a drill swivel head or in the bed of a machine.

tbsp See tablespoonful.

teardrop balloon [AERO ENG] A sounding balloon which, when operationally inflated, resembles an inverted teardrop; this shape was determined primarily by aerodynamic considerations of the problem of obtaining maximum stable rates of a balloon ascension.

tear strength [MECH] The force needed to initiate or to continue tearing a sheet or fabric.

teaspoonful [MECH] A unit of volume used particularly in cookery and pharmacy, equal to 1⅓ fluid drams, or ⅓ tablespoonful; in the United States this is equal to approximately 4.9289 cubic centimeters, in the United Kingdom to approximately 4.7355 cubic centimeters. Abbreviated tsp; tspn.

technical atmosphere [MECH] A unit of pressure in the metric technical system equal to one kilogram-force per square centimeter. Abbreviated at.

teleoperator [CONT SYS] A general-purpose, remotely controlled, cybernetic, dexterous person-machine system.

telescoping gage [DES ENG] An adjustable internal gage with a telescoping plunger that expands under spring tension in the hole to be measured; it is locked into position to allow measurement after being withdrawn from the hole.

telescoping valve [MECH ENG] A valve, with sliding, telescoping members, to regulate water flow in a pipe line with minimum disturbance to stream lines.

television satellite [AERO ENG] An orbiting satellite that relays television signals between ground stations.

telpher [MECH ENG] An electric hoist hanging from and driven by a wheeled cab rolling on a single overhead rail or a rope.

Telsmith breaker [MECH ENG] A type of gyratory crusher, often used for primary crushing; consists of a spindle mounted in a long eccentric sleeve which rotates to impart a gyratory motion to the crushing head, but gives a parallel stroke, that is, the axis of the spindle describes a cylinder rather than a cone, as in the suspended spindle gyratory.

Telstar satellite [AERO ENG] A spherical active repeater communications satellite, first launched on July 10, 1962.

temperature [THERMO] A property of an object which determines the direction of heat flow when the object is placed in thermal contact with another object: heat flows from a region of higher temperature to one of lower temperature; it is measured either by an empirical temperature scale, based on some convenient property of a material or instrument, or by a scale of absolute temperature, for example, the Kelvin scale.

13.0 temperature *See* annealing point.

temperature-actuated pressure relief valve [MECH ENG] A pressure relief valve which operates when subjected to increased external or internal temperature.

temperature bath [THERMO] A relatively large volume of a homogeneous substance held at constant temperature, so that an object placed in thermal contact with it is maintained at the same temperature.

temperature color scale [THERMO] The relation between an incandescent substance's temperature and the color of the light it emits.

temperature gradient [THERMO] For a given point, a vector whose direction is perpendicular to an isothermal surface at the point, and whose magnitude equals the rate of change of temperature in this direction.

temperature scale [THERMO] An assignment of numbers to temperatures in a continuous manner, such that the resulting function is single valued; it is either an empirical temperature scale, based on some convenient property of a substance or object, or it measures the absolute temperature.

temporal decomposition [CONT SYS] The partitioning of the control or decision-making problem associated with a large-scale control system into subproblems based on the different time scales relevant to the associated action functions.

Ten Broecke chart [THERMO] A graphical plot of heat transfer and temperature differences used to calculate the thermal efficiency of a countercurrent cool-fluid–warm-fluid heat-exchange system.

tensile modulus [MECH] The tangent or secant modulus of elasticity of a material in tension.

tensile strength [MECH] The maximum stress a material subjected to a stretching load can withstand without tearing. Also known as hot strength.

tensile stress [MECH] Stress developed by a material bearing a tensile load.

tension [MECH] **1.** The condition of a string, wire, or rod that is stretched between two points. **2.** The force exerted by the stretched object on a support.

tension pulley [MECH ENG] A pulley around which an endless rope passes mounted on a trolley or other movable bearing so that the slack of the rope can be readily taken up by the pull of the weights.

tenthmeter *See* angstrom.

terminal unit [MECH ENG] In an air-conditioning system, a unit at the end of a branch duct through which air is transferred or delivered to the conditioned space.

terminal velocity [FL MECH] The velocity with which a body moves relative to a fluid when the resultant force acting on it (due to friction, gravity, and so forth) is zero.

tertiary air [MECH ENG] Combustion air added to primary and secondary air.

test-bed [AERO ENG] A base, mount, or frame within or upon which a piece of equipment, especially an engine, is secured for testing.

test chamber [AERO ENG] The testof a wind tunnel.

test firing [AERO ENG] The firing of a rocket engine, either live or static, with the purpose of making controlled observations of the engine or of an engine component.

test flight [AERO ENG] A flight to make controlled observations of the operation or performance of an aircraft or a rocket, or of a component of an aircraft or rocket.

test section [AERO ENG] The section of a wind tunnel where objects are tested to determine their aerodynamic characteristics.

test stand [AERO ENG] A stationary platform or table, together with any testing apparatus attached thereto, for testing or proving engines and instruments.

theoretical draft [FL MECH] Draft in a ducted space, neglecting flow losses due to fluid friction.

theoretical relieving capacity [MECH ENG] The capacity of a theoretically perfect nozzle calculated in volumetric or gravimetric units.

therm [THERMO] A unit of heat energy, equal to 10^5 international table British thermal units, or approximately 1.055×10^8 joules.

thermal [THERMO] Of or concerning heat.

thermal barrier [AERO ENG] A limit to the speed of airplanes and rockets in the atmosphere imposed by heat from friction between the aircraft and the air, which weakens and eventually melts the surface of the aircraft. Also known as heat barrier.

thermal capacitance [THERMO] The ratio of the entropy added to a body to the resulting rise in temperature.

thermal capacity *See* heat capacity.

thermal charge *See* entropy.

thermal compressor [MECH ENG] A steam-jet ejector designed to compress steam at pressures above atmospheric.

thermal conductance [THERMO] The amount of heat transmitted by a material divided by the difference in temperature of the surfaces of the material. Also known as conductance.

thermal conductimetry [THERMO] Measurement of thermal conductivities.

thermal conductivity [THERMO] The heat flow across a surface per unit area per unit time, divided by the negative of the rate of change of temperature with distance in a direction perpendicular to the surface. Also known as heat conductivity.

thermal convection *See* heat convection.

thermal coulomb [THERMO] A unit of entropy equal to 1 joule per kelvin.

thermal efficiency *See* efficiency.

thermal emissivity *See* emissivity.

thermal equilibrium [THERMO] Property of a system all parts of which have attained a uniform temperature which is the same as that of the system's surroundings.

thermal farad [THERMO] A unit of thermal capacitance equal to the thermal capacitance of a body for which an increase in entropy of 1 joule per kelvin results in a temperature rise of 1 kelvin.

thermal flame safeguard [MECH ENG] A thermocouple located in the pilot flame of a burner; if the pilot flame is extinguished, an elective circuit is interrupted and the fuel supply is shut off.

thermal flux *See* heat flux.

thermal henry [THERMO] A unit of thermal inductance equal to the product of a temperature difference of 1 kelvin and a time of 1 second divided by a rate of flow of entropy of 1 watt per kelvin.

thermal hysteresis [THERMO] A phenomenon sometimes observed in the behavior of a temperature-dependent property of a body; it is said to occur if the behavior of such a property is different when the body is heated through a given temperature range from when it is cooled through the same temperature range.

thermal inductance [THERMO] The product of temperature difference and time divided by entropy flow.

thermal instability [FL MECH] The instability resulting in free convection in a fluid heated at a boundary.

thermal ohm [THERMO] A unit of thermal resistance equal to the thermal resistance for which a temperature difference of 1 kelvin produces a flow of entropy of 1 watt per kelvin. Also known as fourier.

thermal potential difference [THERMO] The difference between the thermodynamic temperatures of two points.

thermal probe [MECH ENG] A calorimeter in a boiler furnace which measures heat absorption rates.

thermal radiation *See* heat radiation.

thermal resistance [THERMO] A measure of a body's ability to prevent heat from flowing through it, equal to the difference between the temperatures of opposite faces of the body divided by the rate of heat flow. Also known as heat resistance.

thermal resistivity [THERMO] The reciprocal of the thermal conductivity.

thermal Rossby number [FL MECH] The nondimensional ratio of the inertial force due to the thermal wind and the Coriolis force in the flow of a fluid which is heated from below.

thermal stress [MECH] Mechanical stress induced in a body when some or all of its parts are not free to expand or contract in response to changes in temperature.

thermal stress cracking [MECH] Crazing or cracking of materials (plastics or metals) by overexposure to elevated temperatures and sudden temperature changes or large temperature differentials.

thermal value [THERMO] Heat produced by combustion, usually expressed in calories per gram or British thermal units per pound.

thermie [THERMO] A unit of heat energy equal to the heat energy needed to raise 1 tonne of water from 14.5°C to 15.5°C at a constant pressure of 1 standard atmosphere; equal to 10^6 fifteen-degrees calories or $(4.1855 \pm 0.0005) \times 10^6$ joules. Abbreviated th.

thermochemical calorie *See* calorie.

thermocompression evaporator [MECH ENG] A system to reduce the energy requirements for evaporation by compressing the vapor from a single-effect evaporator so that the vapor can be used as the heating medium in the same evaporator.

thermodynamic cycle [THERMO] A procedure or arrangement in which some material goes through a cyclic process and one form of energy, such as heat at an elevated temperature from combustion of a fuel, is in part converted to another form, such as mechanical energy of a shaft, the remainder being rejected to a lower temperature sink. Also known as heat cycle.

thermodynamic equation of state [THERMO] An equation that relates the reversible change in energy of a thermodynamic system to the pressure, volume, and temperature.

thermodynamic equilibrium [THERMO] Property of a system which is in mechanical, chemical, and thermal equilibrium.

thermodynamic function of state [THERMO] Any of the quantities defining the thermodynamic state of a substance in thermodynamic equilibrium; for a perfect gas, the pressure, temperature, and density are the fundamental thermodynamic variables, any two of which are, by the equation of state, sufficient to specify the state. Also known as state parameter; state variable; thermodynamic variable.

thermodynamic potential [THERMO] One of several extensive quantities which are determined by the instantaneous state of a thermodynamic system, independent of its previous history, and which are at a minimum when the system is in thermodynamic equilibrium under specified conditions.

thermodynamic potential at constant volume *See* free energy.

thermodynamic principles [THERMO] Laws governing the conversion of energy from one form to another.

thermodynamic probability [THERMO] Under specified conditions, the number of equally likely states in which a substance may exist; the thermodynamic probability Ω is related to the entropy S by $S = k \ln \Omega$, where k is Boltzmann's constant.

thermodynamic process [THERMO] A change of any property of an aggregation of matter and energy, accompanied by thermal effects.

thermodynamic property [THERMO] A quantity which is either an attribute of an entire system or is a function of position which is continuous and does not vary rapidly over microscopic distances, except possibly for abrupt changes at boundaries between phases of the system; examples are temperature, pressure, volume, concentration, surface tension, and viscosity. Also known as macroscopic property.

thermodynamic system [THERMO] A part of the physical world as described by its thermodynamic properties.

thermodynamic temperature scale [THERMO] Any temperature scale in which the ratio of the temperatures of two reservoirs is equal to the ratio of the amount of heat absorbed from one of them by a heat engine operating in a Carnot cycle to the amount of heat rejected by this engine to the other reservoir; the Kelvin scale and the Rankine scale are examples of this type.

thermodynamic variable *See* thermodynamic function of state.

thermojet [AERO ENG] Air-duct-type engine in which air is scooped up from the surrounding atmosphere, compressed, heated by combustion, and then expanded and discharged at high velocity.

thermometric conductivity *See* diffusivity.

thermometry [THERMO] The science and technology of measuring temperature, and the establishment of standards of temperature measurement.

thermosiphon [MECH ENG] A closed system of tubes connected to a water-cooled engine which permit natural circulation and cooling of the liquid by utilizing the difference in density of the hot and cool portions.

thetagram [THERMO] A thermodynamic diagram with coordinates of pressure and temperature, both on a linear scale.

thickness ratio [AERO ENG] The ratio of the maximum thickness of an airfoil section to the length of its chord.

third law of motion *See* Newton's third law.

third law of thermodynamics [THERMO] The entropy of all perfect crystalline solids is zero at absolute zero temperature.

13.0 temperature *See* annealing point.

Thoma cavitation coefficient [MECH ENG] The equation for measuring cavitation in a hydraulic turbine installation, relating vapor pressure, barometric pressure, runner setting, tail water, and head.

thou *See* mil.

thread [DES ENG] A continuous helical rib, as on a screw or pipe.

thread contour [DES ENG] The shape of thread design as observed in a cross section along the major axis, for example, square or round.

thread cutter [MECH ENG] A tool used to cut screw threads on a pipe, screw, or bolt.

thread gage [DES ENG] A design gage used to measure screw threads.

threading die [MECH ENG] A die which may be solid, adjustable, or spring adjustable, or a self-opening die head, used to produce an external thread on a part.

threading machine [MECH ENG] A tool used to cut or form threads inside or outside a cylinder or cone.

thread plug gage [DES ENG] A thread gage used to measure female screw threads.

thread ring gage [DES ENG] A thread gage used to measure male screw threads.

three-body problem [MECH] The problem of predicting the motions of three objects obeying Newton's laws of motion and attracting each other according to Newton's law of gravitation.

three-dimensional flow [FL MECH] Any fluid flow which is not a two-dimensional flow.

three-jaw chuck [DES ENG] A drill chuck having three serrated-face movable jaws that can grip and hold fast an inserted drill rod.

threshold value [CONT SYS] The minimum input that produces a corrective action in an automatic control system.

throat [DES ENG] The narrowest portion of a constricted duct, as in a diffuser or a venturi tube; specifically, a nozzle throat.

throatable [DES ENG] Of a nozzle, designed to allow a change in the velocity of the exhaust stream by changing the size and shape of the throat of the nozzle.

throat velocity *See* critical velocity.

throttled flow [FL MECH] Flow which is forced to pass through a restricted area, where the velocity must increase.

throttle valve [MECH ENG] A choking device to regulate flow of a liquid, for example, in a pipeline, to an engine or turbine, from a pump or compressor.

throttling [AERO ENG] The varying of the thrust of a rocket engine during powered flight. [CONT SYS] Control by means of intermediate steps between full on and full off.

through-feed centerless grinding [MECH ENG] A metal cutting process by which the external surface of a cylindrical workpiece of uniform diameter is ground by passing the workpiece between a grinding and regulating wheel.

throwout [MECH ENG] In automotive vehicles, the mechanism or assemblage of mechanisms by which the driven and driving plates of a clutch are separated.

throw-out spiral *See* lead-out groove.

thrust [MECH] The force exerted in any direction by a fluid jet or by a powered screw. [MECH ENG] The weight or pressure applied to a bit to make it cut.

thrust augmentation [AERO ENG] The increasing of the thrust of an engine or power plant, especially of a jet engine and usually for a short period of time, over the thrust normally developed.

thrust augmenter [AERO ENG] Any contrivance used for thrust augmentation, as a venturi used in a rocket.

thrust axis [AERO ENG] A line or axis through an aircraft or a rocket, along which the thrust acts; an axis through the longitudinal center of a jet or rocket engine, along which the thrust of the engine acts. Also known as axis of thrust; center of thrust.

thrust bearing [MECH ENG] A bearing which sustains axial loads and prevents axial movement of a loaded shaft.

thrust coefficient *See* nozzle thrust coefficient.

thruster [AERO ENG] A control jet employed in spacecraft; an example would be one utilizing hydrogen peroxide.

thrust horsepower [AERO ENG] **1.** The force-velocity equivalent of the thrust developed by a jet or rocket engine. **2.** The thrust of an engine-propeller combination expressed in horsepower; it differs from the shaft horsepower of the engine by the amount the propeller efficiency varies from 100%.

thrust load [MECH ENG] A load or pressure parallel to or in the direction of the shaft of a vehicle.

Thrustor [MECH ENG] Trademark for a hydraulic device for applying a controllable force, as to a brake.

thrust output [AERO ENG] The net thrust delivered by a jet engine, rocket engine, or rocket motor.

thrust-pound [AERO ENG] A unit of measurement for the thrust produced by a jet engine or rocket.

thrust power [AERO ENG] The power usefully expended on thrust, equal to the thrust (or net thrust) times airspeed.

thrust reverser [AERO ENG] A device or apparatus for reversing thrust, especially of a jet engine.

thrust section [AERO ENG] A section in a rocket vehicle that houses or incorporates the combustion chamber or chambers and nozzles.

thrust terminator [AERO ENG] A device for ending the thrust in a rocket engine, either through propellant cutoff (in the case of a liquid) or through diverting the flow of gases from the nozzle.

thrust-weight ratio [AERO ENG] A quantity used to evaluate engine performance, obtained by dividing the thrust output by the engine weight less fuel.

thrust yoke [MECH ENG] The part connecting the piston rods of the feed mechanism on a hydraulically driven diamond-drill swivel head to the thrust block, which forms the connecting link between the yoke and the drive rod, by means of which link the longitudinal movements of the feed mechanism are transmitted to the swivel-head drive rod. Also known as back end.

thumbscrew [DES ENG] A screw with a head flattened in the same axis as the shaft so that it can be gripped and turned by the thumb and forefinger.

Thunderchief [AERO ENG] A United States supersonic, single-engine, turbojet-powered tactical fighter capable of delivering nuclear weapons as well as nonnuclear bombs and rockets; an all-weather attack fighter, it is also capable of close support for ground forces, and its range can be extended by in-flight refueling; it is equipped with the Sidewinder missile. Designated F-105.

Thunderstreak [AERO ENG] A United States one-pilot fighter also used for reconnaissance; it has a range of over 2000 miles (3200 kilometers) and a speed of over 600 miles (970 kilometers) per hour; the bomb load is 6000 pounds (2700 kilograms) of conventional or nuclear bombs, incendiary gel, or rockets, and there are six .50-caliber machine guns. Designated F-84.

tick [THERMO] An adiabatic, irreversible process in which a gas expands by passing from one chamber to another chamber which is at a lower pressure than the first chamber.

tie plate [MECH ENG] A plate used in a furnace to connect tie rods.

tie rod [MECH ENG] A rod used as a mechanical or structural support between elements of a machine.

Tiger [AERO ENG] A United States single-engine, single-seat, supersonic jet fighter designed for operating from aircraft carriers for the interception and destruction of enemy aircraft, and the support of troops ashore; armament consists of Sidewinders, cannons, and rocket packs. Designated F-11.

tight [MECH ENG] 1. Inadequate clearance or the barest minimum of clearance between working parts. 2. The absence of leaks in a pressure system.

tight fit [DES ENG] A fit between mating parts with slight negative allowance, requiring light to moderate force to assemble.

tilt [AERO ENG] The inclination of an aircraft, winged missile, or the like from the horizontal, measured by reference to the lateral axis or to the longitudinal axis.

tilting dozer [MECH ENG] A bulldozer whose blade can be pivoted on a horizontal center pin to cut low on either side.

tilting idlers [MECH ENG] An arrangement of idler rollers in which the top set is mounted on vertical arms which pivot on spindles set low down on the frame of the roller stool.

tilting mixer [MECH ENG] A small-batch mixer consisting of a rotating drum which can be tilted to discharge the contents; used for concrete or mortar.

tilt rotor [AERO ENG] An assembly of rapidly rotating blades on a vertical takeoff and landing aircraft, whose plane of rotation can be continuously varied from the horizontal to the vertical, permitting performance as helicopter blades or as propeller blades.

time curve *See* time front.

time-distance graph [AERO ENG] A graph used to determine the ground distance for air-route legs of a specified time interval; time-distance relationships are often simplified by considering air, wind, and ground distances for flight legs of 1-hour duration.

time front [AERO ENG] A locus of points representing the maximum ground distances from a departure point that can be covered by an aircraft in a prescribed time interval. Also known as hour-out line; time curve.

time-invariant system [CONT SYS] A system in which all quantities governing the system's behavior remain constant with time, so that the system's response to a given input does not depend on the time it is applied.

time of flight [MECH] Elapsed time in seconds from the instant a projectile or other missile leaves a gun or launcher until the instant it strikes or bursts.

timer [MECH ENG] A device that controls timing of the ignition spark of an internal combustion engine at the correct time.

time separation [AERO ENG] The time interval between adjacent aircraft flying approximately the same path.

time-varying system [CONT SYS] A system in which certain quantities governing the system's behavior change with time, so that the system will respond differently to the same input at different times.

timing [MECH ENG] Adjustment in the relative position of the valves and crankshaft of an automobile engine in order to produce the largest effective output of power.

timing belt [DES ENG] A power transmission belt with evenly spaced teeth on the bottom side which mesh with grooves cut on the periphery of the pulley to produce a positive, no-slip, constant-speed drive. Also known as cogged belt; synchronous belt. [MECH ENG] A positive drive belt that has axial cogs molded on the underside of the belt which fit into grooves on the pulley; prevents slip, and makes accurate timing possible; combines the advantages of belt drives with those of chains and gears. Also known as positive drive belt.

timing belt pulley [MECH ENG] A pulley that is similar to an uncrowned flat-belt pulley, except that the grooves for the belt's teeth are cut in the pulley's face parallel to the axis.

timing gears [MECH ENG] The gear train of reciprocating engine mechanisms for relating camshaft speed to crankshaft speed.

tinner's rivet [DES ENG] A special-purpose rivet that has a flat head, used in sheet metal work.

tip [DES ENG] A piece of material secured to and differing from a cutter tooth or blade.

tipped bit [DES ENG] A drill bit in which the cutting edge is made of especially hard material.

tipped solid cutters [DES ENG] Cutters made of one material and having tips or cutting edges of another material bonded in place.

tire iron [DES ENG] A single metal bar having bladelike ends of various shapes to insert between the rim and the bead of a pneumatic tire to remove or replace the tire.

Tiros satellite [AERO ENG] Television infrared observation satellite; a meteorological satellite that takes television pictures of cloud cover, using radiation sensors and cameras; it stores and transmits this information on ground command.

TME *See* metric-technical unit of mass.

to-and-fro ropeway *See* jig back.

toe-in [MECH ENG] The degree (usually expressed in fractions of an inch) to which the forward part of the front wheels of an automobile are closer together than the rear part, measured at hub height with the wheels in the normal "straight ahead" position of the steering gear.

toe-out [MECH ENG] The outward inclination of the wheels of an automobile at the front on turns due to setting the steering arms at an angle.

toggle [MECH ENG] A form of jointed mechanism for the amplification of forces.

toggle bolt [DES ENG] A bolt having a nut with a pair of pivotal wings that close against a spring; wings open after emergence through a hole or passage in a thin or hollow wall to fasten the unit securely.

toggle press [MECH ENG] A mechanical press in which a toggle mechanism actuates the slide.

tolerance [DES ENG] The permissible variations in the dimensions of machine parts.

tolerance chart [DES ENG] A chart indicating graphically the sequence in which dimensions must be produced on a part so that the finished product will meet the prescribed tolerance limits.

tolerance limits [DES ENG] The extreme values (upper and lower) that are permitted by the tolerance.

tolerance unit [DES ENG] A unit of length used to express the degree of tolerance allowed in fitting cylinders into cylindrical holes, equal, in micrometers, to $0.45\ D^{1/3} + 0.001\ D$, where D is the cylinder diameter in millimeters.

ton [MECH] **1.** A unit of weight in common use in the United States, equal to 2000 pounds or 907.18474 kilogram-force. Also known as just ton; net ton; short ton. **2.** A unit of mass in common use in the United Kingdom equal to 2240 pounds, or to 1016.0469088 kilogram-force. Also known as gross ton; long ton. **3.** A unit of weight in troy measure, equal to 2000 troy pounds, or to 746.4834432 kilogram-force. **4.** *See* tonne. [MECH ENG] A unit of refrigerating capacity, that is, of rate of heat flow, equal to the rate of extraction of latent heat when one short ton of ice of specific latent heat 144 international table British thermal units per pound is produced from water at the same temperature in 24 hours; equal to 200 British thermal units per minute, or to approximately 3516.85 watts. Also known as standard ton.

tondal [MECH] A unit of force equal to the force which will impart an acceleration of 1 foot per second to a mass of 1 long ton; equal to approximately 309.6911 newtons.

tongs [DES ENG] Any of various devices for holding, handling, or lifting materials and consisting of two legs joined eccentrically by a pivot or spring.

tongue and groove [DES ENG] A joint in which a projecting rib on the edge of one board fits into a groove in the edge of another board.

tonne [MECH] A unit of mass in the metric system, equal to 1000 kilograms or to approximately 2204.62 pound mass. Also known as metric ton; millier; ton; tonneau.

tonneau *See* tonne.

tool design [DES ENG] The division of mechanical design concerned with the design of tools.

tool-dresser [MECH ENG] A tool-stone-grade diamond inset in a metal shank and used to trim or form the face of a grinding wheel.

toolhead [MECH ENG] The adjustable tool-carrying part of a machine tool.

tool post [MECH ENG] A device to clamp and position a tool holder on a machine tool.

tooth [DES ENG] **1.** One of the regular projections on the edge or face of a gear wheel. **2.** An angular projection on a tool or other implement, such as a rake, saw, or comb.

tooth point [DES ENG] The chamfered cutting edge of the blade of a face mill.

top [MECH] A rigid body, one point of which is held fixed in an inertial reference frame, and which usually has an axis of symmetry passing through this point; its motion is usually studied when it is spinning rapidly about the axis of symmetry.

top dead center [MECH ENG] The dead-center position of an engine piston and its crankshaft arm when at the top or outer end of its stroke.

topping governor *See* limit governor.

topple [MECH] In gyroscopes for marine or aeronautical use, the condition of a sudden upset gyroscope or a gyroscope platform evidenced by a sudden and rapid precession of the spin axis due to large torque disturbances such as the spin axis striking the mechanical stops. Also known as tumble.

topple axis [MECH] Of a gyroscope, the horizontal axis, perpendicular to the horizontal spin axis, around which topple occurs. Also known as tumble axis.

topside sounder [AERO ENG] A satellite designed to measure ion concentration in the ionosphere from above the ionosphere.

toromatic transmission [MECH ENG] A semiautomatic transmission; it contains a compound planetary gear train with a torque converter.

torque [MECH] **1.** For a single force, the cross product of a vector from some reference point to the point of application of the force with the force itself. Also known as moment of force; rotation moment. **2.** For several forces, the vector sum of the torques (first definition) associated with each of the forces.

torque arm [MECH ENG] In automotive vehicles, an arm to take the torque of the rear axle.

torque converter [MECH ENG] A device for changing the torque speed or mechanical advantage between an input shaft and an output shaft.

torque reaction [MECH ENG] On a shaft-driven vehicle, the reaction between the bevel pinion with its shaft (which is supported in the rear axle housing) and the bevel ring gear (which is fastened to the differential housing) that tends to rotate the axle housing around the axle instead of rotating the axle shafts alone.

torque-winding diagram [MECH ENG] A diagram showing how the winding load on a winch drum varies and is used to decide the method of balancing needed; made by plotting the turning moment in pounds per foot on the vertical axis against time, or revolutions or depth on the horizontal axis.

torr [MECH] A unit of pressure, equal to 1/760 atmosphere; it differs from 1 millimeter of mercury by less than one part in seven million; approximately equal to 133.3224 pascals.

Torricellian vacuum [FL MECH] The space enclosed above a column of mercury when a tube, closed at one end, is filled with mercury and then placed, open end downward, in a well of mercury; this space is evacuated except for mercury vapor.

Torricelli's law of efflux [FL MECH] The velocity of efflux of liquid from an orifice in a container is equal to that which would be attained by a body falling freely from rest at the free surface of the liquid to the orifice.

torsiometer [MECH ENG] An instrument which measures power transmitted by a rotating shaft; consists of angular scales mounted around the shaft from which twist of the loaded shaft is determined. Also known as torsionmeter.

torsion [MECH] A twisting deformation of a solid body about an axis in which lines that were initially parallel to the axis become helices.

torsional angle [MECH] The total relative rotation of the ends of a straight cylindrical bar when subjected to a torque.

torsional modulus [MECH] The ratio of the torsional rigidity of a bar to its length. Also known as modulus of torsion.

torsional pendulum [MECH] A device consisting of a disk or other body of large moment of inertia mounted on one end of a torsionally flexible elastic rod whose other end is held fixed; if the disk is twisted and released, it will undergo simple harmonic motion, provided the torque in the rod is proportional to the angle of twist.

torsional rigidity [MECH] The ratio of the torque applied about the centroidal axis of a bar at one end of the bar to the resulting torsional angle, when the other end is held fixed.

torsional vibration [MECH] A periodic motion of a shaft in which the shaft is twisted about its axis first in one direction and then in the other; this motion may be superimposed on rotational or other motion.

torsion bar [MECH ENG] A spring flexed by twisting about its axis; found in the spring suspension of truck and passenger car wheels, in production machines where space limitations are critical, and in high-speed mechanisms where inertia forces must be minimized.

torsion damper [MECH ENG] A damper used on automobile internal combustion engines to reduce torsional vibration.

torsionmeter *See* torsiometer.

total heat *See* enthalpy.

total impulse [AERO ENG] The product of the thrust and the time over which the thrust is produced, expressed in pounds (force)–seconds; used especially in reference to a rocket motor or a rocket engine.

total lift [AERO ENG] The upward force produced by the gas in a balloon; it is equal to the sum of the free lift, the weight of the balloon, and the weight of auxiliary equipment carried by the balloon.

total pressure [FL MECH] *See* dynamic pressure. [MECH] The gross load applied on a given surface.

total vorticity [FL MECH] Usually, the magnitude of the vorticity vector, all components included, as opposed to the vertical (component of the) vorticity.

toughness [MECH] A property of a material capable of absorbing energy by plastic deformation; intermediate between softness and brittleness.

towed load [MECH] The weight of a carriage, trailer, or other equipment towed by a prime mover.

tower bolt *See* barrel bolt.

tracer milling [MECH ENG] Cutting a duplicate of a three-dimensional form by using a mastic form to direct the tracer-controlled cutter.

track [AERO ENG] The actual line of movement of an aircraft or a rocket over the surface of the earth; it is the projection of the history of the flight path on the surface. Also known as flight track. [MECH ENG] 1. The slide or rack on which a diamond-drill swivel head can be moved to positions above and clear of the collar of a borehole. 2. A crawler mechanism for earth-moving equipment.

Tracker [AERO ENG] A United States twin-reciprocating-engine, antisubmarine aircraft capable of operating from carriers, and designed primarily for the detection, location, and destruction of submarines. Designated S-2.

tracking problem [CONT SYS] The problem of determining a control law which when applied to a dynamical system causes its output to track a given function; the performance index is in many cases taken to be of the integral square error variety.

track made good [AERO ENG] The actual path of an aircraft over the surface of the earth, or its graphic representation.

traction [MECH] Pulling friction of a moving body on the surface on which it moves.

tractional force [FL MECH] The force exerted on particles under flowing water by the current; it is proportional to the square of the velocity.

tractor [MECH ENG] 1. An automotive vehicle having four wheels or a caterpillar tread used for pulling agricultural or construction implements. 2. The front pulling section of a semitrailer. Also known as truck-tractor.

tractor drill [MECH ENG] A drill having a crawler mounting to support the feed-guide bar on an extendable arm.

tractor loader [MECH ENG] A tractor equipped with a tipping bucket which can be used to dig and elevate soil and rock fragments to dump at truck height. Also known as shovel dozer; tractor shovel.

tractor shovel *See* tractor loader.

traffic pattern [AERO ENG] The traffic flow that is prescribed for aircraft landing at, taxiing on, and taking off from an airport; the usual components of a traffic pattern are upwind leg, crosswind leg, downwind leg, base leg, and final approach.

trail angle [AERO ENG] The angle at an aircraft between the vertical and the line of sight to an object over which the aircraft has passed.

trailer [MECH ENG] The section of a semitrailer that is pulled by the tractor.

trail formation [AERO ENG] Aircraft flying singly or in elements in such manner that each aircraft or element is in line behind the preceding aircraft or element.

trailing edge [AERO ENG] The rear section of a multipiece airfoil, usually that portion aft of the rear spar.

trailing-edge tab [AERO ENG] One of the devices on the aircraft elevator that reduce or eliminate hinge movements required to deflect the elevator during flight.

trajectory [MECH] The curve described by an object moving through space, as of a meteor through the atmosphere, a planet around the sun, a projectile fired from a gun, or a rocket in flight.

tramway [MECH ENG] An overhead rail, rope, or cable on which wheeled cars run to convey a load.

transfer caliper [DES ENG] A caliper having one leg which can be opened (or closed) to remove the instrument from the piece being measured; used to measure inside recesses or over projections.

transfer ellipse *See* transfer orbit.

transfer function [CONT SYS] The mathematical relationship between the output of a control system and its input: for a linear system, it is the Laplace transform of the output divided by the Laplace transform of the input under conditions of zero initial-energy storage.

transfer matrix [CONT SYS] The generalization of the concept of a transfer function to a multivariable system; it is the matrix whose product with the vector representing the input variables yields the vector representing the output variables.

transfer orbit [AERO ENG] In interplanetary travel, an elliptical trajectory tangent to the orbits of both the departure planet and the target planet. Also known as transfer ellipse.

transient problem *See* initial-value problem.

transition [THERMO] A change of a substance from one of the three states of matter to another.

transitional fit [DES ENG] A fit with varying clearances due to specified tolerances on the shaft and sleeve or hole.

transitional flow [FL MECH] A flow in which the viscous and Reynolds stresses are of approximately equal magnitude; it is transitional between laminar flow and turbulent flow.

transition altitude [AERO ENG] The altitude in the vicinity of an aerodrome at or below which the vertical position of an aircraft is controlled by reference to true altitude.

transition flow [AERO ENG] A flow of fluid about an airfoil that is changing from laminar flow to turbulent flow.

transition point [THERMO] Either the temperature at which a substance changes from one state of aggregation to another (a first-order transition), or the temperature of culmination of a gradual change, such as the lambda point, or Curie point (a second-order transition). Also known as transition temperature.

transition temperature *See* transition point.

transition zone [FL MECH] Those conditions of fluid flow in which the nature of the flow is changing from laminar to turbulent.

Transit satellite [AERO ENG] One of a system of passive, low-orbiting satellites which provide high-accuracy fixes using the Doppler technique several times a day at every point on earth, for navigation and geodesy.

translation [MECH] The linear movement of a point in space without any rotation.

translational motion [MECH] Motion of a rigid body in such a way that any line which is imagined rigidly attached to the body remains parallel to its original direction.

transmissibility [MECH] A measure of the ability of a system either to amplify or to suppress an input vibration, equal to the ratio of the response amplitude of the system in steady-state forced vibration to the excitation amplitude; the ratio may be in forces, displacements, velocities, or accelerations.

transmission [MECH ENG] The gearing system by which power is transmitted from the engine to the live axle in an automobile. Also known as gearbox.

transonic flight [AERO ENG] Flight of vehicles at speeds near the speed of sound (660 miles per hour or 1060 kilometers per hour, at 35,000 feet or 10,700 meters altitude), characterized by great increase in drag, decrease in lift at any altitude, and abrupt changes in the moments acting on the aircraft; the vehicle may shake or buffet.

transonic flow [FL MECH] Flow of a fluid over a body in the range just above and just below the acoustic velocity.

transonic range [FL MECH] The range of speeds between the speed at which one point on a body reaches supersonic speed, and the speed at which all points reach supersonic speed.

transonic speed [FL MECH] The speed of a body relative to the surrounding fluid at which the flow is in some places on the body subsonic and in other places supersonic.

transpiration cooling *See* sweat cooling.

transportation lag *See* distance/velocity lag.

transporter crane [MECH ENG] A long lattice girder supported by two lattice towers which may be either fixed or moved along rails laid at right angles to the girder; a crab with a hoist suspended from it travels along the girder.

transport lag *See* distance/velocity lag.

transport vehicle [MECH ENG] Vehicle primarily intended for personnel and cargo carrying.

transverse vibration [MECH] Vibration of a rod in which elements of the rod move at right angles to the axis of the rod.

trap [AERO ENG] That part of a rocket motor that keeps the propellant grain in place. [MECH ENG] A device which reduces the effect of the vapor pressure of oil or mercury on the high-vacuum side of a diffusion pump.

trapezoidal excavator [MECH ENG] A digging machine which removes earth in a trapezoidal cross-section pattern for canals and ditches.

traveling block [MECH ENG] The movable unit, consisting of sheaves, frame, clevis, and hook, connected to, and hoisted or lowered with, the load in a block-and-tackle system. Also known as floating block; running block.

traveling-grate stoker [MECH ENG] A type of furnace stoker; coal feeds by gravity into a hopper located on top of one end of a moving (traveling) grate; as the grate passes under the hopper, it carries a bed of fresh coal toward the furnace.

tray elevator [MECH ENG] A device for lifting drums, barrels, or boxes; a parallel pair of vertical-mounted continuous chains turn over upper and lower drive gears, and spaced trays on the chains cradle and lift the objects to be moved.

trencher *See* trench excavator.

trench excavator [MECH ENG] A digging machine, usually on crawler tracks, and having either a movable wheel or a continuous chain on which buckets are mounted. Also known as bucket-ladder excavator; ditcher; trencher; trenching machine.

trenching machine *See* trench excavator.

trepanning tool [MECH ENG] A cutting tool in the form of a circular tube, having teeth on the end; the workpiece or tube, or both, are rotated and the tube is fed axially into the workpiece, leaving behind a narrow grooved surface in the workpiece.

tricycle landing gear [AERO ENG] A landing-gear arrangement that places the nose gear well forward of the center of gravity on the fuselage and the two main gears slightly aft of the center of gravity, with sufficient distance between them to provide stability against rolling over during a yawed landing in a crosswind, or during ground maneuvers.

trigger pull [MECH] Resistance offered by the trigger of a rifle or other weapon; force which must be exerted to pull the trigger.

trim [AERO ENG] The orientation of an aircraft relative to the airstream, as indicated by the amount of control pressure required to maintain a given flight performance.

trimmer conveyor [MECH ENG] A self-contained, lightweight portable conveyor, usually of the belt type, for use in unloading and delivering bulk materials from trucks to domestic storage places, and for trimming bulk materials in bins or piles.

trip hammer [MECH ENG] A large power hammer whose head is tripped and falls by cam or lever action.

triple thread [DES ENG] A multiple screw thread having three threads or starts equally spaced around the periphery; the lead is three times the pitch.

triplex chain block [MECH ENG] A geared hoist using an epicyclic train.

tripod [DES ENG] An adjustable, collapsible three-legged support, as for a camera or surveying instrument.

tripod drill [MECH ENG] A reciprocating rock drill mounted on three legs and driven by steam or compressed air; the drill steel is removed and a longer drill inserted about every 2 feet (60 centimeters).

tripper [MECH ENG] A device that snubs a conveyor belt causing the load to be discharged.

trochoidal wave [FL MECH] A progressive oscillatory wave whose form is that of a prolate cycloid or trochoid; it is approximated by waves of small amplitude.

trolley [MECH ENG] 1. A wheeled car running on an overhead track, rail, or ropeway. 2. An electric streetcar.

trolley locomotive [MECH ENG] A locomotive operated by electricity drawn from overhead conductors by means of a trolley pole.

troughed belt conveyor [MECH ENG] A belt conveyor with the conveyor belt edges elevated on the carrying run to form a trough by conforming to the shape of the troughed carrying idlers or other supporting surface.

troughed roller conveyor [MECH ENG] A roller conveyor having two rows of rolls set at an angle to form a trough over which objects are conveyed.

troughing idler [MECH ENG] A belt idler having two or more rolls arranged to turn up the edges of the belt so as to form the belt into a trough.

troughing rolls [MECH ENG] The rolls of a troughing idler that are so mounted on an incline as to elevate each edge of the belt into a trough.

Trouton's rule [THERMO] The rule that, for a nonassociated liquid, the latent heat of vaporization in calories is equal to approximately 22 times the normal boiling point on the Kelvin scale.

trowel [DES ENG] Any of various hand tools consisting of a wide, flat or curved blade with a short wooden handle; used by gardeners, plasterers, and bricklayers.

troy ounce *See* ounce.

troy pound   *See* pound.

troy system   [MECH] A system of mass units used primarily to measure gold and silver; the ounce is the same as that in the apothecaries' system, being equal to 480 grains or 31.1034768 grams. Abbreviated t. Also known as troy weight.

troy weight   *See* troy system.

truck   [MECH ENG] A self-propelled wheeled vehicle, designed primarily to transport goods and heavy equipment; it may be used to tow trailers or other mobile equipment.

truck crane   [MECH ENG] A crane carried on the bed of a motortruck.

truck-mounted drill rig   [MECH ENG] A drilling rig mounted on a lorry or caterpillar tracks.

truck-tractor   *See* tractor.

true airspeed   [AERO ENG] The actual speed of an aircraft relative to the air through which it flies, that is, the calibrated airspeed corrected for temperature, density, or compressibility.

true-airspeed indicator   [AERO ENG] An instrument for measuring true airspeed. Also known as true-airspeed meter.

true-airspeed meter   *See* true-airspeed indicator.

true rake   [MECH ENG] The angle, measured in degrees, between a plane containing a tooth face and the axial plane through the tooth point in the direction of chip flow.

truing   [MECH ENG] 1. Cutting a grinding wheel to make its surface run concentric with the axis. 2. Aligning a wheel to be concentric and in one plane.

trunnion   [DES ENG] 1. Either of two opposite pivots, journals, or gudgeons, usually cylindrical and horizontal, projecting one from each side of a piece of ordnance, the cylinder of an oscillating engine, a molding flask, or a converter, and supported by bearings to provide a means of swiveling or turning. 2. A pin or pivot usually mounted on bearings for rotating or tilting something.

Tschudi engine   [MECH ENG] A cat-and-mouse engine in which the pistons, which are sections of a torus, travel around a toroidal cylinder; motion of the pistons is controlled by two cams which bear against rollers attached to the rotors.

tsi   [MECH] A unit of force equal to 1 ton-force per square inch; equal to approximately 1.54444×10^7 pascals.

T slot   [DES ENG] A recessed slot, in the form of an inverted T, in the table of a machine tool, to receive the square head of a T-slot bolt.

tsp   *See* teaspoonful.

tspn   *See* teaspoonful.

tube cleaner   [MECH ENG] A device equipped with cutters or brushes used to clean tubes in heat transfer equipment.

tube core   [AERO ENG] One type of sandwich configuration used in structural materials in aircraft; aluminum, steel, and titanium have been used for face materials with cores of wood, rubber, plastics, steel, and aluminum in the form of tubes.

tube door   [MECH ENG] A door in a boiler furnace wall which facilitates the removal or installation of tubes.

tube mill   [MECH ENG] A revolving cylinder used for fine pulverization of ore, rock, and other such materials; the material, mixed with water, is fed into the chamber from one end, and passes out the other end as slime.

tube turbining [MECH ENG] Cleaning tubes by passing a power-driven rotary device through them.

tumble *See* topple.

tumble axis *See* topple axis.

tumbler feeder *See* drum feeder.

tumbler gears [MECH ENG] Idler gears interposed between spindle and stud gears in a lathe gear train; used to reverse rotation of lead screw or feed rod.

tumbling [AERO ENG] An attitude situation in which the vehicle continues on its flight, but turns end over end about its center of mass.   [MECH ENG] Loss of control in a two-frame free gyroscope, occurring when both frames of reference become coplanar.

tumbling mill [MECH ENG] A grinding and pulverizing machine consisting of a shell or drum rotating on a horizontal axis.

tunnel borer [MECH ENG] Any boring machine for making a tunnel; often a ram armed with cutting faces operated by compressed air.

tunnel carriage [MECH ENG] A machine used for rapid tunneling, consisting of a combined drill carriage and manifold for water and air so that immediately the carriage is at the face, drilling may commence with no lost time for connecting up or waiting for drill steels; the air is supplied at pressures of 95 to 100 pounds per square inch (6.55 to 6.89 × 10^5 newtons per square meter).

turbine [MECH ENG] A fluid acceleration machine for generating rotary mechanical power from the energy in a stream of fluid.

turbine propulsion [MECH ENG] Propulsion of a vehicle or vessel by means of a steam or gas turbine.

turbine pump *See* regenerative pump.

turbining [MECH ENG] The removal of scale or other foreign material from the internal surface of a metallic cylinder.

turboblower [MECH ENG] A centrifugal or axial-flow compressor.

turbofan [AERO ENG] An air-breathing jet engine in which additional propulsive thrust is gained by extending a portion of the compressor or turbine blades outside the inner engine case.

turbojet [AERO ENG] A jet engine incorporating a turbine-driven air compressor to take in and compress the air for the combustion of fuel (or for heating by a nuclear reactor), the gases of combustion (or the heated air) being used both to rotate the turbine and to create a thrust-producing jet.

turbosupercharger [MECH ENG] A centrifugal air compressor, gas-turbine driven, usually used to increase induction system pressure in an internal combustion reciprocating engine.

turbulence *See* turbulent flow.

turbulence energy *See* eddy kinetic energy.

turbulent boundary layer [FL MECH] The layer in which the Reynolds stresses are much larger than the viscous stresses.

turbulent diffusion *See* eddy diffusion.

turbulent flow [FL MECH] Motion of fluids in which local velocities and pressures fluctuate irregularly, in a random manner. Also known as turbulence.

turbulent flux *See* eddy flux.

turbulent Schmidt number [FL MECH] A dimensionless number used in the study of mass transfer in turbulent flow, equal to the ratio of the eddy viscosity to the eddy mass diffusivity. Symbolized Sc_T.

turbulent shear force [FL MECH] A shear force in a fluid which arises from turbulent flow.

turnbuckle [DES ENG] A sleeve with a thread at one end and a swivel at the other, or with threads of opposite hands at each end so that by turning the sleeve connected rods or wire rope will be drawn together and tightened.

turning [MECH ENG] Shaping a member on a lathe.

turning-block linkage [MECH ENG] A variation of the sliding-block mechanical linkage in which the short link is fixed and the frame is free to rotate. Also known as the Wentworth quick-return motion.

turning error *See* northerly turning error.

turret lathe [MECH ENG] A semiautomatic lathe differing from the engine lathe in having the tailstock replaced with a multisided, indexing tool holder or turret designed to hold several tools.

twin-cable ropeway [MECH ENG] An aerial ropeway which has parallel track cables with carriers running in opposite directions; both rows of carriers are pulled by the same traction rope.

twin-geared press [MECH ENG] A crank press having the drive gears attached to both ends of the crankshaft.

twist [DES ENG] In a fiber, rope, yarn, or cord, the turns about its axis per unit length; usually expressed as TPI (turns per inch).

twist drill [DES ENG] A tool having one or more helical grooves, extending from the point to the smooth part of the shank, for ejecting cuttings and admitting a coolant.

two-body problem [MECH] The problem of predicting the motions of two objects obeying Newton's laws of motion and exerting forces on each other according to some specified law such as Newton's law of gravitation, given their masses and their positions and velocities at some initial time.

two-cycle engine [MECH ENG] A reciprocating internal combustion engine that requires two piston strokes or one revolution to complete a cycle.

two-degrees-of-freedom gyro [MECH] A gyro whose spin axis is free to rotate about two orthogonal axes, not counting the spin axis.

two-dimensional flow [FL MECH] Fluid flow in which all flow occurs in a set of parallel planes with no flow normal to them, and the flow is identical in each of these parallel planes.

two-lip end mill [MECH ENG] An end-milling cutter having two cutting edges and straight or helical flutes.

two-phase flow [FL MECH] Cocurrent movement of two phases (for example, gas and liquid) through a closed conduit or duct (for example, a pipe).

two-point press [MECH ENG] A mechanical press in which the slide is actuated at two points.

two-port system [CONT SYS] A system which has only one input or excitation and only one response or output.

two-position propeller [AERO ENG] An airplane propeller whose blades are limited to two angles, one for take off and climb and the other for cruising.

two-stroke cycle [MECH ENG] An internal combustion engine cycle completed in two strokes of the piston.

U-bend die [MECH ENG] A die with a square or rectangular cross section which provides two edges over which metal can be drawn.

U blades [DES ENG] Curved bulldozer blades designed to increase moving capacity of tractor equipment.

U bolt [DES ENG] A U-shaped bolt with threads at the ends of both arms to receive nuts.

ullage rocket [AERO ENG] A small rocket used in space to impart an acceleration to a tank system to ensure that the liquid propellants collect in the tank in such a manner as to flow properly into the pumps or thrust chamber.

ultimate-load design [DES ENG] Design of a beam that is proportioned to carry at ultimate capacity the design load multiplied by a safety factor. Also known as limit-load design; plastic design; ultimate-strength design.

ultimate strength [MECH] The tensile stress, per unit of the original surface area, at which a body will fracture, or continue to deform under a decreasing load.

ultimate-strength design *See* ultimate-load design.

ultragravity waves [FL MECH] Gravity waves which are characterized by periods in the 0.1–1.0 second range.

ultrasonic atomizer [MECH ENG] An atomizer in which liquid is fed to, or caused to flow over, a surface which vibrates at an ultrasonic frequency; uniform drops may be produced at low feed rates.

ultrasonic drill [MECH ENG] A drill in which a magnetostrictive transducer is attached to a tapered cone serving as a velocity transformer; with an appropriate tool at the end of the transformer, practically any shape of hole can be drilled in hard, brittle materials such as tungsten carbide and gems.

ultrasonic drilling [MECH ENG] A vibration drilling method in which ultrasonic vibrations are generated by the compression and extension of a core of electrostrictive or magnetostrictive material in a rapidly alternating electric or magnetic field.

ultrasonic machining [MECH ENG] The removal of material by abrasive bombardment and crushing in which a flat-ended tool of soft alloy steel is made to vibrate at a frequency of about 20,000 hertz and an amplitude of 0.001–0.003 inch (0.0254–0.0762 millimeter) while a fine abrasive of silicon carbide, aluminum oxide, or boron carbide is carried by a liquid between tool and work.

umbilical connections [AERO ENG] Electrical and mechanical connections to a launch vehicle prior to lift off; the umbilical tower adjacent to the vehicle on the launch pad supports these connections which supply electrical power, control signals, data links, propellant loading, high pressure gas transfer, and air conditioning.

umbilical cord [AERO ENG] Any of the servicing electrical or fluid lines between the ground or a tower and an uprighted rocket vehicle before the launch. Also known as umbilical.

umbilical tower [AERO ENG] A vertical structure supporting the umbilical cords running into a rocket in launching position.

unavailable energy [THERMO] That part of the energy which, when an irreversible process takes place, is initially in a form completely available for work and is converted to a form completely unavailable for work.

undercarriage [AERO ENG] The landing gear assembly for an aircraft.

underdrive press [MECH ENG] A mechanical press having the driving mechanism located within or under the bed.

underhung crane [MECH ENG] An overhead traveling crane in which the end trucks carry the bridge suspended below the rails.

underplate [DES ENG] An unfinished plate which forms part of an armored front for a mortise lock, and which is fastened to the case.

undershot wheel [MECH ENG] A water wheel operated by the impact of flowing water against blades attached around the periphery of the wheel, the blades being partly or totally submerged in the moving stream of water.

underspin [MECH] Property of a projectile having insufficient rate of spin to give proper stabilization.

unfinished bolt [DES ENG] One of three degrees of finish in which standard hexagon wrench-head bolts and nuts are available; only the thread is finished.

uniaxial stress [MECH] A state of stress in which two of the three principal stresses are zero.

unified screw thread [DES ENG] Three series of threads: coarse (UNC), fine (UNF), and extra fine (UNEF); a ¼-inch (0.006 millimeter) diameter thread in the UNC series has 20 threads per inch, while in the UNF series it has 28.

uniflow engine [MECH ENG] A steam engine in which steam enters the cylinder through valves at one end and escapes through openings uncovered by the piston as it completes its stroke.

uniform circular motion [MECH] Circular motion in which the angular velocity remains constant.

uniform load [MECH] A load distributed uniformly over a portion or over the entire length of a beam; measured in pounds per foot.

unilateral tolerance method [DES ENG] Method of dimensioning and tolerancing wherein the tolerance is taken as plus or minus from an explicitly stated dimension; the dimension represents the size or location which is nearest the critical condition (that is maximum material condition), and the tolerance is applied either in a plus or minus direction, but not in both directions, in such a way that the permissible variation in size or location is away from the critical condition.

union [DES ENG] A screwed or flanged pipe coupling usually in the form of a ring fitting around the outside of the joint.

union joint [DES ENG] A threaded assembly used for the joining of ends of lengths of installed pipe or tubing where rotation of neither length is feasible.

unitary air conditioner [MECH ENG] A small self-contained electrical unit enclosing a motor-driven refrigeration compressor, evaporative cooling coil, air-cooled condenser, filters, fans, and controls.

United States standard dry seal thread [DES ENG] A modified pipe thread used for pressure-tight connections that are to be assembled without lubricant or sealer in refrigeration pipes, automotive and aircraft fuel-line fittings, and gas and chemical shells.

unit heater [MECH ENG] A heater consisting of a fan for circulating air over a heat-exchange surface, all enclosed in a common casing.

unit strain [MECH] **1.** For tensile strain, the elongation per unit length. **2.** For compressive strain, the shortening per unit length. **3.** For shear strain, the change in angle between two lines originally perpendicular to each other.

unit stress [MECH] The load per unit of area.

univariant system [THERMO] A system which has only one degree of freedom according to the phase rule.

universal grinding machine [MECH ENG] A grinding machine having a swivel table and headstock, and a wheel head that can be rotated on its base.

universal joint [MECH ENG] A linkage that transmits rotation between two shafts whose axes are coplanar but not coinciding.

unloader [MECH ENG] A power device for removing bulk materials from railway freight cars or highway trucks; in the case of railway cars, the car structure may aid the unloader; a transitional device between interplant transportation means and intraplant handling equipment.

unloading conveyor [MECH ENG] Any of several types of portable conveyors adapted for unloading bulk materials, packages, or objects from conveyances.

Unsin engine [MECH ENG] A type of rotary engine in which the trochoidal rotors of eccentric-rotor engines are replaced with two circular rotors, one of which has a single gear tooth upon which gas pressure acts, and the second rotor has a slot which accepts the gear tooth.

unsprung axle [MECH ENG] A rear axle in an automobile in which the housing carries the right and left rear-axle shafts and the wheels are mounted at the outer end of each shaft.

unsprung weight [MECH ENG] The weight of the various parts of a vehicle that are not carried on the springs such as wheels, axles, brakes, and so forth.

unsteady flow [FL MECH] Fluid flow in which properties of the flow change with respect to time.

unsteady-state flow [FL MECH] A condition of fluid flow in which the volumetric ratios of two or more phases (liquid-gas, liquid-liquid, and so on) vary along the course of flow; can be the result of changes in temperature, pressure, or composition.

updraft carburetor [MECH ENG] For a gasoline engine, a fuel-air mixing device in which both the fuel jet and the airflow are upward.

updraft furnace [MECH ENG] A furnace in which volumes of air are supplied from below the fuel bed or supply.

upmilling [MECH ENG] Milling a workpiece by rotating the cutter against the direction of feed of the workpiece.

upper consolute temperature *See* consolute temperature.

upper critical solution temperature *See* consolute temperature.

utilization rate [AERO ENG] The amount of flying time produced in a specific period expressed in hours per period per aircraft. Also known as flying hour rate.

V

vac *See* millibar.

vacuum brake [MECH ENG] A form of air brake which operates by maintaining low pressure in the actuating cylinder; braking action is produced by opening one side of the cylinder to the atmosphere so that atmospheric pressure, aided in some designs by gravity, applies the brake.

vacuum cleaner [MECH ENG] An electrically powered mechanical appliance for the dry removal of dust and loose dirt from rugs, fabrics, and other surfaces.

vacuum heating [MECH ENG] A two-pipe steam heating system in which a vacuum pump is used to maintain a suction in the return piping, thus creating a positive return flow of air and condensate.

vacuum pump [MECH ENG] A compressor for exhausting air and noncondensable gases from a space that is to be maintained at subatmospheric pressure.

vacuum support [MECH ENG] That portion of a rupture disk device which prevents deformation of the disk resulting from vacuum or rapid pressure change.

valve [MECH ENG] A device used to regulate the flow of fluids in piping systems and machinery.

valve follower [MECH ENG] A linkage between the cam and the push rod of a valve train.

valve guide [MECH ENG] A channel which supports the stem of a poppet valve for maintenance of alignment.

valve head [MECH ENG] The disk part of a poppet valve that gives a tight closure on the valve seat.

valve-in-head engine *See* overhead-valve engine.

valve lifter [MECH ENG] A device for opening the valve of a cylinder as in an internal combustion engine.

valve positioner [CONT SYS] A pneumatic servomechanism which is used as a component in process control systems to improve operating characteristics of valves by reducing hysteresis. Also known as pneumatic servo.

valve seat [DES ENG] The circular metal ring on which the valve head of a poppet valve rests when closed.

valve stem [MECH ENG] The rod by means of which the disk or plug is moved to open and close a valve.

valve train [MECH ENG] The valves and valve-operating mechanism for the control of fluid flow to and from a piston-cylinder machine, for example, steam, diesel, or gasoline engine.

van der Waals surface tension formula [THERMO] An empirical formula for the dependence of the surface tension on temperature: $\gamma = K p_c^{2/3} \mathbf{T}_c^{1/3} (1 - T/T_c)^n$, where

γ is the surface tension, T is the temperature, T_c and p_c are the critical temperature and pressure, K is a constant, and n is a constant equal to approximately 1.23.

vane [AERO ENG] A device that projects ahead of an aircraft to sense gusts or other actions of the air so as to create impulses or signals that are transmitted to the control system to stabilize the aircraft. [MECH ENG] A flat or curved surface exposed to a flow of fluid so as to be forced to move or to rotate about an axis, to rechannel the flow, or to act as the impeller; for example, in a steam turbine, propeller fan, or hydraulic turbine.

vane motor rotary actuator [MECH ENG] A type of rotary motor actuator which consists of a rotor with several spring-loaded sliding vanes in an elliptical chamber; hydraulic fluid enters the chamber and forces the vanes before it as it moves to the outlets.

Vanguard satellite [AERO ENG] One of three artificial satellites launched by the United States in 1958 and 1959, using a modified Viking rocket, as a part of the International Geophysical Year program; *Vanguard 1* was the first spacecraft to use solar cells.

vapor [THERMO] A gas at a temperature below the critical temperature, so that it can be liquefied by compression, without lowering the temperature.

vapor-compression cycle [MECH ENG] A refrigeration cycle in which refrigerant is circulated through a machine which allows for successive boiling (or vaporization) of liquid refrigerant as it passes through an expansion valve, thereby producing a cooling effect in its surroundings, followed by compression of vapor to liquid.

vapor cycle [THERMO] A thermodynamic cycle, operating as a heat engine or a heat pump, during which the working substance is in, or passes through, the vapor state.

vaporization *See* volatilization.

vapor lock [FL MECH] Interruption of the flow of fuel in a gasoline engine caused by formation of vapor or gas bubbles in the fuel-feeding system.

vapor pressure [THERMO] For a liquid or solid, the pressure of the vapor in equilibrium with the liquid or solid.

variable-area exhaust nozzle [AERO ENG] On a jet engine, an exhaust nozzle of which the exhaust exit opening can be varied in area by means of some mechanical device, permitting variation in the jet velocity.

variable-cycle engine [AERO ENG] A type of gas turbine jet engine whose cycle parameters, such as pressure ratio, temperature, gas flow paths, and air-handling characteristics, can be varied between those of a turbojet and a turbofan, enabling it to combine the advantages of both.

variable flow [FL MECH] Fluid flow in which the velocity changes both with time and from point to point.

variable force [MECH] A force whose direction or magnitude or both change with time.

variable geometry aircraft [AERO ENG] Aircraft with variable profile geometry, such as variable sweep wings.

variable-speed drive [MECH ENG] A mechanism transmitting motion from one shaft to another that allows the velocity ratio of the shafts to be varied continuously.

Varignon's theorem [MECH] The theorem that the moment of a force is the algebraic sum of the moments of its vector components acting at a common point on the line of action of the force.

V belt [DES ENG] An endless power-transmission belt with a trapezoidal cross section which runs in a pulley with a V-shaped groove; it transmits higher torque at less width and tension than a flat belt. [MECH ENG] A belt, usually endless, with a trapezoidal cross section which runs in a pulley with a V-shaped groove, with the top surface of the belt approximately flush with the top of the pulley.

V-bend die [MECH ENG] A die with a triangular cross-sectional opening to provide two edges over which bending is accomplished.

V-bucket carrier [MECH ENG] A conveyor consisting of two strands of roller chain separated by V-shaped steel buckets; used for elevating and conveying nonabrasive materials, such as coal.

vector steering [AERO ENG] A steering method for rockets and spacecraft wherein one or more thrust chambers are gimbal-mounted so that the direction of the thrust force (thrust vector) may be tilted in relation to the center of gravity of the vehicle to produce a turning movement.

vehicle [AERO ENG] 1. A structure, machine, or device, such as an aircraft or rocket, designed to carry a burden through air or space. 2. More restrictively, a rocket vehicle. [MECH ENG] A self-propelled wheeled machine that transports people or goods on or off roads; automobiles and trucks are examples.

vehicle control system [AERO ENG] A system, incorporating control surfaces or other devices, which adjusts and maintains the altitude and heading, and sometimes speed, of a vehicle in accordance with signals received from a guidance system. Also known as flight control system.

vehicle mass ratio [AERO ENG] The ratio of the final mass of a vehicle after all propellant has been used, to the initial mass.

velocity [MECH] 1. The time rate of change of position of a body; it is a vector quantity having direction as well as magnitude. Also known as linear velocity. 2. The speed at which the detonating wave passes through a column of explosives, expressed in meters or feet per second.

velocity analysis [MECH] A graphical technique for the determination of the velocities of the parts of a mechanical device, especially those of a plane mechanism with rigid component links.

velocity coefficient [FL MECH] The ratio of the actual velocity of gas emerging from a nozzle to the velocity calculated under ideal conditions; it is less than 1 because of friction losses. Also known as coefficient of velocity.

velocity constant [CONT SYS] The ratio of the rate of change of the input command signal to the steady-state error, in a control system where these two quantities are proportional.

velocity control *See* rate control.

velocity gradient [FL MECH] The rate of change of velocity of propagation with distance normal to the direction of flow.

velocity head [FL MECH] The square of the speed of flow of a fluid divided by twice the acceleration of gravity; it is equal to the static pressure head corresponding to a pressure equal to the kinetic energy of the fluid per unit volume.

velocity potential [FL MECH] For a fluid flow, a scalar function whose gradient is equal to the velocity of the fluid.

velocity pressure *See* wind pressure.

velocity profile [FL MECH] A graph of the speed of a fluid flow as a function of distance perpendicular to the direction of flow.

velocity ratio [MECH ENG] The ratio of the velocity given to the effort or input of a machine to the velocity acquired by the load or output.

velocity servomechanism [CONT SYS] A servomechanism in which the feedback-measuring device generates a signal representing a measured value of the velocity of the output shaft. Also known as rate servomechanism.

vena contracta [FL MECH] The contraction of a jet of liquid which comes out of an opening in a container to a cross section smaller than the opening.

Venera space program [AERO ENG] A series of unmanned space vehicle flights to probe the conditions near and on the planet Venus; the program was initiated by the Soviet Union.

ventilator [MECH ENG] A mechanical apparatus for producing a current of air, as a blowing or exhaust fan.

Venus probe [AERO ENG] A probe for exploring and reporting on conditions on or about the planet Venus, such as Pioneer and Mariner probes of the United States, and Venera probes of the Soviet Union.

vernier engine [AERO ENG] A rocket engine of small thrust used primarily to obtain a fine adjustment in the velocity and trajectory of a rocket vehicle just after the thrust cutoff of the last sustainer engine, and used secondarily to add thrust to a booster or sustainer engine. Also known as vernier rocket.

vernier rocket *See* vernier engine.

vertical band saw [MECH ENG] A band saw whose blade operates in the vertical plane; ideal for contour cutting.

vertical boiler [MECH ENG] A fire-tube boiler having vertical tubes between top head and tube sheet, connected to the top of an internal furnace.

vertical boring mill [MECH ENG] A large type of boring machine in which a rotating workpiece is fastened to a horizontal table, which resembles a four-jaw independent chuck with extra radial T slots, and the tool has a traverse motion.

vertical broaching machine [MECH ENG] A broaching machine having the broach mounted in the vertical plane.

vertical conveyor [MECH ENG] A materials-handling machine designed to move or transport bulk materials or packages upward or downward.

vertical drop [MECH] The drop of an object in trajectory or along a plumb line, measured vertically from its line of departure to the object.

vertical firing [MECH ENG] The discharge of fuel and air perpendicular to the burner in a furnace.

vertical guide idlers [MECH ENG] Idler rollers about 3 inches (8 centimeters) in diameter so placed as to make contact with the edge of the belt conveyor should it run too much to one side.

vertical gyro [AERO ENG] A two-degree-of-freedom gyro with provision for maintaining its spin axis vertical; output signals are produced by gimbal angular displacements which correspond to components of the angular displacements of the base about two orthogonal axes; used in aircraft to measure both bank angle and pitch attitude.

vertical scale [DES ENG] The ratio of the vertical dimensions of a laboratory model to those of the natural prototype; usually exaggerated in relation to the horizontal scale.

vertical separation [AERO ENG] A specified vertical distance measured in terms of space between aircraft in flight at different altitudes or flight levels.

vertical speed indicator *See* rate-of-climb indicator.

vertical tail [AERO ENG] A part of the tail assembly of an aircraft; consists of a fin (a symmetrical airfoil in line with the center line of the fuselage) fixed to the fuselage or body and a rudder which is movable by the pilot.

vertical takeoff and landing [AERO ENG] A flight technique in which an aircraft rises directly into the air and settles vertically onto the ground. Abbreviated VTOL.

vertical turbine pump *See* deep-well pump.

vertical turret lathe [DES ENG] Similar in principle to the horizontal turret lathe but capable of handling heavier, bulkier workpieces; it is constructed with a rotary, horizontal worktable whose diameter (30–74 inches, or 76–188 centimeters) normally designates the capacity of the machine; a crossrail mounted above the worktable carries a turret, which indexes in a vertical plane with tools that may be fed either across or downward.

vertical vorticity [FL MECH] The vertical component of the vorticity vector.

VFR *See* visual flight rules.

VFR between layers [AERO ENG] A flight condition wherein an aircraft is operated under modified visual flight rules while in flight between two layers of clouds or obscuring phenomena, each of which constitutes a ceiling.

VFR flight *See* visual flight.

VFR on top [AERO ENG] A flight condition wherein an aircraft is operated under modified visual flight rules while in flight above a layer of clouds or an obscuring phenomenon sufficient to constitute a ceiling.

VFR terminal minimums [AERO ENG] A set of operational weather limits at an airport, that is, the minimum conditions of ceiling and visibility under which visual flight rules may be used.

V guide [MECH ENG] A V-shaped groove serving to guide a wedge-shaped sliding machine element.

vibrating feeder [MECH ENG] A feeder for bulk materials (pulverized or granulated solids), which are moved by the vibration of a slightly slanted, flat vibrating surface.

vibrating grizzlies [MECH ENG] Bar grizzlies mounted on eccentrics so that the entire assembly is given a forward and backward movement at a speed of some 100 strokes a minute.

vibrating pebble mill [MECH ENG] A size-reduction device in which feed is ground by the action of vibrating, moving pebbles.

vibrating screen [MECH ENG] A sizing screen which is vibrated by solenoid or magnetostriction, or mechanically by eccentrics or unbalanced spinning weights.

vibrating screen classifier [MECH ENG] A classifier whose screening surface is hung by rods and springs, and moves by means of electric vibrators.

vibration [MECH] A continuing periodic change in a displacement with respect to a fixed reference.

vibration damping [MECH ENG] The processes and techniques used for converting the mechanical vibrational energy of solids into heat energy.

vibration drilling [MECH ENG] Drilling in which a frequency of vibration in the range of 100 to 20,000 hertz is used to fracture rock.

vibration machine [MECH ENG] A device for subjecting a system to controlled and reproducible mechanical vibration. Also known as shake table.

vibration separation [MECH ENG] Classification or separation of grains of solids in which separation through a screen is expedited by vibration or oscillatory movement of the screening mediums.

vibrator [MECH ENG] An instrument which produces mechanical oscillations.

vibratory centrifuge [MECH ENG] A high-speed rotating device to remove moisture from pulverized coal or other solids.

vibratory equipment [MECH ENG] Reciprocating or oscillating devices which move, shake, dump, compact, settle, tamp, pack, screen, or feed solids or slurries in process.

vibratory hammer [MECH ENG] A type of pile hammer which uses electrically activated eccentric cams to vibrate piles into place.

vibroenergy separator [MECH ENG] A screen-type device for classification or separation of grains of solids by a combination of gyratory motion and auxiliary vibration caused by balls bouncing against the lower surface of the screen cloth.

Victaulic coupling [DES ENG] A development in which a groove is cut around each end of pipe instead of the usual threads; two ends of pipe are then lined up and a rubber ring is fitted around the joint; two semicircular bands, forming a sleeve, are placed around the ring and are drawn together with two bolts, which have a ridge on both edges to fit into the groove of the pipe; as the bolts are tightened, the rubber ring is compressed, making a watertight joint, while the ridges fitting in the grooves make it strong mechanically.

Vigilante [AERO ENG] A United States supersonic, twin-engine, turbojet tactical, all-weather attack aircraft designed to operate from aircraft carriers, and capable of delivering nuclear or nonnuclear weapons; it has electronic countermeasures equipment, long-range radar, automatic pilot guidance features, inflight refueling capabilities, and a crew of pilot and bombardier. Designated A-5.

Viking spacecraft [AERO ENG] A series of two United States spacecraft, each consisting of one module which remained in orbit around Mars and another which landed on the planet's surface in 1976.

virial coefficients [THERMO] For a given temperature T, one of the coefficients in the expansion of P/RT in inverse powers of the molar volume, where P is the pressure and R is the gas constant.

Virmel engine [MECH ENG] A cat-and-mouse engine that employs vanelike pistons whose motion is controlled by a gear-and-crank system; each set of pistons stops and restarts when a chamber reaches the spark plug.

virtual displacement [MECH] 1. Any change in the positions of the particles forming a mechanical system. 2. An infinitesimal change in the positions of the particles forming a mechanical system, which is consistent with the geometrical constraints on the system.

virtual entropy [THERMO] The entropy of a system, excluding that due to nuclear spin. Also known as practical entropy.

virtual work [MECH] The work done on a system during any displacement which is consistent with the constraints on the system.

virtual work principle *See* principle of virtual work.

viscoelastic fluid [FL MECH] A fluid that displays viscoelasticity.

viscoelasticity [MECH] Property of a material which is viscous but which also exhibits certain elastic properties such as the ability to store energy of deformation, and in which the application of a stress gives rise to a strain that approaches its equilibrium value slowly.

viscoelastic theory [MECH] The theory which attempts to specify the relationship between stress and strain in a material displaying viscoelasticity.

viscometric analysis [FL MECH] Measurement of the flow properties of substances by viscometry.

viscosity [FL MECH] The resistance that a gaseous or liquid system offers to flow when it is subjected to a shear stress. Also known as flow resistance.

viscosity coefficient [FL MECH] An empirical number used in equations of fluid mechanics to account for the effects of viscosity.

viscosity curve [FL MECH] A graph showing the viscosity of a liquid or gaseous material as a function of temperature.

viscous damping [MECH ENG] A method of converting mechanical vibrational energy of a body into heat energy, in which a piston is attached to the body and is arranged to move through liquid or air in a cylinder or bellows that is attached to a support.

viscous dissipation function [FL MECH] A quadratic function of spatial derivatives of components of fluid velocity which gives the rate at which mechanical energy is converted into heat in a viscous fluid per unit volume. Also known as dissipation function.

viscous drag [FL MECH] That part of the rearward force on an aircraft that results from the aircraft carrying air forward with it through viscous adherence.

viscous fillers [MECH ENG] A packaging machine that fills viscous product into cartons; there are two basic types, straight-line and rotary plunger; the former operates intermittently on a given number of containers, while the latter fills and discharges containers continuously.

viscous flow [FL MECH] 1. The flow of a viscous fluid. 2. The flow of a fluid through a duct under conditions such that the mean free path is small in comparison with the smallest, transverse section of the duct.

viscous fluid [FL MECH] A fluid whose viscosity is sufficiently large to make the viscous forces a significant part of the total force field in the fluid.

viscous force [FL MECH] The force per unit volume or per unit mass arising from viscous effects in fluid flow.

vise [DES ENG] A tool consisting of two jaws for holding a workpiece; opened and closed by a screw, lever, or cam mechanism.

visual flight [AERO ENG] An aircraft flight occurring under conditions which allow navigation by visual reference to the earth's surface at a safe altitude and with sufficient horizontal visibility, and operating under visual flight rules. Also known as VFR flight.

visual flight rules [AERO ENG] A set of regulations set down by the U.S. Civil Aeronautics Board (in Civil Air Regulations) to govern the operational control of aircraft during visual flight. Abbreviated VFR.

vitrified wheel [DES ENG] A grinding wheel with a glassy or porcelanic bond.

vixen file [DES ENG] A flat file with curved teeth; used for filing soft metals.

V jewels [DES ENG] Jewel bearings used in conjunction with a conical pivot, the bearing surface being a small radius located at the apex of a conical recess; found primarily in electric measuring instruments.

Voigt notation [MECH] A notation employed in the theory of elasticity in which elastic constants and elastic moduli are labeled by replacing the pairs of letters xx, yy, zz, yz, zx, and xy by the number 1, 2, 3, 4, 5, and 6 respectively.

volatility [THERMO] The quality of having a low boiling point or subliming temperature at ordinary pressure or, equivalently, of having a high vapor pressure at ordinary temperatures.

volatilization [THERMO] The conversion of a chemical substance from a liquid or solid state to a gaseous or vapor state by the application of heat, by reducing pressure, or by a combination of these processes. Also known as vaporization.

volume flow rate [FL MECH] The volume of the fluid that passes through a given surface in a unit time.

volumetric efficiency [MECH ENG] In describing an engine or gas compressor, the ratio of volume of working substance actually admitted, measured at a specified temperature and pressure, to the full piston displacement volume; for a liquid-fuel engine, such as a diesel engine, volumetric efficiency is the ratio of the volume of air drawn into a cylinder to the piston displacement.

volumetric strain [MECH] One measure of deformation; the change of volume per unit of volume.

volute [DES ENG] A spiral casing for a centrifugal pump or a fan designed so that speed will be converted to pressure without shock.

volute pump [MECH ENG] A centrifugal pump housed in a spiral casing.

vortex [FL MECH] 1. Any flow possessing vorticity; for example, an eddy, whirlpool, or other rotary motion. 2. A flow with closed streamlines, such as a free vortex or line vortex. 3. See vortex tube.

vortex distribution method [FL MECH] An analytic method used in ideal aerodynamics which ignores the thickness of the profile of the aerodynamic figure being studied.

vortex filament [FL MECH] The line of concentrated vorticity in a line vortex. Also known as vortex line.

vortex line [FL MECH] 1. A line drawn through a fluid such that it is everywhere tangent to the vorticity. 2. See vortex filament.

vortex ring [FL MECH] A line vortex in which the line of concentrated vorticity is a closed curve. Also known as collar vortex; ring vortex.

vortex shedding [FL MECH] In the flow of fluids past objects, the shedding of fluid vortices periodically downstream from the restricting object (for example, smokestacks, pipelines, or orifices).

vortex sheet [FL MECH] A surface across which there is a discontinuity in fluid velocity, such as in slippage of one layer of fluid over another; the surface may be regarded as being composed of vortex filaments.

vortex street [FL MECH] A series of vortices which are systematically shed from the downstream side of a body around which fluid is flowing rapidly. Also known as vortex trail; vortex train.

vortex trail See vortex street.

vortex train See vortex street.

vortex tube [FL MECH] A tubular surface consisting of the collection of vortex lines which pass through a small closed curve. Also known as vortex.

vorticity [FL MECH] For a fluid flow, a vector equal to the curl of the velocity of flow.

vorticity equation [FL MECH] An equation of fluid mechanics describing horizontal circulation in the motion of particles around a vertical axis: $(d/dt) (S + f) = - (S + f) \operatorname{div}_h c$, where $(S + f)$ is the absolute vorticity and $\operatorname{div}_h c$ is the horizontal divergence of the fluid velocity.

vorticity-transport hypothesis [FL MECH] The hypothesis that, owing to the existence of pressure fluctuations, vorticity, and not momentum, is conservative in turbulent eddy flux.

Voskhod program [AERO ENG] The multi-crew member spaceflight program of the Soviet Union which began with the flight of *Voskhod 1*, October 12, 1964.

Vostok spacecraft [AERO ENG] One of a series of crewed artificial satellites launched by the Soviet Union; *Vostok 1* launched on April 12, 1961, was the first crewed spacecraft.

votator   [MECH ENG] Efficient heat-exchange units for chilling and mechanically working a continuous stream of emulsion; used in food industries in preparation of margarine.

VTOL   *See* vertical takeoff and landing.

wagon drill [MECH ENG] 1. A vertically mounted, pneumatic, percussive-type rock drill supported on a three- or four-wheeled wagon. 2. A wheel-mounted diamond drill machine.

wake [FL MECH] The region behind a body moving relative to a fluid in which the effects of the body on the fluid's motion are concentrated.

wake flow [FL MECH] Turbulent eddying flow that occurs downstream from bluff bodies.

walking beam [MECH ENG] A lever that oscillates on a pivot and transmits power in a manner producing a reciprocating or reversible motion; used in rock drilling and oil well pumping.

walking dragline [MECH ENG] A large-capacity dragline built with moving feet; disks 20 feet (6 meters) in diameter support the excavator while working.

walking machine [MECH ENG] A machine designed to carry its operator over various types of terrain; the operator sits on a platform carried on four mechanical legs, and movements of his arms control the front legs of the machine while movements of his legs control the rear legs of the machine.

wall-attachment amplifier [FL MECH] A bistable fluidic device utilizing two walls set back from the supply jet port, control ports, and channels to define two downstream outputs. Also known as flip-flop amplifier.

wall crane [MECH ENG] A jib crane mounted on a wall.

Walley engine [MECH ENG] A multirotor engine employing four approximately elliptical rotors that turn in the same clockwise sense, leading to excessively high rubbing velocities.

wall friction [FL MECH] The drag created in the flow of a liquid or gas because of contact with the wall surfaces of its conductor, such as the inside surfaces of a pipe.

wall ratio [DES ENG] Ratio of the outside radius of a gun, a tube, or jacket to the inside radius; or ratio of the corresponding diameters.

Walter engine [MECH ENG] A multirotor rotary engine that uses two different-sized elliptical rotors.

Wankel engine [MECH ENG] An eccentric-rotor-type internal combustion engine with only two primary moving parts, the rotor and the eccentric shaft; the rotor moves in one direction around the trochoidal chamber containing peripheral intake and exhaust ports.

Ward-Leonard speed-control system [CONT SYS] A system for controlling the speed of a direct-current motor in which the armature voltage of a separately excited direct-current motor is controlled by a motor-generator set.

warm-air heating [MECH ENG] Heating by circulating warm air; system contains a direct-fired furnace surrounded by a bonnet through which air circulates to be heated.

warpage [MECH] The action, process, or result of twisting or turning out of shape.

wash [AERO ENG] The stream of air or other fluid sent backward by a jet engine or a propeller. [FL MECH] The surge of disturbed air or other fluid resulting from the passage of something through the fluid.

wash boring *See* jet drilling.

washer [DES ENG] A flattened, ring-shaped device used to improve the tightness of a screw fastener.

water column [MECH ENG] A tubular column located at the steam and water space of a boiler to which protective devices such as gage cocks, water gage, and level alarms are attached.

water-cooled condenser [MECH ENG] A steam condenser which is for the maintenance of vacuum, and in which water is the heat-receiving fluid.

water-cooled furnace [MECH ENG] A fuel-fired furnace containing tubes in which water is circulated to limit heat loss to the surroundings, control furnace temperature, and generate steam.

water hammer [FL MECH] Pressure rise in a pipeline caused by a sudden change in the rate of flow or stoppage of flow in the line.

water heater [MECH ENG] A tank for heating and storing hot water for domestic use.

waterpower [MECH] Power, usually electric, generated from an elevated water supply by the use of hydraulic turbines.

water swivel [DES ENG] A device connecting the water hose to the drill-rod string and designed to permit the drill string to be rotated in the borehole while water is pumped into it to create the circulation needed to cool the bit and remove the cuttings produced. Also known as gooseneck; swivel neck.

water-tube boiler [MECH ENG] A steam boiler in which water circulates within tubes and heat is applied from outside the tubes to generate steam.

water tunnel [AERO ENG] A device similar to a wind tunnel, but using water as the working fluid instead of air or other gas.

waterwall [MECH ENG] The side of a boiler furnace consisting of water-carrying tubes which absorb radiant heat and thereby prevent excessively high furnace temperatures.

waterwheel [MECH ENG] A vertical wheel on a horizontal shaft that is made to revolve by the action or weight of water on or in containers attached to the rim.

Watt's law [THERMO] A law which states that the sum of the latent heat of steam at any temperature of generation and the heat required to raise water from 0°C to that temperature is constant; it has been shown to be substantially in error.

wave [FL MECH] A disturbance which moves through or over the surface of a liquid, as of a sea.

wave motor [MECH ENG] A motor that depends on the lifting power of sea waves to develop its usable energy.

ways [MECH ENG] Bearing surfaces used to guide and support moving parts of machine tools; may be flat, V-shaped, or dovetailed.

weathercocking [AERO ENG] The aerodynamic action causing alignment of the longitudinal axis of a rocket with the relative wind after launch. Also known as weather vaning.

weathercock stability *See* directional stability.

weather vaning *See* weathercocking.

web [MECH ENG] For twist drills and reamers, the central portion of the tool body that joins the loads.

Weber number 1 [FL MECH] A dimensionless number used in the study of surface tension waves and bubble formation, equal to the product of the square of the velocity of the wave or the fluid velocity, the density of the fluid, and a characteristic length, divided by the surface tension. Symbolized N_{We1}, We.

Weber number 2 [FL MECH] A dimensionless number, equal to the square root of Weber number 1. Symbolized N_{We2}.

wedge [DES ENG] A piece of resistant material whose two major surfaces make an acute angle.

wedge bit [DES ENG] A tapered-nose noncoring bit, used to ream out the borehole alongside the steel deflecting wedge in hole-deflection operations. Also known as bull-nose bit; wedge reaming bit; wedging bit.

wedge core lifter [MECH ENG] A core-gripping device consisting of a series of three or more serrated-face, tapered wedges contained in slotted and tapered recesses cut into the inner surface of a lifter case or sleeve; the case is threaded to the inner tube of a core barrel, and as the core enters the inner tube, it lifts the wedges up along the case taper; when the barrel is raised, the wedges are pulled tight, gripping the core.

wedge reaming bit *See* wedge bit.

wedging bit *See* wedge bit.

weight [MECH] 1. The gravitational force with which the earth attracts a body. 2. By extension, the gravitational force with which a star, planet, or satellite attracts a nearby body.

weight and balance sheet [AERO ENG] A sheet which records the distribution of weight in an aircraft and shows the center of gravity of an aircraft at takeoff and landing.

weightlessness [MECH] A condition in which no acceleration, whether of gravity or other force, can be detected by an observer within the system in question. Also known as zero gravity.

well drill [MECH ENG] A drill, usually a churn drill, used to drill water wells.

well-regulated system [CONT SYS] A system with a regulator whose action, together with that of the environment, prevents any disturbance from permanently driving the system from a state in which it is stable, that is, a state in which it retains its structure and survives.

Wentworth quick-return motion *See* turning-block linkage.

wet cooling tower [MECH ENG] A structure in which water is cooled by atomization into a stream of air; heat is lost through evaporation. Also known as evaporative cooling tower.

wet drill [MECH ENG] A percussive drill with a water feed either through the machine or by means of a water swivel, to suppress the dust produced when drilling.

wet emplacement [AERO ENG] A launch emplacement that provides a deluge of water for cooling the flame bucket, the rocket engines, and other equipment during the launch of a missile.

wet engine [MECH ENG] An engine with its oil, liquid coolant (if any), and trapped fuel inside.

wet grinding [MECH ENG] 1. The milling of materials in water or other liquid. 2. The practice of applying a coolant to the work and the wheel to facilitate the grinding process.

wet mill [MECH ENG] 1. A grinder in which the solid material to be ground is mixed with liquid. 2. A mill in which the grinding energy is developed by a fast-flowing liquid stream; for example, a jet pulverizer.

wet sleeve [MECH ENG] A cylinder liner which is exposed to the coolant over 70% or more of its surface.

wetting angle [FL MECH] A contact angle which lies between 0 and 90°.

wet well [MECH ENG] A chamber which is used for collecting liquid, and to which the suction pipe of a pump is attached.

wheel [DES ENG] A circular frame with a hub at the center for attachment to an axle, about which it may revolve and bear a load.

wheel base [DES ENG] The distance in the direction of travel from front to rear wheels of a vehicle, measured between centers of ground contact under each wheel.

wheeled crane [MECH ENG] A self-propelled crane that rides on a rubber-tired chassis with power for transportation provided by the same engine that is used for hoisting.

wheel sleeve [DES ENG] A flange used as an adapter on precision grinding machines where the hole in the wheel is larger than the machine arbor.

Whitworth screw thread [DES ENG] A British screw thread standardized to form and dimension.

whole range point [AERO ENG] The point vertically below an aircraft at the moment of impact of a bomb released from that aircraft, assuming that the aircraft's velocity has remained unchanged.

Willans line [MECH ENG] The line (nearly straight) on a graph showing steam consumption (pounds per hour) versus power output (kilowatt or horsepower) for a steam engine or turbine; frequently extended to show total fuel consumed (pounds per hour) for gas turbines, internal combustion engines, and complete power plants.

Williams-Hazen formula [FL MECH] In a liquid-flow system, a method for calculation of head loss due to the friction in a pipeline.

winch [MECH ENG] A machine having a drum on which to coil a rope, cable, or chain for hauling, pulling, or hoisting.

windage [MECH] 1. The deflection of a bullet or other projectile due to wind. 2. The correction made for such deflection.

wind deflection [MECH] Deflection caused by the influence of wind on the course of a projectile in flight.

windmill [MECH ENG] Any of various mechanisms, such as a mill, pump, or electric generator, operated by the force of wind against vanes or sails radiating about a horizontal shaft.

window [AERO ENG] An interval of time during which conditions are favorable for launching a spacecraft on a specific mission.

wind pressure [MECH] The total force exerted upon a structure by wind. Also known as velocity pressure.

wind triangle [AERO ENG] A vector diagram showing the effect of the wind on the flight of an aircraft; it is composed of the wind direction and wind speed vector, the true heading and true airspeed vector, and the resultant track and ground speed vector.

wind-tunnel balance [AERO ENG] A device or apparatus that measures the aerodynamic forces and moments acting upon a body tested in a wind tunnel.

wing [AERO ENG] 1. A major airfoil. 2. An airfoil on the side of an airplane's fuselage or cockpit, paired off by one on the other side, the two providing the principal lift for the airplane.

wing assembly [AERO ENG] An aeronautical structure designed to maintain a guided missile in stable flight; it consists of all panels, sections, fastening devices, chords, spars, plumbing accessories, and electrical components necessary for a complete wing assembly.

wing axis [AERO ENG] The locus of the aerodynamic centers of all the wing sections of an airplane.

wing loading [AERO ENG] A measure of the load carried by an airplane wing per unit of wing area; commonly used units are pounds per square foot and kilograms per square meter.

wing nut [DES ENG] An internally threaded fastener with wings to permit it to be tightened or loosened by finger pressure only. Also known as butterfly nut.

wing panel [AERO ENG] That portion of a multipiece wing section that usually lies between the front and rear spars; it may be designed to include either the leading edge or the trailing edge as an integral part, but never both, and excludes control surfaces.

wing profile [AERO ENG] The outline of a wing section.

wing rib [AERO ENG] A chordwise member of the wing structure of an airplane, used to give the wing section its form and to transmit the load from the fabric to the spars.

wing section *See* airfoil profile.

wing structure [AERO ENG] In an aircraft, the combination of outside fairing panels that provide the aerodynamic lifting surfaces and the inside supporting members that transmit the lifting force to the fuselage; the primary load-carrying portion of a wing is a box beam (the prime box) made up usually of two or more vertical webs, plus a major portion of the upper and lower skins of the wing, which serve as chords of the beam.

wing-tip rake [AERO ENG] The shape of the wing when the tip edge is straight in plan but not parallel to the plane of symmetry; the amount of rake is measured by the acute angle between the straight portion of the wing tip and the plane of symmetry; the rake is positive when the trailing edge is longer than the leading edge.

wire cloth [DES ENG] Screen composed of wire crimped or woven into a pattern of squares or rectangles.

wire gage [DES ENG] 1. A gage for measuring the diameter of wire or thickness of sheet metal. 2. A standard series of sizes arbitrarily indicated by numbers, to which the diameter of wire or the thickness of sheet metal is usually made, and which is used in describing the size or thickness.

wire line [DES ENG] 1. Any cable or rope made of steel wires twisted together to form the strands. 2. A steel wire rope $\frac{5}{16}$ inch or less in diameter.

wire nail [DES ENG] A nail made of wire and having a circular cross section.

wire saw [MECH ENG] A machine employing one- or three-strand wire cable, up to 16,000 feet (4900 meters) long, running over a pulley as a belt; used in quarries to cut rock by abrasion.

wire tack [DES ENG] A tack made from wire stock.

Wobbe index [THERMO] A measure of the amount of heat released by a gas burner with a constant orifice, equal to the gross calorific value of the gas in British thermal units per cubic foot at standard temperature and pressure divided by the square root of the specific gravity of the gas.

wobble wheel roller [MECH ENG] A roller with freely suspended pneumatic tires used in soil stabilization.

Woodruff key [DES ENG] A self-aligning machine key made by a side-milling cutter in the form of a segment of a disk.

wood screw [DES ENG] A threaded fastener with a pointed shank, a slotted or recessed head, and a sharp tapered thread of relatively coarse pitch for use only in wood.

woodstave pipe [DES ENG] A pipe made of narrow strips of wood placed side by side and banded with wire, metal collars, and inserted joints, used largely for municipal water supply, outfall sewers, and mining irrigation.

work [MECH] The transference of energy that occurs when a force is applied to a body that is moving in such a way that the force has a component in the direction of the body's motion; it is equal to the line integral of the force over the path taken by the body.

work function *See* free energy.

work-kinetic energy theorem [MECH] The theorem that the change in the kinetic energy of a particle during a displacement is equal to the work done by the resultant force on the particle during this displacement.

work of adhesion *See* adhesional work.

worm [DES ENG] A shank having at least one complete tooth (thread) around the pitch surface; the driver of a worm gear.

worm conveyor *See* screw conveyor.

worm gear [DES ENG] A gear with teeth cut on an angle to be driven by a worm; used to connect nonparallel, nonintersecting shafts.

worm wheel [DES ENG] A gear wheel with curved teeth that meshes with a worm.

wrap forming *See* stretch forming.

wrapper sheet [MECH ENG] 1. The outer plate enclosing the firebox in a fire-tube boiler. 2. The thinner sheet of a boiler drum having two sheets.

wrench [MECH] The combination of a couple and a force which is parallel to the torque exerted by the couple.

wrench-head bolt [DES ENG] A bolt with a square or hexagonal head designed to be gripped between the jaws of a wrench.

wringing fit [DES ENG] A fit of zero-to-negative allowance.

wrist pin *See* piston pin.

X engine [MECH ENG] An in-line engine with the cylinder banks so arranged around the crankshaft that they resemble the letter X when the engine is viewed from the end.

X frame [DES ENG] An automotive frame which either has side rails bent in at the center of the vehicle, making the overall form that of an X, or has an X-shaped member which joins the side rails with diagonals for added strength and resistance to torsional stresses.

Y

yard [MECH] A unit of length in common use in the United States and United Kingdom, equal to 0.9144 meter, or 3 feet. Abbreviated yd.

yardage [MECH] An amount expressed in yards.

yard crane *See* crane truck.

yaw [MECH] **1.** The rotational or oscillatory movement of a ship, aircraft, rocket, or the like about a vertical axis. Also known as yawing. **2.** The amount of this movement, that is, the angle of yaw. **3.** To rotate or oscillate about a vertical axis.

yaw acceleration [MECH] The angular acceleration of an aircraft or missile about its normal or Z axis.

yaw axis [MECH] A vertical axis through an aircraft, rocket, or similar body, about which the body yaws; it may be a body, wind, or stability axis. Also known as yawing axis.

yaw damper [AERO ENG] A control system or device that reduces the yaw of an aircraft, guided missile, or the like.

yaw indicator [AERO ENG] A device that measures the angular direction of the airflow relative to the longitudinal vertical plane of the aircraft; this may be accomplished by a balanced vane or by a differential pressure sensor that aligns the detector to the airflow, and in so doing transmits the measured angle between the normal axis and the detector as the yaw angle.

yawing *See* yaw.

yaw simulator [CONT SYS] A test instrument used to derive and thereby permit study of probable aerodynamic behavior in controlled flight under specific initial conditions; certain components of the missile guidance system, such as the receiver or servo loop, are connected into the simulator circuitry; also, certain aerodynamic parameters of the specific missile must be known and set into the simulator; applicable to the yaw plane.

yd *See* yard.

yield [MECH] That stress in a material at which plastic deformation occurs.

yield point [MECH] The lowest stress at which strain increases without increase in stress.

yield strength [MECH] The stress at which a material exhibits a specified deviation from proportionality of stress and strain.

yield stress [MECH] The lowest stress at which extension of the tensile test piece increases without increase in load.

yoke [DES ENG] A clamp or similar device to embrace and hold two other parts. [MECH ENG] A slotted crosshead used instead of a connecting rod in some steam engines.

Young's modulus [MECH] The ratio of a simple tension stress applied to a material to the resulting strain parallel to the tension.

z

zero bevel gears [DES ENG] A special form of bevel gear having curved teeth with a zero-degree spiral angle.

zero gravity *See* weightlessness.

zero length [AERO ENG] In rocket launchers, zero length indicates that the launcher is designed to hold the rocket in position for launching but not to give it guidance.

zero-lift angle [AERO ENG] The angle of attack of an airfoil when its lift is zero.

zero-lift chord [AERO ENG] A chord taken through the trailing edge of an airfoil in the direction of the relative wind when the airfoil is at zero-lift angle of attack.

zero-order hold [CONT SYS] A device which converts a sampled output into an output which is held constant between samples at the last sampled value.

zipper conveyor [MECH ENG] A type of conveyor belt with zipperlike teeth that mesh to form a closed tube; used to handle fragile materials.

Zond spacecraft [AERO ENG] One of a series of Soviet space probes which have photographed the moon and made observations in interplanetary space.